Applied Anatomy in Liver Resection and Liver Transplantation

W. Y. Lau

Applied Anatomy in Liver Resection and Liver Transplantation

PMPH PEOPLE'S MEDICAL PUBLISHING HOUSE

Springer

W. Y. Lau
Prince of Wales Hospital
Chinese University of Hong Kong
Hong Kong
China

ISBN 978-981-16-0799-8 ISBN 978-981-16-0800-1 (eBook)
https://doi.org/10.1007/978-981-16-0800-1

This Springer imprint is published by the registered company Springer Nature Singapore Pte Ltd.
The registered company address is: 152 Beach Road, #21-01/04 Gateway East, Singapore 189721, Singapore

This book is dedicated to my three most respected mentors who have the greatest influence on my professional career:

- *To Dr. Shu Hong Wong who inspired me to take up surgery as a professional career*
- *To Professor Arthur K.C. Li who brought me into the challenging field of academic surgery and the international scene of surgery*
- *To the late Professor Qiu Fazu who set a role model for me to follow and showed me the virtues of a great surgeon:*
  - *Self-discipline and lifelong pursuit of excellence and knowledge*
  - *Self-sacrifice for the betterment of others*
  - *Willingness and eagerness to teach*

Foreword

Learning different techniques in liver resection is similar to learning Chinese Kung Fu.

	Chinese Kung Fu (Bruce Lee)	Liver resection (W.Y. Lau)
Styles	Different styles, e.g., Shaolin, Wudang	Different approaches, e.g., conventional, anterior
First step of learning	Before learning how to attack (active), first learn how to defend (passive)	Before learning actual cutting (active), first learn how to control and prevent bleeding (and be passively led by bleeding)
Elementary and foundation steps	Stance	Anatomy, physiology
Techniques	Punch, block, kick	Dissection, transection, ligation, coagulation
Combinations	Combinations of punches, blocks, and kicks	Different techniques of liver resection
Chinese weapons	Sword, staff, spear, etc.	Various instruments for liver parenchymal transection and hemostasis
Real combat	To apply different free-style boxing techniques to different opponents to achieve the best result	Clinical application of the most suitable approach in liver resection to achieve the best result
Final goal to achieve	"The highest technique is to have no technique."	Same

Hong Kong, China

W. Y. Lau

Preface to the Second Edition

Six years have passed since the publication of the first edition of this book. As clearly stated by Rajasekaran S in his editorial for SICOT e-Newsletter published in March 2012 "We are in an era of knowledge explosion with new ideas and surgical techniques being described every day." The knowledge and the new advances in liver surgery necessitate the publication of this second edition.

About 20 years ago, many surgeons from all parts of the world came to Hong Kong to learn liver surgery from me. It was the handouts, hand-written notes, diagrams, and figures which I used to teach these surgeons liver anatomy and surgery that I collected and based on that came up with the draft for the first edition of this book.

The first edition became so popular among liver surgeons in China that I have been invited to conduct 12 workshops in 12 different cities in China (Shanghai, Wuhan, Chongqing, Guangzhou, Nanning, Hangzhou, Beijing, Chengdu, Zhengzhou, Xi'an, Tianjin, and Changchun). On average, there were two such workshops per year and each workshop lasted for 2.5 days. Lectures on applied liver anatomy were given in the morning of the first day of the workshop followed by hands-on dissection of human cadaveric livers preserved in formalin in the afternoon. Lectures on various techniques of liver resection were given on the morning of the second day followed by hands-on application of liver instruments on cow's and pig's livers. The evening on the second day of the workshop was spent on going through and selecting cases for live demonstration of liver resection which was done on the third day. Such workshops were so popular that each workshop was attended by over 100 surgeons who had varying degrees of experience in liver surgery—from the novice to the very experienced. During these workshops which were designed to be interactive, I learned as much from the delegates as they learned from me. All these form the basis for the second edition.

This book has expanded from the 12 chapters in the first edition to the 20 chapters in this second edition. I have also taken this opportunity to correct mistakes which I have made in the first edition.

I have tried to keep the particular feature of this book—the heavy use of diagrams which makes reading easy. I hope this book will remain an important textbook for those who have some experience with liver surgery and as a reference book for the most experienced liver surgeons.

Hong Kong, China W. Y. Lau

Acknowledgments of the Second Edition

I am deeply indebted to the advisers and deputy editors of this book for their valuable advice and help in drafting, in proof-reading, and in the production of the operative photographs and pictures of the specimens. I am grateful to the People's Medical Publishing House in agreeing to publish the second edition of this book. I appreciate the dedication and professionalism of the production team to this project.

I am grateful to my secretary Ms. Helena Lee for her dedication in typing the manuscript and for her support in compiling this book.

Finally, I am thankful to my wife Corinna, my daughter Stephanie, and my son Declan for their patience, support, encouragement, and love, and for the time stolen from them so that I can pursue an academic career.

W. Y. Lau

Preface to the First Edition

This book is written based on my 30 years of experience in liver surgery, extensive research in medical literature on applied liver anatomy, and meticulous studies on liver specimens and liver casts.

I strongly believe that surgery should be based on a solid foundation of anatomy, for without anatomy, there is no surgery. A thorough understanding of the relevant anatomy is a prerequisite to good surgery as a surgeon needs to know with confidence what needs to be removed and what needs to be preserved. When a surgeon encounters unexpected difficulties in an operation, it is the knowledge of anatomy which guides him from rough sea back to safe land. A change in surgical strategy/approach, and the design of a new operation/instrument, can only be done scientifically and logically with adequate knowledge in anatomy.

This book has 12 chapters which cover a full range of knowledge for what one needs to know before one embarks on carrying out a liver operation on a patient. The knowledge ranges from external to internal anatomy of the liver, from pure anatomy to its application in liver operations, from vascular inflow/outflow of the liver to techniques used in reducing intraoperative blood loss, from Couinaud's liver segments to segment-based liver resection, and from the different approaches to liver resectional techniques to the different types of liver transplantation.

The particular feature of this book is the heavy use of diagrams which makes reading easier. It is a must-read for all those surgeons who wish to embark on, and is an important textbook for those who have some experience with, liver surgery. Even the most experienced liver surgeons will find this book of value as a reference book.

Hong Kong, China

W. Y. Lau

Acknowledgments to the First Edition

Acknowledgments are always difficult to write as it is never possible to acknowledge all of those who have helped contribute to a book.

I am deeply indebted to the advisors of this book Professor Wu Mengchao and Professor Zhong Shizhen for their valuable advice. I am honored and thankful to my deputy editors Professor David Yew, Professor Zhou Weiping, Dr. Eric C.H. Lai, Dr. Stephanie H.Y. Lau, and Dr. S.S. Ho for their help in redrafting, in proof-reading, and in the production of the operative photographs and pictures of the specimens. I am grateful to Jia Xiaowei for his coordination and Li Xiaojie for the time she spent on the production of the beautiful medical illustrations.

From the production side at the People's Medical Publishing House, I am grateful to Mr. Hu Guochen, President and Editor in Chief, and Dr. Du Xian, Vice-Editor in Chief, who believed that a book on applied anatomy in liver resection and transplantation was needed. I am especially grateful to them in agreeing to publish this book in an English version for the international market, and a Chinese version for the local market. The production team has been a joy to work with. I appreciate their dedication and professionalism to this project.

I am grateful to my secretary Ms. Helena Lee for her dedication in typing the manuscript and for her support in compiling this book.

W. Y. Lau

Contents

Contributors

Advisors

Wu Mengchao, MD Chinese Academy of Sciences, Eastern Hepatobiliary Surgry Hospital, Second Military Medical University, Shanghai, People's Republic of China

Zhong Shizhen, MD Chinese Academy of Engineering, Clinical Anatomy Research Institute, Southern Medical University, Guangzhou, People's Republic of China

Editor in Chief

Joseph W. Y. Lau, MD, DSc, FRCS (Edin) Chinese Academy of Sciences, Choh-Ming Li Research Professor of Surgery, The Chinese University of Hong Kong, Shatin, New Teritories, Hong Kong SAR

Editors

N. H. Chia, MBBS (HK), FRCS (Edin) Department of Surgery, Queen Elizabeth Hospital, Kowloon, Hong Kong SAR

David W. Chu, LMC HK, FCSHK, FRCS (Edin) Department of Surgery, Queen Elizabeth Hospital, Kowloon, Hong Kong SAR

Thomas K. M. Chung, MBBS (HK), FRCS (Glasgow), Department of Surgery, Queen Elizabeth Hospital, Kowloon, Hong Kong SAR

S. S. Ho, MBBS (HK), MRCP (UK), FHKCA Honorary Clinical Tutor, The Chinese University of Hong Kong, Shatin, New Territories, Hong Kong SAR

Eric C. H. Lai, MBChB, MRCS (Ed), FRACS Department of Surgery, Pamela Youde Nethersole Eastern Hospital, Hong Kong, SAR, China

Stephanie H. Y. Lau, MBChB, MRCS (Ed), FRCS (Edin) Department of Surgery, Queen Elizabeth Hospital, Kowloon, Hong Kong SAR

Zhou Weiping, MD, PhD The 3rd Department of Hepatic Surgery, Eastern Hepatobiliary Surgery Hospital, Second Military Medical University, Shanghai, People's Republic of China

David T. W. Yew, PhD, DSc, Dr Med (Habil) Faculty of Medicine, The Chinese University of Hong Kong, Shatin, New Territories, Hong Kong SAR

1 Applied Anatomy of the Liver

1.1 Surface Anatomy (Fig. 1.1)

When viewed from the front, the normal liver surface markings are:

Upper margin: approximately level with the xiphisternal joint, arching slightly upwards on each side. On the left, it reaches the fifth intercostal space 7–8 cm from the midline, and on the right to the fifth rib.

Right border: curving down to the right border from ribs 7 to 11 in the midaxillary line.

Inferior border: along a line that joins the right lower and upper left extremities. On the right side, the inferior border lines approximately level with the right costal margin while centrally it crosses behind the right upper abdominal wall between the costal margins.

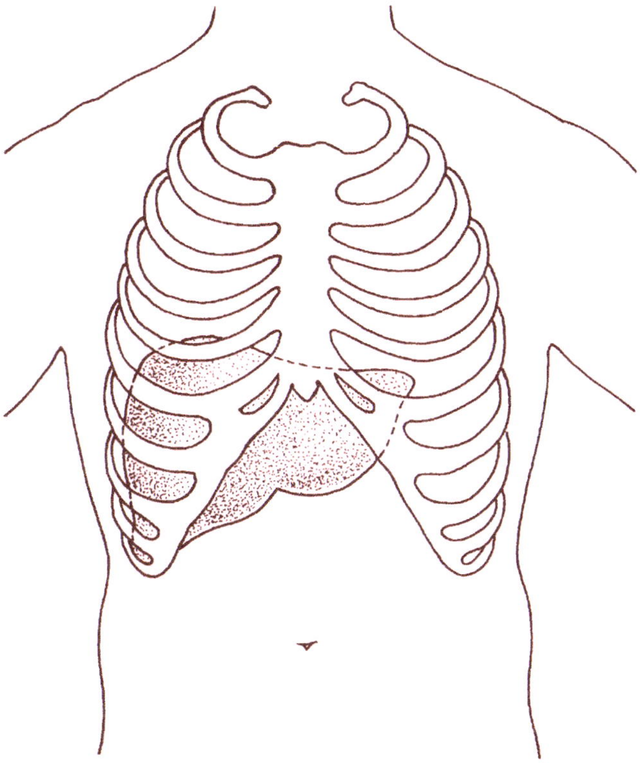

Fig. 1.1 Surface anatomy of the liver

1.1.1 Clinical Applications

1. As most of the liver is undercover of the right rib cage, a tumour in the liver is difficult to detect clinically unless the tumour is very large.
2. Surgical access to the liver requires a bilateral subcostal incision with a midline upward extension with good retraction of the rib cages in a forward, outward and upward direction, or a right thoracoabdominal incision with division of the diaphragm.
3. Trauma to the right lower chest can result in injury to the liver and vice versa.
4. Ultrasound-guided needle puncture for biopsy or interventional procedure is technically more difficult for liver lesions immediately underneath the diaphragm, as the needle might have to go through the right pleural cavity. Approaches include the intercostal and the subcostal routes.

1.2 Gross Anatomy

The liver has three surfaces: diaphragmatic, visceral and posterior surfaces.

1.2.1 Diaphragmatic Surface (Fig. 1.2)

The diaphragmatic surface is covered for the most part in the peritoneum, which forms a sheath around the liver, except in places where the ligaments reflect to join the adjacent diaphragm. In the midline of the abdomen and over the anterior convexity of the liver, the falciform ligament is attached, and

W. Y. Lau, *Applied Anatomy in Liver Resection and Liver Transplantation*, https://doi.org/10.1007/978-981-16-0800-1_1

divides the liver into the anatomical right lobe and left lobe. The ligamentum teres, a remnant of the left umbilical vein, runs from the umbilicus in between the two leaves of the falciform ligament to the visceral surface of the liver, where it disappears behind a bridge of either fibrous or liver tissue which connects the left lobe with the quadrate lobe to end in the left portal vein at the junction between the branches to segments 3 and 4 (Fig. 1.3). The fundus of the gallbladder peeps below the inferior border of the liver. This paragraph is reprinted from: Lunan Yan, Operative Techniques in Liver Resection. Springer, 2016. ISBN 978-94-017-7409-3.

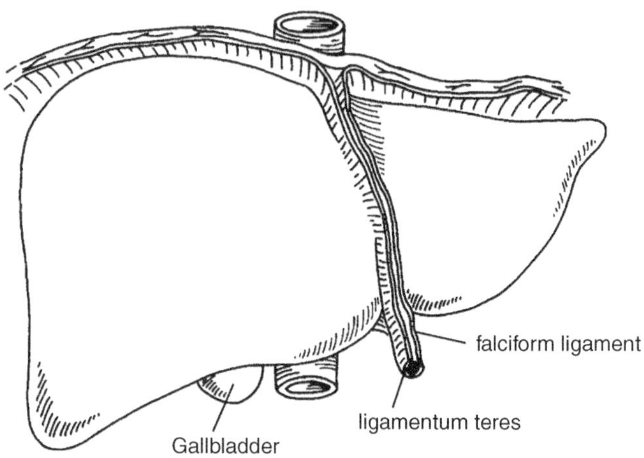

Fig. 1.2 Diaphragmatic surface of the liver

1.2.2 Visceral Surface (Fig. 1.3)

The sharp inferior border of the liver joins the diaphragmatic surface with the visceral surface of the liver. The main structures here are arranged in an H-shaped pattern. The cross-piece of the H is made by the porta hepatis (the hilum of the liver). The right limb of the H is made incompletely by the inferior vena cava posteriorly, and the gallbladder anteriorly. The left limb of the H is made by the continuity of the fissures for the ligamentum teres anteriorly and the ligamentum venosum posteriorly. The vena cava lies in a deep groove and it is crossed over by the hepatocaval ligament. On its right side is the bare area and its left side the caudate lobe. This paragraph is reprinted from: Lunan Yan, Operative Techniques in Liver Resection. Springer, 2016. ISBN 978-94-017-7409-3.

1.2.3 Posterior Surface (Fig. 1.4)

The inferior vena cava (IVC) runs in the centre of the posterior surface of the liver. A fibrous band called the ligamentum venae cavae (hepatocaval ligament) covers part of the inferior vena cava posteriorly. This fibrous band, sometimes replaced by a bridge of liver tissue, is attached to the bare area on the right side and the caudate lobe on the left side. The ligamentum venosum runs in a groove just to the left of

Fig. 1.3 Visceral surface of the liver

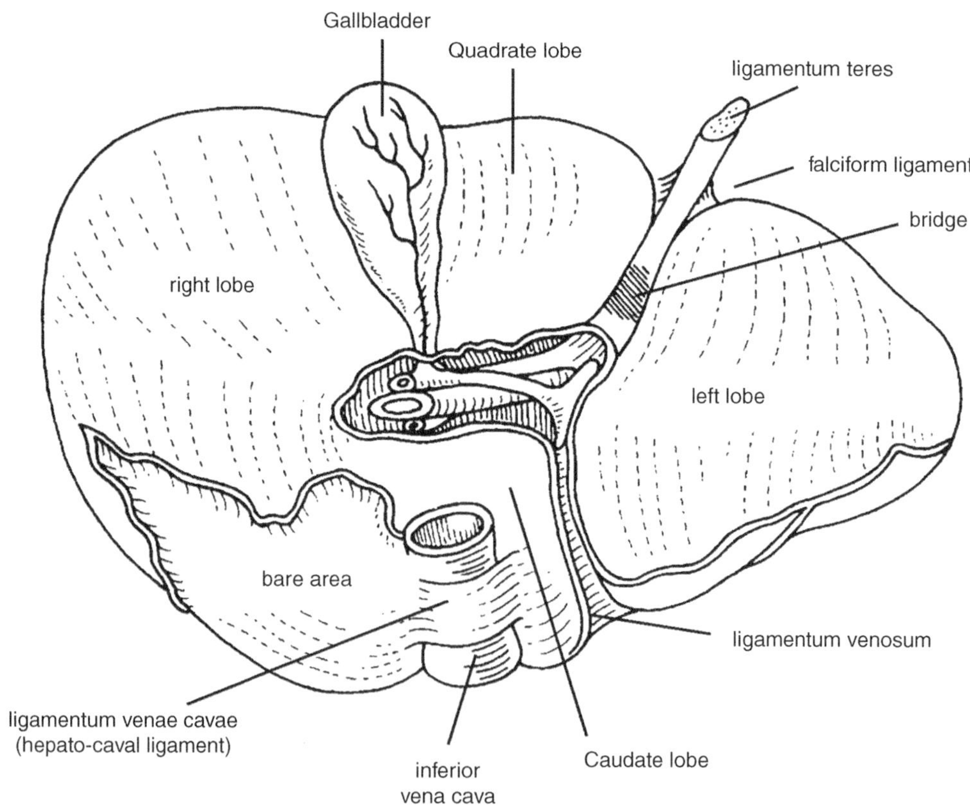

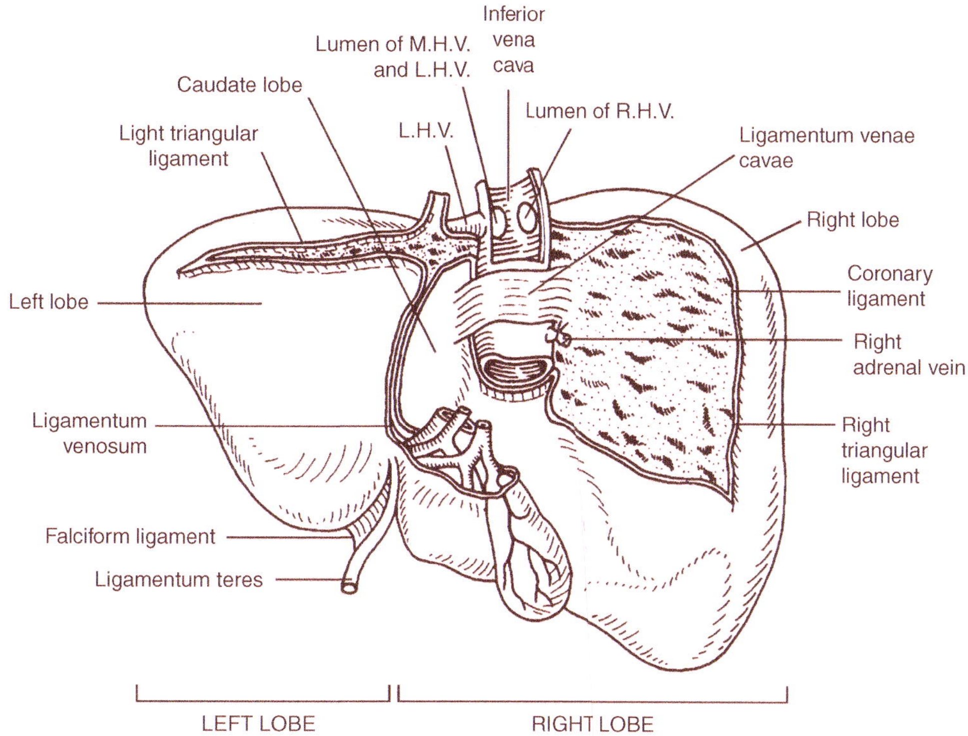

Fig. 1.4 Posterior surface of the liver. *RHV* right hepatic vein, *MHV* middle hepatic vein, *LHV* left hepatic vein

the caudate lobe. The rest of the posterior surface of the liver is made up of the ligaments (the left triangular ligament, the coronary ligament and the right triangular ligament) which attach the liver to the diaphragm. This paragraph is reprinted from: Lunan Yan, Operative Techniques in Liver Resection. Springer, 2016. ISBN 978-94-017-7409-3.

1.2.4 Clinical Applications

The gross external anatomy of the liver is important. It serves as an important anatomical landmark for a more in-depth study of the internal anatomy of the liver.

The internal anatomy of the liver can be seen in a living human body through modern imaging techniques with ultrasound (USG), computed tomography (CT) or magnetic resonance imaging (MRI). The hurdle that one has to overcome on looking at the two-dimensional (2D) image is to reconstruct a three-dimensional (3D) image in one's mind. This is now overcome by the use of 3D CT and MRI images.

1.3 Ligaments of the Liver

The falciform ligament is a sickle-shaped fold, consisting of two closely applied layers of peritoneum which connects the liver to the diaphragm and to the supra-umbilical part of the anterior abdominal wall. The ligamentum teres runs on its free edge, together with the small paraumbilical veins. At the upper end the two layers of the falciform ligament separate from each other (Fig. 1.5).

On the right, it forms the upper layer of the coronary ligament, which continues inferiorly to form the right triangular ligament, then the lower layer of the coronary ligament. In between these ligaments is the bare area of the liver. At its left extremity, the lower layer of the coronary ligament passes in front of the lower end of the groove for the inferior vena cava and becomes continuous with the line of peritoneal reflexion from the right border of the caudate lobe (Fig. 1.6). This paragraph is reprinted from: Lunan Yan, Operative Techniques in Liver Resection. Springer, 2016. ISBN 978-94-017-7409-3.

On the left, the falciform ligament forms the anterior layer of the left triangular ligament, which turns backward to form the posterior layer. At the upper end of the fissure for the ligamentum venosum, it becomes the anterior layer of the lesser omentum. The posterior layer of the lesser omentum is the line of reflexion of the peritoneum from the upper end of the right border of the caudate lobe. This layer then goes around the caudate lobe to join the lower layer of the coronary ligament (Figs. 1.6 and 1.7). This paragraph is reprinted from: Lunan Yan, Operative Techniques in Liver Resection. Springer, 2016. ISBN 978-94-017-7409-3.

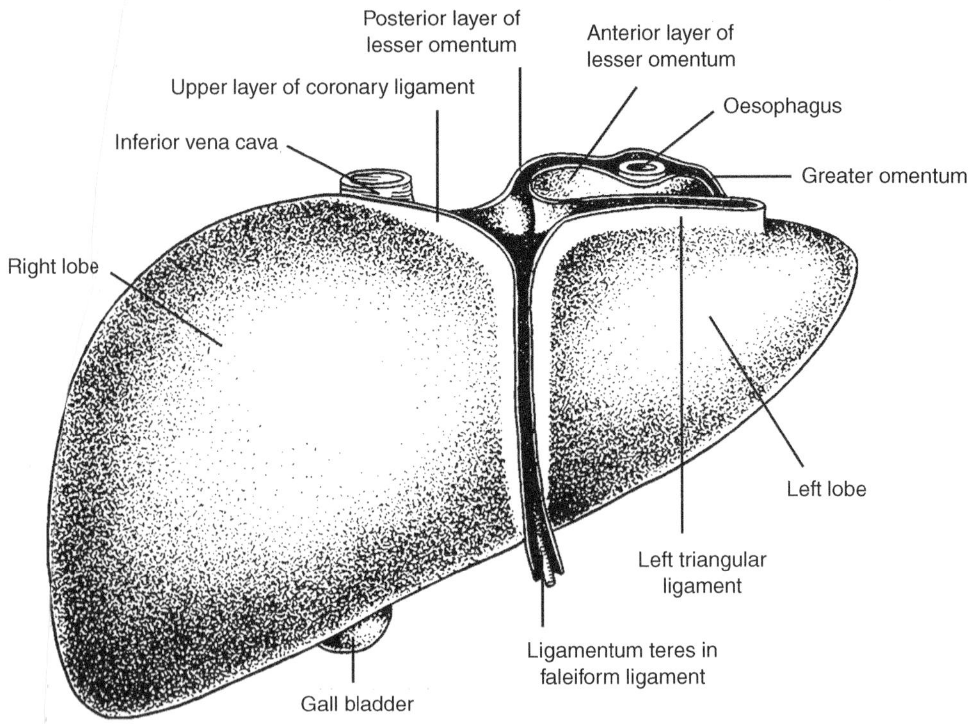

Fig. 1.5 Diaphragmatic surface of the liver and its ligaments

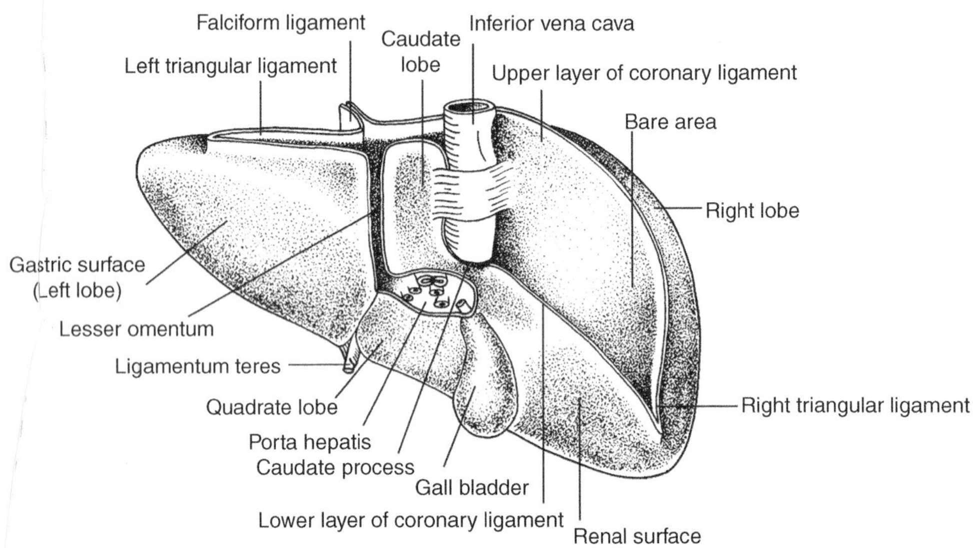

Fig. 1.6 Posterior surface of the liver and its ligaments

1.3.1 Clinical Applications

1. Division of all the ligaments that connect the liver to the diaphragm and the anterior abdominal wall (i.e. division of the falciform ligament, ligamentum teres, coronary ligament, right triangular ligament and left triangular ligament) leaves the liver attached to the body by three structures: the porta hepatis, the major hepatic veins (right, middle/left trunk) and the short hepatic veins which run directly from the inferior vena cava to the liver.
2. In the division of the lesser omentum, particular attention should be paid to avoid damaging the anterior and posterior vagus nerves and their gastric branches, and the biliary branch of the anterior vagus nerve. In patients with the left hepatic artery arising from the left gastric artery, the left hepatic artery may inadvertently be divided if the anomaly is not looked for.

1.4 Shapes of the Liver

The liver comes with many shapes (Figs. 1.8 and 1.9). The average weight is 1500 g and it receives 1500 mL of blood/min.

The liver has good regenerative power. Atrophy in one part of the liver can result in an atrophy–hypertrophy complex with the normal principal plane between the right and left hemilivers rotated either clockwise or anticlockwise depending on which part of the liver hypertrophies.

1.4.1 Clinical Applications

1. Although it is often said that the right hemiliver is larger than the left hemiliver in the ratio of 55:45, how much of the liver is really going to be resected in a hemihepatectomy would depend on the shape of the liver and its size. A CT volumetric study helps to determine the expected

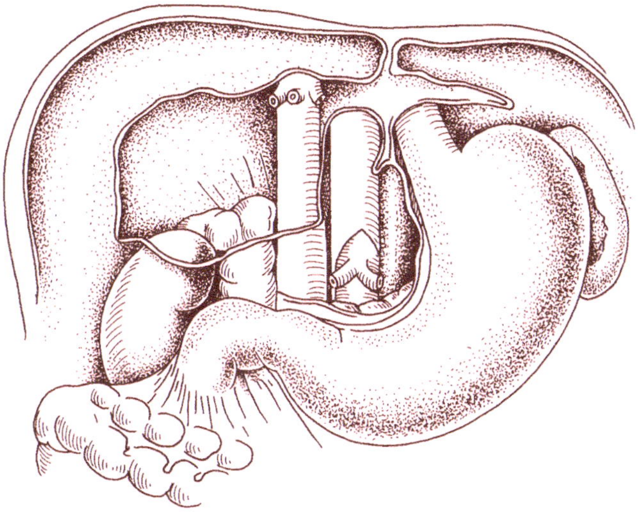

Fig. 1.7 Ligaments and bare area left after removal of the liver

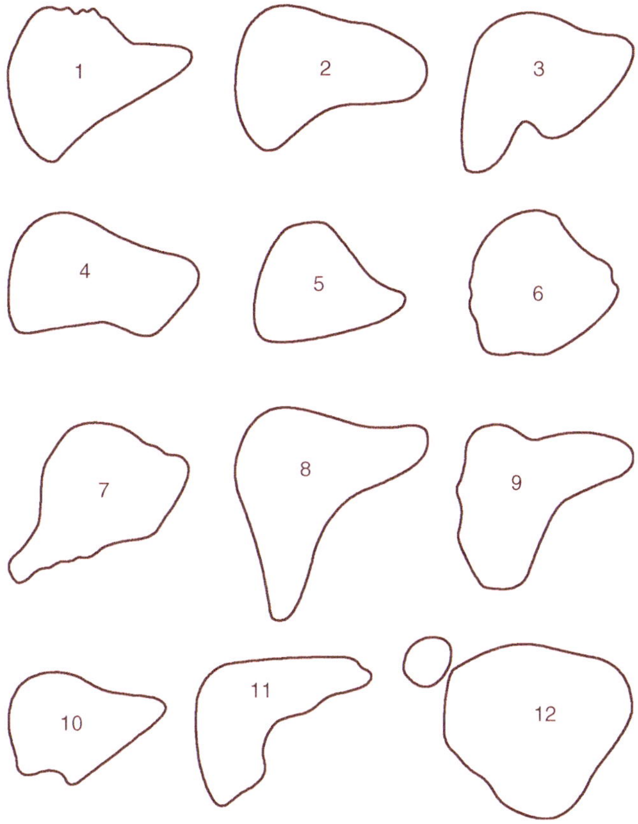

Fig. 1.8 Normal variations in the shapes of the liver as shown on CT scans

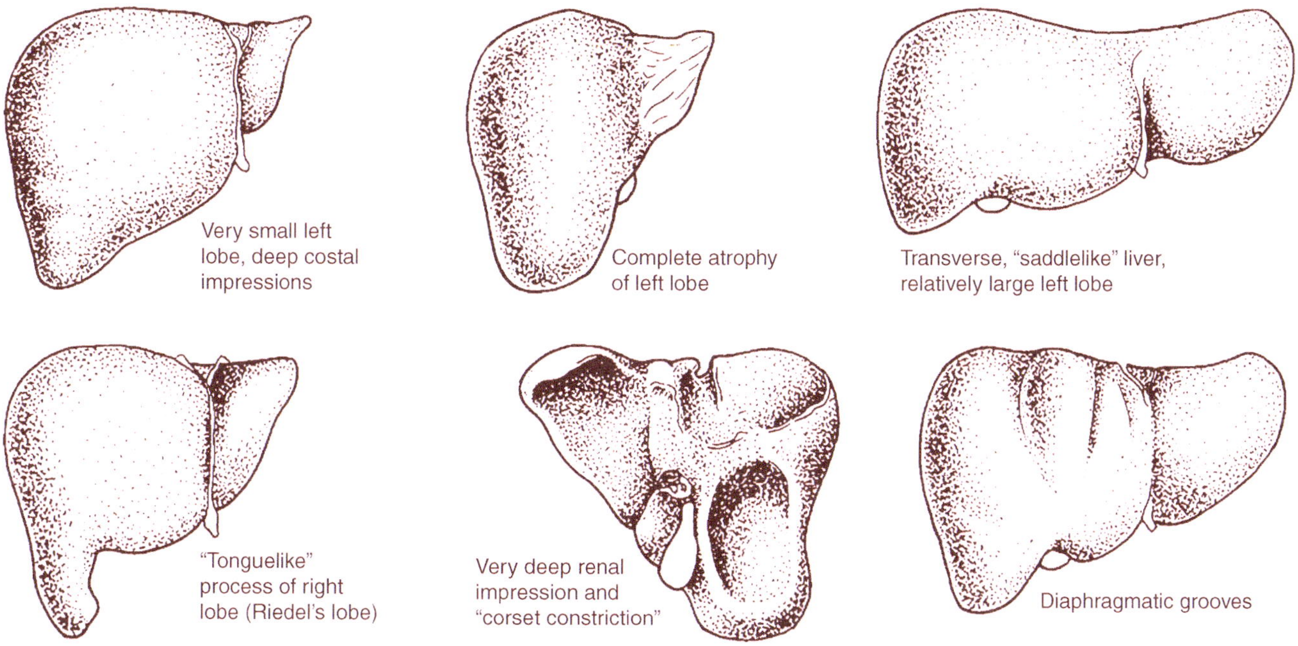

Fig. 1.9 Variations in form of liver on gross examination

liver volume to be resected and the volume of the future liver remnant (FLR) to be left behind.

2. Similarly for living-related liver transplantation, CT volumetric is necessary to determine the exact volume of liver to be transplanted to the recipient.
3. In patients with atrophy–hypertrophy complex, the axis of the liver can be rotated clockwise or anticlockwise, and the left hemiliver may be larger than the right hemiliver.

Further Reading

Jamieson G, Launois B. Chapter 2: Liver resections and liver transplantation: the anatomy of the liver and associated structures. In: Jamieson GG, editor. The anatomy of general surgical operation. 2nd ed. Edinburgh: Elsevier Churchill Livingstone; 2006. p. 8–23.

Lau WY. The history of liver surgery. J R Coll Surg Edinb. 1997;42:303–9.

Hemilivers, Sections (Sectors), Segments

2

2.1 Concept of Liver Sections

The concept of segmental anatomy of the liver was first described by Francis Glisson from Cambridge in 1654. This was to be the foundation of modern liver surgery but was forgotten for 300 years. Rex in 1888 reported on the 'new' arrangements of the right and left lobes of the liver and widely described the lobar anatomy. From the external gross anatomy of the liver, it is easy to understand why our forefathers divided the liver into the right and the left lobes along the plane of the falciform ligament/the umbilical fissure/the fissure for ligamentum venosum, i.e. the left limb of the H as seen on the visceral surface of the liver. This left limb of the H, together with the right limb of the H formed incompletely by the gallbladder anteriorly and the inferior vena cava posteriorly, defined the boundaries of the quadrate lobe and the caudate lobe. The transverse hilar fissure where the porta hepatis enters into the liver separates the quadrate lobe anteriorly from the caudate lobe posteriorly (Fig. 2.1).

The old concept of the liver is divided into right and left lobes by the falciform ligament was disproved by Cantlie, who studied the liver of a cadaver to determine that the main lobar fissure (= midplane of the liver) is oblique, and extends from right to left and from the visceral to the diaphragmatic surface at about a 70° angle. Thus, it was ascertained that the main division between the right and the left lobes extends from approximately the bed of the gallbladder antero-inferiorly to the right side of the inferior vena cava postero-superiorly. This work was further extended by Hjorstjo, and by Healey and Schroy, who demonstrated by using casts that the right lobe was further divided into an anterior and posterior section (which Healey and Schroy called segment), and the left lobe was divided into a medial and lateral section (which Healey and Schroy called segment) by the line of the falciform ligament. These anatomical divisions were based on topography and the intrahepatic artery/biliary duct systems (i.e. by arteriobiliary segmentation). Healey described third-order 'areas' which are referred to as 'subsegments' by others. The areas of Healey and the segments of Couinaud

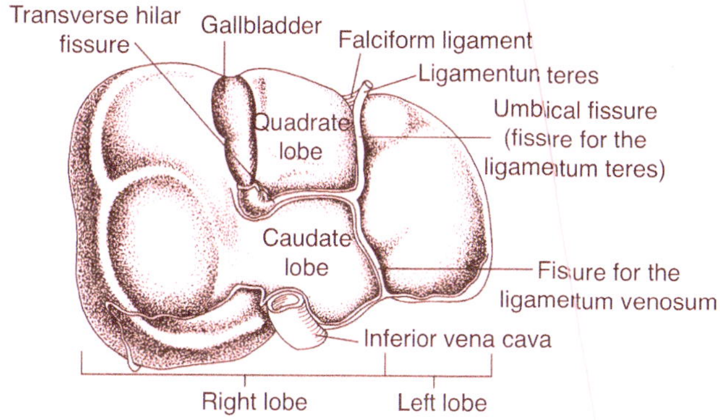

Fig. 2.1 Old concept of the liver lobes

seem to correspond with one exception: Couinaud described a single segment IV to describe the mass of liver tissue between the midplane and the umbilical fissure, but Healey described two 'areas'. It has generally been agreed upon that there is only one anatomical mass of tissue in this portion of the liver and that subdividing it further as proposed by Healey and Schroy is artificial. Some surgeons have divided segment IV into IVa and IVb. Although they recognise that the division is artificial, they found it useful to describe the location of lesions within segment IV (Fig. 2.2).

2.1.1 Clinical Applications

An application of this more-refined anatomy led to the development of surgical planes of the liver and to the recommended action of using these intersectional planes for performing liver resections (Fig. 2.3). The first left lateral sectionectomy was performed by Keen in 1899. At that time, because of the lack of knowledge, the operation was considered as a left hepatic lobectomy. This operation is also called by some authors as a left lateral segmentectomy (Healey's left lateral segment) which we now know involves the resection of two Couinaud liver segments. In this procedure carried out for cancer of the liver, blood vessels were tied

W. Y. Lau, *Applied Anatomy in Liver Resection and Liver Transplantation*, https://doi.org/10.1007/978-981-16-0800-1_2

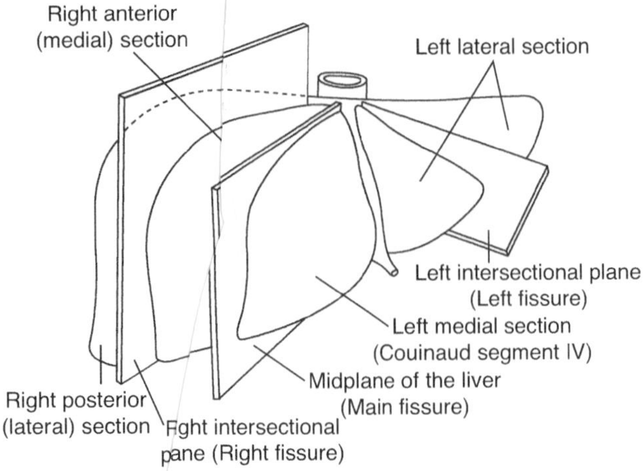

Fig. 2.2 Newer concept of lobes and sections

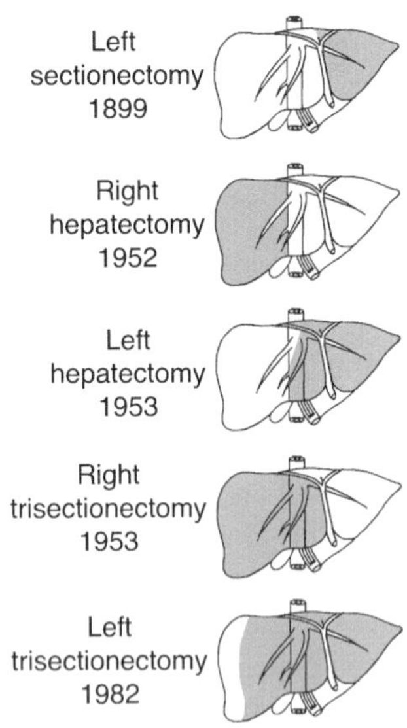

Fig. 2.3 Liver resections based on liver sections

individually with catgut intrahepatically and cautery was applied. In 1948, Raven reported a left lateral sectionectomy (called by him only as partial hepatectomy) for metastatic colon carcinoma. An anatomical resection was carried out in which the triangular and coronary ligaments were divided, branches of the left portal vein, the left hepatic artery and left hepatic duct were ligated within the hepatoduodenal ligament. The left hepatic vein was then isolated extrahepatically and divided, following which the liver parenchyma was transected. Lortat-Jacob and Robert advanced the procedure by performing a right hepatectomy (which was called a right hepatic lobectomy) in 1952, using a technique designed to control haemorrhage with ligation of the blood vessels and bile ducts to the right liver in the hepatoduodenal ligament followed by extrahepatic ligation of the right hepatic vein prior to transection of the liver parenchyma. Seneque and his associates reported in 1953 a left hepatectomy for hydatid cysts. In the same year, Quattlebaum carried out the first right trisectionectomy. The procedure of the left trisectionectomy took much longer to realise. Although Couinaud in 1957 described the anatomical basis for this operation, Leslie Blumgart in 1978 and Joishy and Balasegaram in 1980 reported on the operation of extended left hepatectomy with incomplete resection of the right anterior section of the liver. It was not until Starzl et al. in 1982 described the necessary hilar dissection that an anatomical left trisectionectomy was performed in four patients.

2.2 Concept of Liver Sectors and Segments

The concept of functional liver anatomy based on the distribution of the portal pedicles and the location of the hepatic veins (portal segmentation), instead of Healey's arteriobiliary segmentation, evolved from Couinaud's study of casts made by plastic injection into portal and hepatic veins followed by corrosion of the surrounding parenchyma. The liver is divided by the three hepatic veins into sectors (called suprahepatic segmentation by Couinaud). The middle hepatic vein runs in the main scissura (= midplane of the liver) which divides the liver into the right and the left livers (or hemilivers). On the right side, the right hepatic vein runs in the right scissura (= right fissure, or right intersectoral plane) which divides the right liver into the right anterior sector (= right paramedian sector) and the right posterior sector (= right lateral sector). It should be noted that in the right liver, the Healey's liver sections (he called them segments) are exactly the same as the Couinaud's sectors. On the left side, the left hepatic vein runs in the left scissura (= left fissure) which divides the left liver into a left medial sector (= left paramedian sector) and a left lateral sector (= left posterior sector). Thus, in the left liver, the Healey's liver sections (he called them segments) are not the same as the Couinaud's sectors. Couinaud further subdivided the liver into eight segments (subhepatic segmentation) by using the branches of the portal vein (Fig. 2.4). This paragraph is reprinted from: Lunan Yan, Operative Techniques in Liver Resection. Springer, 2016. ISBN 978-94-017-7409-3.

In the right liver, as section is the same as sector, the right anterior section (= sector) can be divided into segment VII superiorly and segment V inferiorly. The right posterior section (= sector) consists of segment VII superiorly and segment VI inferiorly.

In the left liver, section is not the same as sector. The left medial section lies between the main scissura (= main fissure, or the midplane of the liver) and the falciform ligament, and it consists of only the segment IV, while the left lateral section consists of segments III and II, being separated by

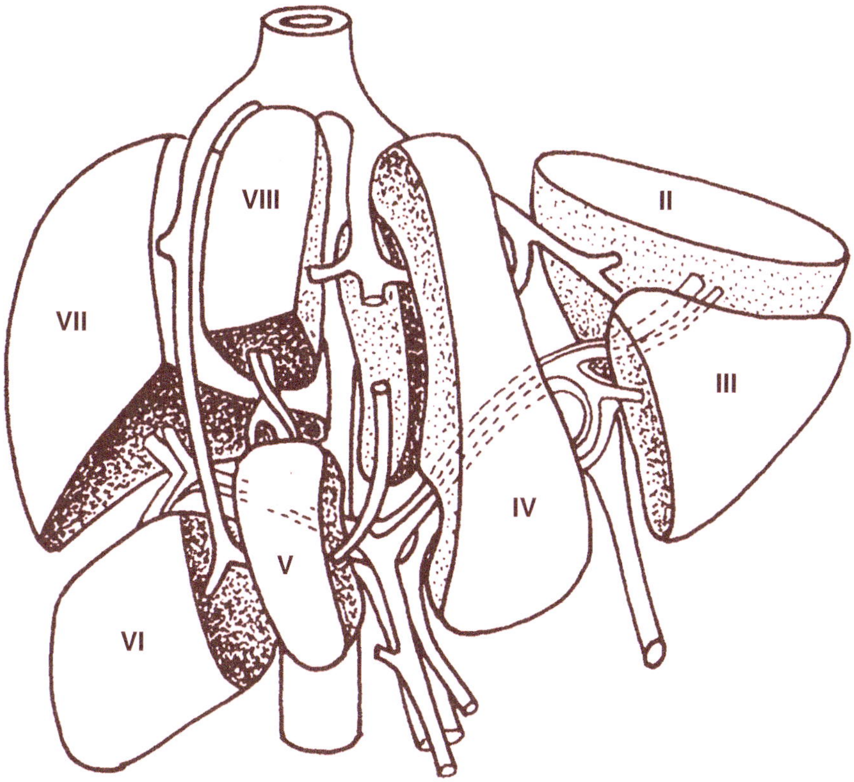

Fig. 2.4 Concepts of liver sectors and segments

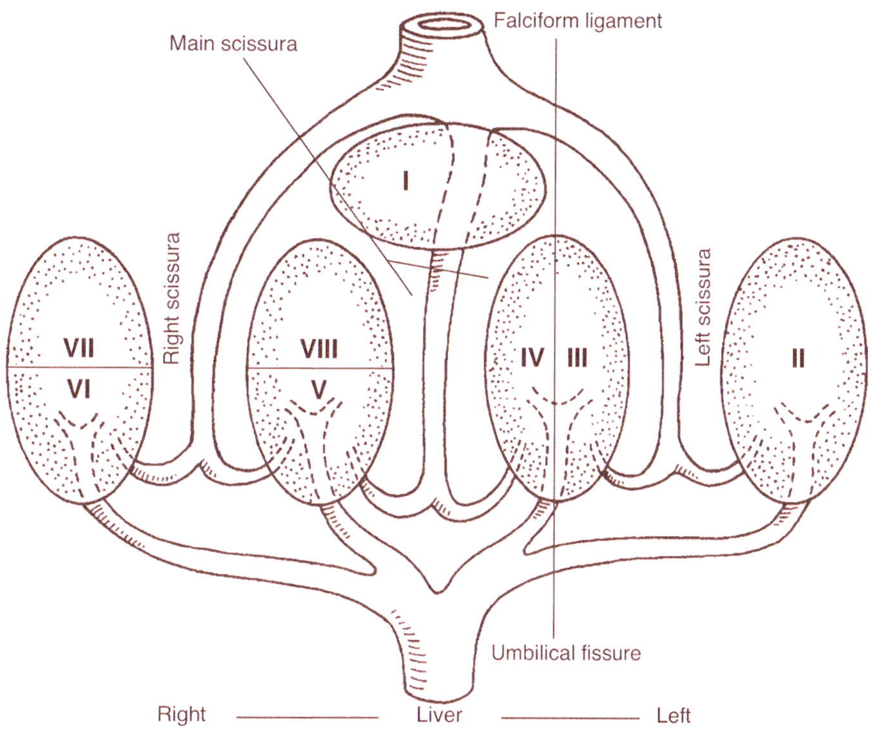

Fig. 2.5 Liver sectors and segments

the left hepatic vein which runs in the left scissura (= left fissure). For the left medial sector, it consists of segments III and IV, lying between the middle hepatic vein in the main scissura, and the left hepatic vein in the left scissura. The falciform ligament/umbilical fissure divides the liver segment IV from III. The left lateral sector, which lies on the left of the left hepatic vein, consists of liver segment II only (Fig. 2.5).

The liver segment I is almost the same as the caudate lobe in both the Healey's arteriobiliary, and the Couinaud's portal segmentations (See Chap. 3, Caudate Lobe).

Our readers of this book must be very clever if up to this point that we have failed to confuse him/her with all the terminologies used by the different anatomists who tried to divide the liver in their own ways, using different terminologies while they do so. The American terminology is largely that proposed by Healey, while the European terminology by Couinaud. It must be clearly pointed out that the Healey's segment is not the same as Couinaud's segment; and the term 'section' which we use to represent Healey's segment can be the same, or different from, Couinaud's sector. To add things more confusing, Takasaki, a Japanese surgeon, divided the liver into three almost equal parts, and he called these parts the right segment (equivalent to Couinaud segments 6, 7, or the right posterior sector), the middle segment (Couinaud segments 5, 8, or the right anterior sector), and the left segment (Couinaud segments 4, 3, 2, or the left medial and lateral sectors, or the left hemiliver) (Fig. 2.6). Furthermore, the term 'lobes' may have different meanings to different people. On the other hand, there are many terms that have been used to mean one thing, e.g. the midplane of the liver which divides the liver into the right and the left hemilivers, can also be called Cantlie's line, midline, principal plane, main scissura, main fissure, main sulcus, main portal scissura (by Couinaud) and interlobar plane (in American terminology).

It is therefore desirable to have a uniform, internationally agreed-upon terminology for liver anatomy and liver resection.

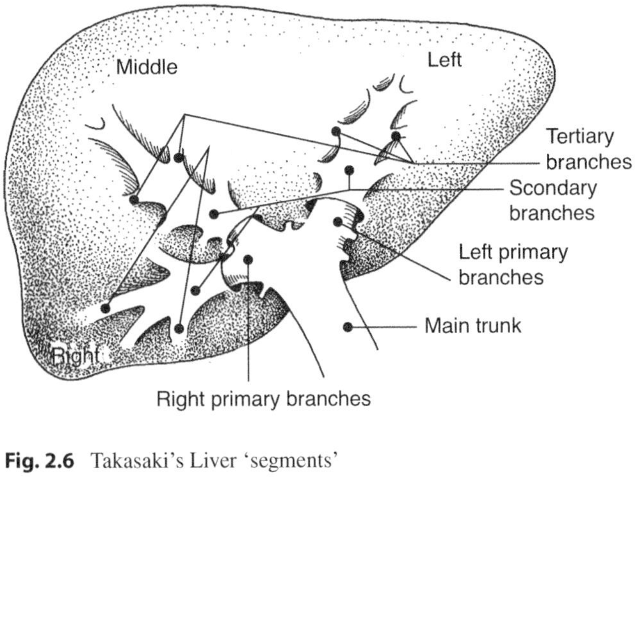

Fig. 2.6 Takasaki's Liver 'segments'

2.3 The Brisbane 2000 Terminology of Liver Anatomy and Resections

The Scientific Committee of the International Hepato-Pancreato-Biliary Association (IHPBA), at a meeting held in Berne, Switzerland in December 1998 decided to create a Terminology Committee of international experts to deal with the confusion in the nomenclature of hepatic anatomy and liver resections. A terminology was sought which was anatomically correct in which anatomical and surgical terms agreed, and which was consistent, self-explanatory, linguistically correct, translatable, precise and concise. After 18 months, the Committee presented a terminology which was endorsed by the IHPBA at the World Congress of the IHPBA held in Brisbane, Australia, in May 2000 (Fig. 2.7). This paragraph is reprinted from: Lunan Yan, Operative Techniques in Liver Resection. Springer, 2016. ISBN 978-94-017-7409-3.

To summarise this Terminology, the liver is divided into two parts: the main liver, and the caudate lobe (called dorsal

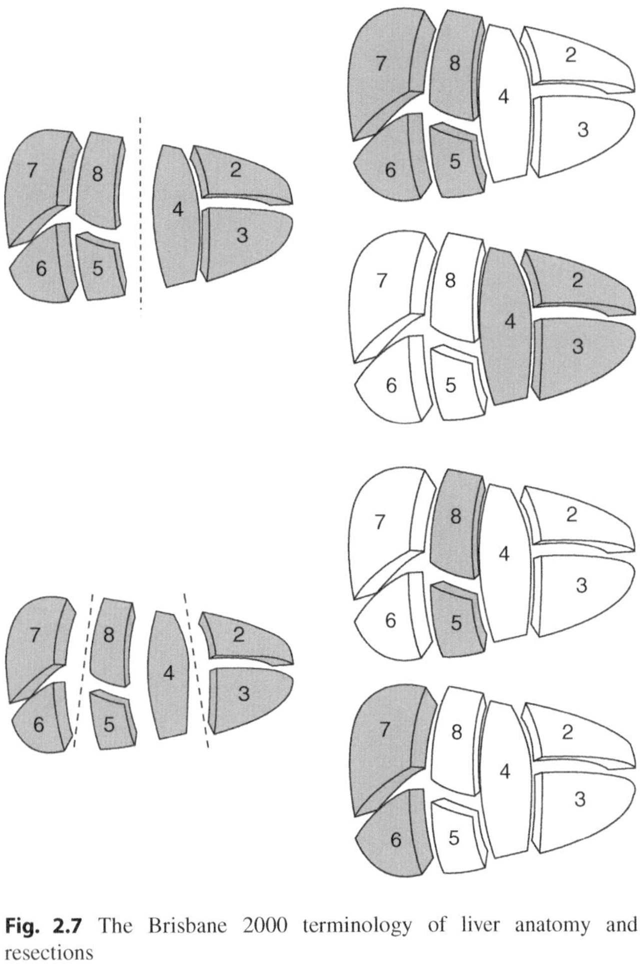

Fig. 2.7 The Brisbane 2000 terminology of liver anatomy and resections

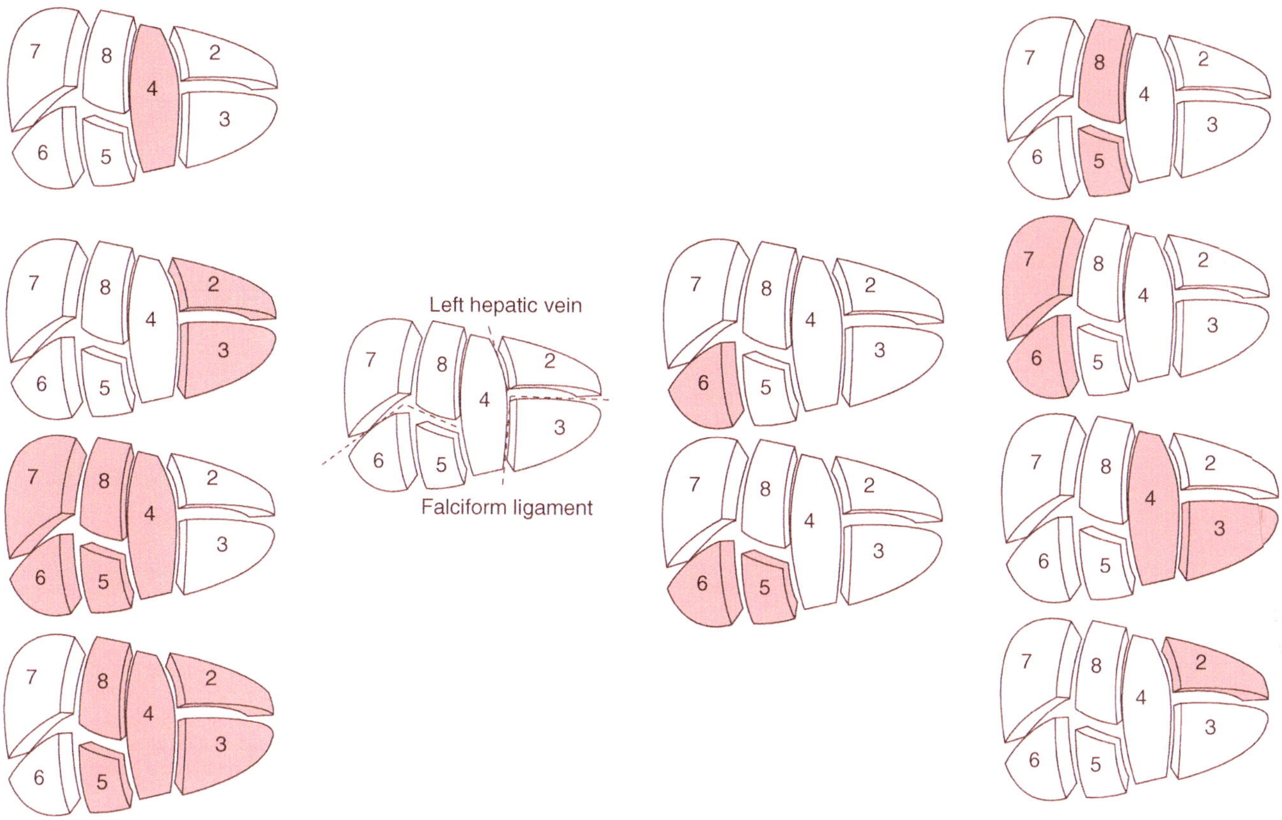

Fig. 2.7 (continued)

sector by Couinaud). There are still some controversies on the terminology of the caudate lobe or the dorsal sector as called by Couinaud (see Chap. 3 Caudate Lobe). This paragraph is reprinted from: Lunan Yan, Operative Techniques in Liver Resection. Springer, 2016. ISBN 978-94-017-7409-3.

The main liver is divided by three orders of division into the hemilivers (or livers), sections and segments, respectively. Each segment is an independent unit, with a separate arteriobiliary and portal venous supply, and a separate hepatic venous drainage. Thus, each segment can be resected individually, or together with an adjacent segment. This paragraph is reprinted from: Lunan Yan, Operative Techniques in Liver Resection. Springer, 2016. ISBN 978-94-017-7409-3.

The first-order division which separates the right and the left liver is a plane that intersects the gallbladder fossa and the fossa for the IVC and is referred to as the midplane of the liver. Within this plane runs the middle hepatic vein. This paragraph is reprinted from: Lunan Yan, Operative Techniques in Liver Resection. Springer, 2016. ISBN 978-94-017-7409-3.

Anatomical term	Couinaud segments referred to	Term for surgical resection	Diagram (pertinent area is in grey colour)
First-Order Division			
Right hemiliver **OR** *Right liver*	Sg 5-8 (+/–Sg 1)	*Right hepatectomy* **OR** *Right hemihepatectomy (stipulate +/– segment 1)*	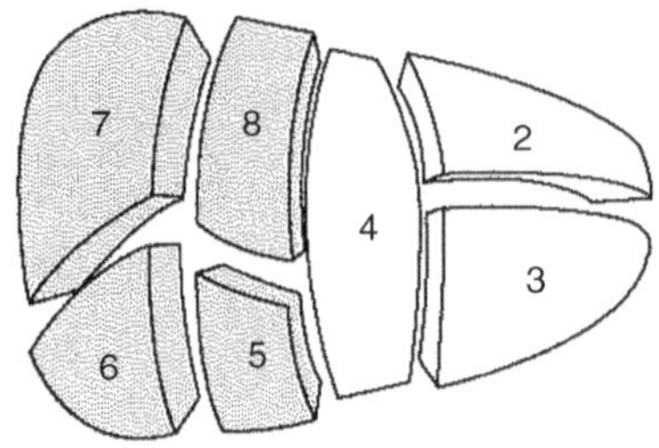
Left hemiliver **OR** *Left liver*	Sg 2-4 (+/–Sg 1)	*Left hepatectomy* **OR** *Left hemihepatectomy (stipulate +/– segment 1)*	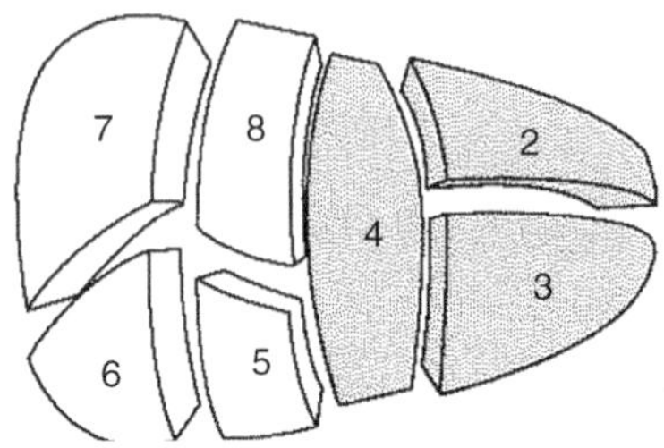

Second-Order Division

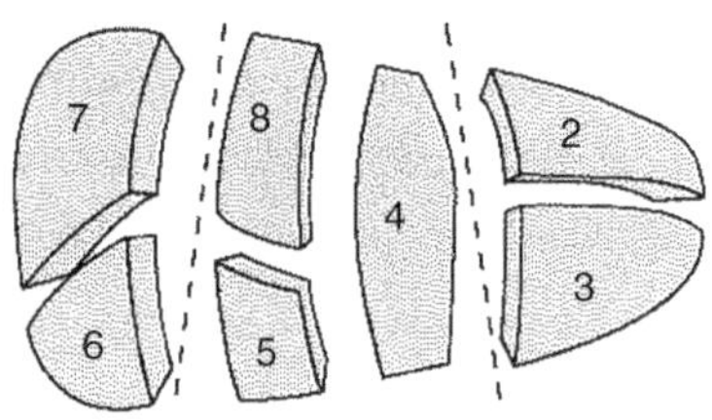

The second order divisions are the right and left intersectional planes. The right intersectional plane has no surface marking, but in this plane runs the right hepatic vein. The left intersectional plane passes through the umbilical fissure and the attachment of the falciform ligament.

Anatomical term	Couinaud segments referred to	Term for surgical resection	Diagram
Right anterior section	Sg 5,8	*Add ' -ectomy' to any of the anatomical terms as in Right anterior sectionectomy*	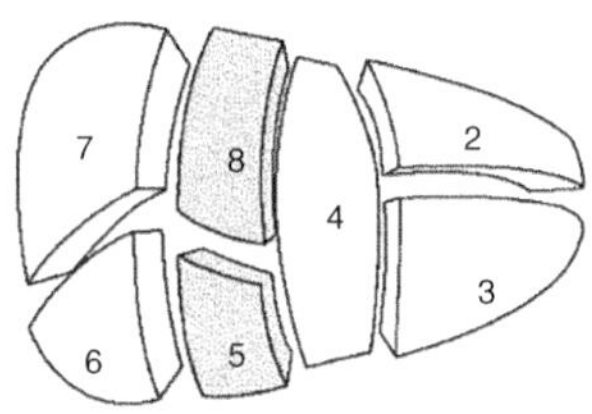
Right posterior section	Sg 6,7	*Right posterior sectionectomy*	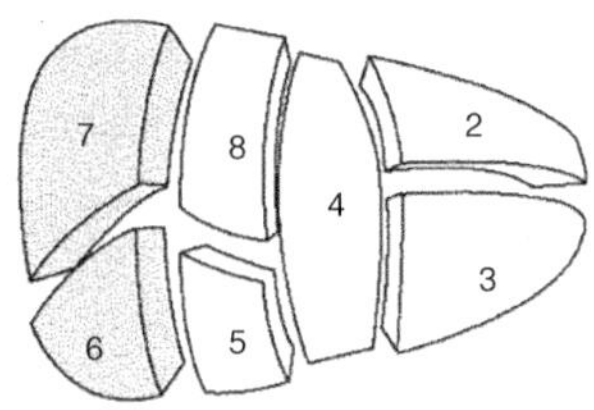
Left medial section	Sg 4	*Left medial sectionectomy* **OR** *Resection segment 4 (also see under third order)* **OR** *Segmentectomy 4 (also see under third order)*	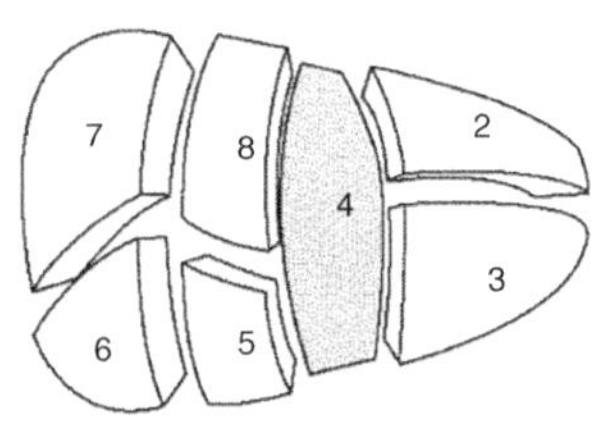

Left lateral section	Sg 2,3	*Left lateral sectionectomy* **OR** *Bisegmentectomy 2, 3 (also see under third order)*	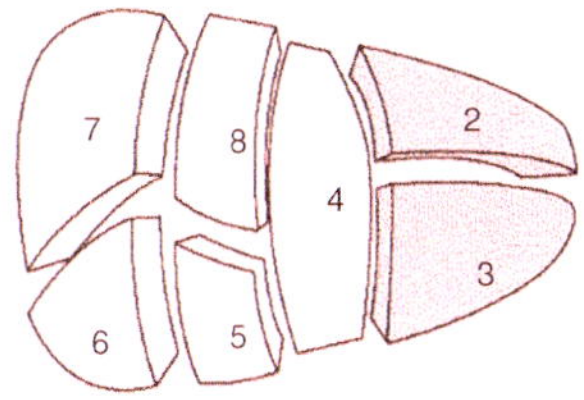
Right hemiliver plus left medial section	Sg 4-8 (+/– Sg 1)	*Right trisectionectomy* **or** *Extended right hepatectomy* **or** *Extended right hernihepatectomy*	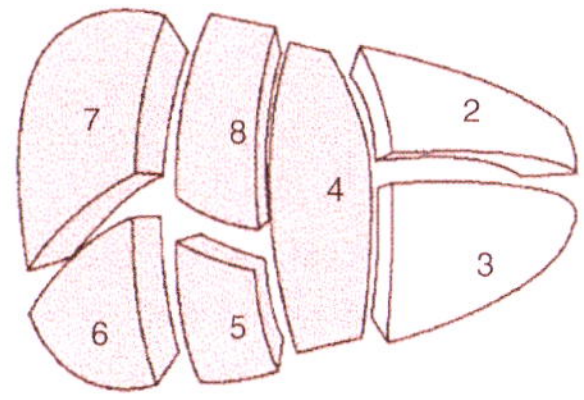
Left hemiliver plus right anterior section	Sg 2-5, 8 (+/– Sg 1)	*Left trisectionectomy* **or** *Extended left hepatectomy* **or** *Extended left hemihepatectomy*	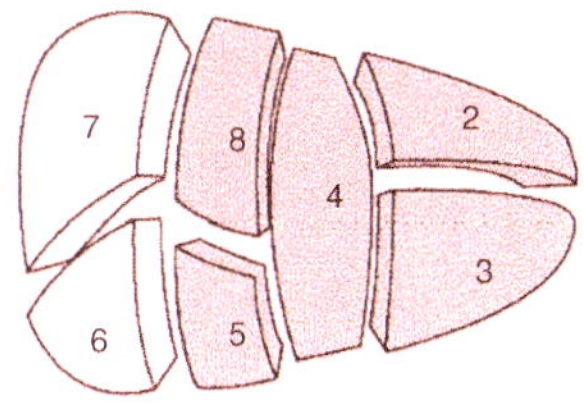

Third-Order Division

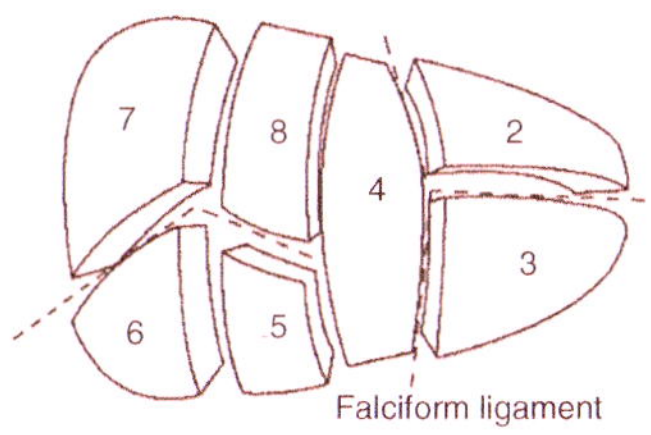

The third order division is the planes between the liver segments and they are referred to as intersegmental planes. There are no surface markings, and no major structures run in these planes. Please note that the left hepatic vein runs between segments 2 and 4 and then segments 2 and 3.

Anatomical term	Couinaud segments referred to	Term for surgical resection	Diagram
Segments 1-9 (Segmentectomy 6)	Any one of Sg 1 to Sg 9	*Segrnentectomy 6*	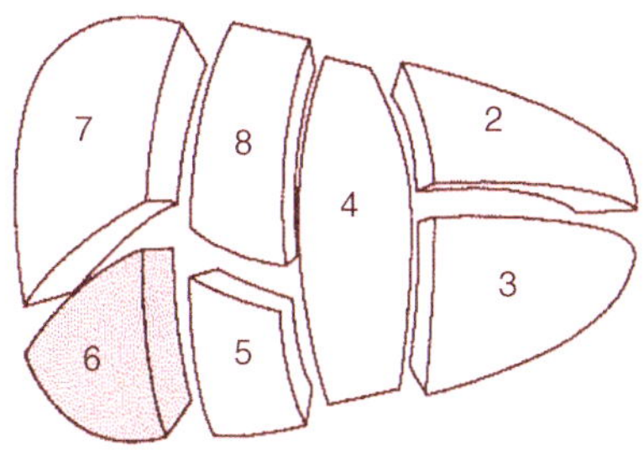
2 contiguous segments	Any two of Sg 1 to Sg 9 in continuity	*Bisegrnentectorny 5, 6*	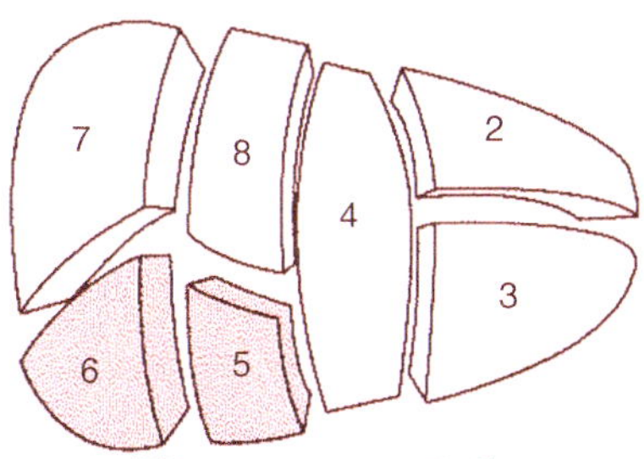 (Bisegrnentectomy 5, 6)

Note: The above four tables are reprinted with permission from: S.M. Strasberg, J. Belghiti, P.-A. Clavien, E. Gadzijev, J.O. Garden, W.-Y. Lau, M. Makuuchi, R.W. Strong. "The Brisbane 2000 Terminology of Liver Anatomy and Resections", HPB, 2000.

Please note that the above terminology is based on Healey's second-order division (arteriobiliary segmentation).

An alternative but equally correct terminology for the second-order division is contained in the addendum to the Brisbane 2000 Terminology. This addendum is based on Couinaud's concept of portal segmentation. This addendum is included because it preserves the ability to name certain uncommon resections on the left side according to Couinaud's concepts of the portal and hepatic vein, e.g. left paramedian sectorectomy (i.e. resection of liver segments 3 and 4).

Alternative second-order division (second-order division based on portal vein)

Anatomical term	**Couinaud segments referred to**	**Term for surgical resection**	**Diagram**
Right anterior sector **OR** *Right p aramedian sector*	Sg 5,8	*Add '-ectomy' to any of the anatomical terms, as in* *Right anterior sectorectomy* **OR** *Right paramedian sectorectomy*	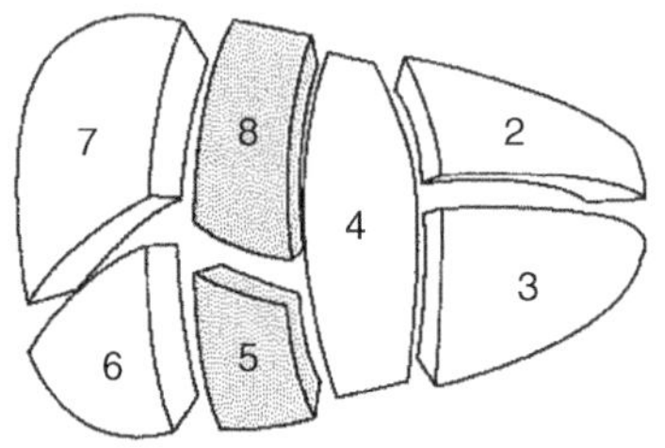
Right posterior sector **OR** *Right lateral sector*	Sg 6,7	*Right posterior sectorectomy* **OR** *Right lateral sectorectomy*	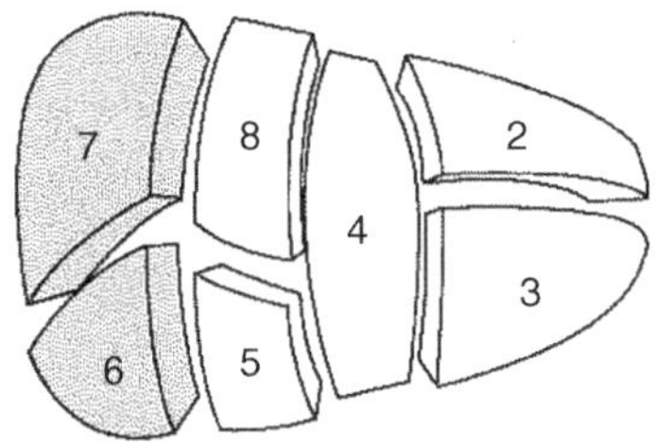
Left medial sector **OR** *Left paramedian sector*	Sg 3,4	*Left medial sectorectomy* **OR** *Left paramedian sectorectomy* **OR** *Bisegmentectomy 3, 4*	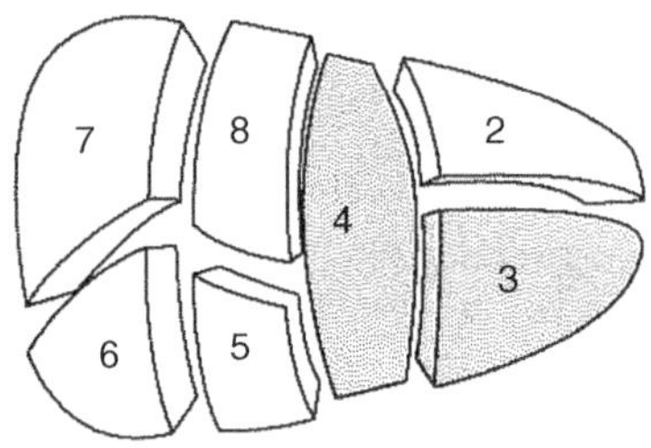
Left lateral sector **OR** *Left posterior sector*	Sg 2	*Left lateral sectorectomy* **OR** *Left posterior sectorectomy* **OR** *Segmentectomy 2*	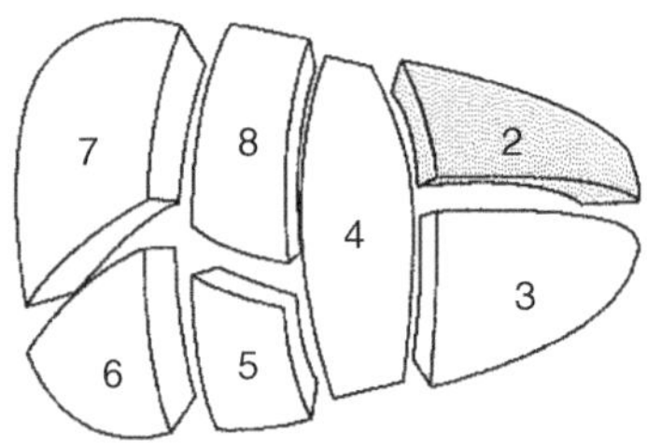

Addendum

Right anterior sector and right anterior section are synonyms. Right posterior sector and right posterior section are synonyms.

Left medial sector and left medial section are NOT synonyms and are NOT exchangeable terms. They do not describe the same anatomic areas.

Left lateral sector and left lateral section are NOT synonyms and are NOT exchangeable terms.

Borders or watersheds:

The border or watersheds of second-order division based on the portal vein are referred to as right and left intersectoral planes. These have no surface markings. This paragraph is reprinted with permission from: S.M. Strasberg, J. Belghiti, P.-A. Clavien, E. Gadzijev, J.O. Garden, W.-Y. Lau, M. Makuuchi, R.W. Strong. "The Brisbane 2000 Terminology of Liver Anatomy and Resections", HPB, 2000.

Additional points that are important in this Brisbane 2000 Terminology are:

1. Couinaud segments are indicated in short forms as Sg 1–9 (e.g. Sg 6). Sg is chosen rather than S, to avoid confusion of segment with section or sector. Arabic numerals are chosen rather than Roman numerals because many non-western nations do not use Roman numerals. This paragraph is reprinted with permission from: S.M. Strasberg, J. Belghiti, P.-A. Clavien, E. Gadzijev, J.O. Garden, W.-Y. Lau, M. Makuuchi, R.W. Strong. "The Brisbane 2000 Terminology of Liver Anatomy and Resections", HPB, 2000.
2. Wherever the word ('OR' uppercase, bold) appears in the table, it indicates equally acceptable terminology, e.g. 'right hemiliver "OR" right liver'. The choice is that of the user. Wherever the word 'or' (lowercase, bold) appears in the table, it indicates that the first choice is preferred but that the second is acceptable, e.g. 'right trisectionectomy "or" extended right hepatectomy'. The choice is up to the user; but the first term is the preferred term. The reason in this case is as follow: as some use the adjective 'extended' to indicate any degree of extension of a resection over the midplane (which in some cases, is less than a whole section), the terms in the table containing the word 'extended', which acceptable, are less preferred. This paragraph is reprinted with permission from: S.M. Strasberg, J. Belghiti, P.-A. Clavien, E. Gadzijev, J.O. Garden, W.-Y. Lau, M. Makuuchi, R.W. Strong. "The Brisbane 2000 Terminology of Liver Anatomy and Resections", HPB, 2000.

2.3.1 Clinical Applications

1. A unified terminology has been advocated by the IHPBA. Hopefully, with wide-spread acceptance and usage, the confusion caused in the nomenclature of hepatic anatomy and liver resections can be avoided in the future.
2. As each liver segment can be resected individually and in combination, liver resection can be segment based.
3. For segment 4, some authors suggest that the quadrate lobe is only the anterior part of segment 4, calling it segment 4b, although there is no particular reason why the whole segment should not be called the quadrate lobe. However, there is a practical point here. If the area of the quadrate lobe on the visceral surface of the liver is resected vertically, then not all of segment 4 is removed. On the diaphragmatic surface of the liver, segment 4 extends posteriorly to the region of the inferior vena cava. The studies by Couinaud in 1957 suggested that there was no useful purpose in removing the posterior part of segment 4 (segment 4a). In fact, there are several reasons why segment 4a is often not resected: (a) segment 4a represents less than 20% of segment 4; (b) it is the most difficult part to resect. Draining into the middle hepatic vein are several moderate-sized veins which transverse segment 4a, and there is no clear demarcation between segments 1 and 4; (c) in some operations, conservation of functioning liver tissue may be important; (d) if segment 4 is being resected to gain access to the confluence of the bile duct (see Sect. 4.3), it is not necessary to resect segment 4a; (e) resecting segment 4a is putting the blood supply and/or biliary drainage of segments 2 and 3 at considerable risk. The pedicles to segment 4 have more variations than any other segments in the liver. Thus, the portal pedicles commonly are between three and ten in numbers and maybe many more. The arterial and biliary pedicles are even more variable (see Chap. 6) and when the structures are dissected individually for the removal of the whole segment 4, the vasculobiliary supply to segments 2 and 3 can be compromised.
4. The caudate lobe is referred to by Couinaud as the dorsal sector. To some clinicians, the caudate lobe is equivalent to segment 1, thus making the liver have a total of eight liver segments only. To other clinicians, the caudate lobe consists of two liver segments, segment 1 (equivalent to Spigelian lobe) and segment 9 (equivalent to paracaval portion and caudate process), making a total of nine liver segments. Although some clinicians named the paracaval portion of the caudate lobe as segment 9 and the caudate process segment 10 to make a

total of ten liver segments because of their independent blood supplies, this view is generally considered to be too extreme to be acceptable.

2.4 Terms with the Same Meaning but in Different Terminologies

We hope that Table 2.1 helps to clarify some of the confusion caused by different anatomists who used different terms to describe the same anatomical part of the liver. It is hoped that the Brisbane 2000 Terminology of Liver Anatomy and Liver Resections will be adopted by anatomists and surgeons all around the world.

2.4.1 Commonly Used Terms

There are several terms that are quite commonly encountered but have not been defined clearly:

2.4.1.1 The Umbilical Fissure

This fissure separates segment 3 from segment 4 and it is marked morphologically on the inferior liver surface at the point where the falciform ligament reaches the anterior border of the liver. There usually is a bridge of liver tissue over the ligamentum teres as it lies in the depths of this fissure. On the superior surface of the liver, the fissure lies deep to the line of attachment of the falciform ligament.

There is often a vein of some surgical importance which lies in this fissure-best called the umbilical portion of the left portal vein to avoid confusing it with the umbilical vein which becomes obliterated after birth to form the ligamentum teres (see Sect. 6.9).

2.4.1.2 The Fissure Venosum

This fissure is also marked morphologically on the liver surface and it is the continuation of the umbilical fissure on the posterior surface of the liver. The lesser omentum passes into the fissure and in its depth lies a fibrous cord that connects the left branch of the portal vein with the trunk of the middle/

Table 2.1 Different terminologies used

	Brisbane 2000	European (Couinaud)	American (Healey & Schroy)	Others
1.	Midplane	Main portal scissura, main scissura	Interlobar plane	Cantlie's line, Cantlie's fissure, midline, main fissure, main sulcus, principal plane, watershed between right and left livers
2.	Right liver, right hemiliver	Right liver, right hemiliver	Right lobe	Right lobe can be anatomical lobe, i.e. right liver = segments 4–8; or physiological lobe, i.e. right liver = segments 5–8
3.	Left liver, left hemiliver	Left liver, left hemiliver	Left lobe	Left lobe can be anatomical lobe, i.e. left liver = segments 2, 3; or physiological lobe, i.e. left liver = segments 2–4; left hemiliver = Takasaki's left liver segment
4.	Right intersectional plane, right intersectoral plane	Right portal scissura, right scissura	Right intersegmental plane	Right fissure, right sulcus
5.	Right anterior section, Sg 5, 8	Right anterior sector, right paramedian sector	Right anterior segment	Takasaki's middle liver segment
6.	Right posterior section, Sg 6, 7	Right posterior sector, right lateral sector	Right posterior segment	Takasaki's right liver segment
7.	Left intersectional plane	–	Left intersegmental plane	Umbilical fissure
8.	Left medial section, Sg 4	–	Left medial segment	Quadrate lobe
9.	Left lateral section, Sg 2, 3	–	Left lateral segment	Lateral lobe
10.	Left intersectoral plane	Left portal scissura, left scissura	–	–
11.	Sg 3, 4	Left medial sector, left paramedian sector	–	–
12.	Sg 2	Left lateral sector, left posterior sector	–	–
13.	Intersegmental plane	Plane between segments	Plane between areas	Plane between subsegments
14.	Segments, Sg 1–9	Segments I to IX	Areas	Subsegments
15.	Caudate lobe, Sg 1, 9	Dorsal sector, Sg I, IX	Caudate lobe	Spiegelian lobe + paracaval portion + caudate process
16.	Segment 9	Segment IX	–	Paracaval portion of caudate lobe + caudate process

left hepatic veins—this was the ductus venosum in foetal life. The caudate process lies behind this fissure, separating from it only by the lesser omentum. The fissure lies more or less in a coronal plane and the plane leads to the plane of the dorsal fissure (see Sect. 6.10).

2.4.1.3 The Dorsal Fissure

The dorsal fissure separates the posterior aspect of segment 4 from segment 1. It is in the same plane as the fissure venosum, so that when the surgeon places the fingers in this fissure, they point to the division between segment 1 behind and segment 4 in front (Fig. 2.8).

Couinaud calls the portion of the caudate lobe lying posterior to segment 8 and 4 the segment 9. Other people call this portion the paracaval portion and the caudate process of the caudate lobe (see Chap. 3, Caudate Lobe) (Fig. 2.9).

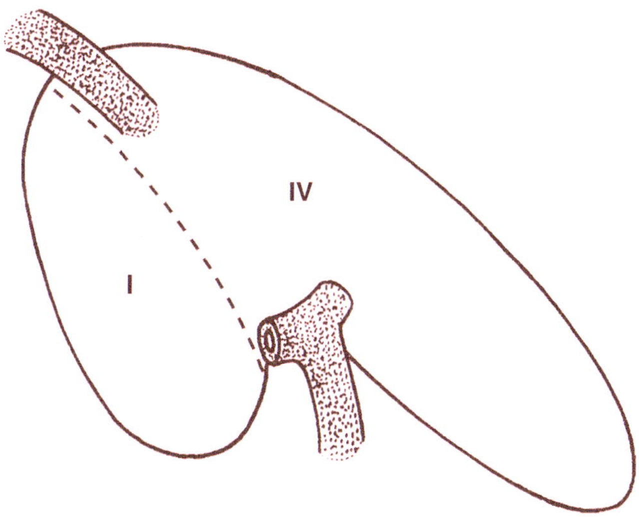

Fig. 2.8 Lateral view showing the plane of the dorsal fissure

2.4.1.4 Lobe and Segment

There are two terms that are worth further discussion: the terms 'lobe' and 'segment' are not precise, and they mean different things to different people.

For the word 'lobe', it has one meaning in the European terminology and another meaning in the American terminology. It can mean physiological right/left lobes, or anatomical right/left lobes. Segment 4 is also called the quadrate lobe. The Spigelian lobe forms part of the caudate lobe (see Chap. 3). Occasionally the lower border of the right liver to the right of the gallbladder may project downwards for a considerable distance as a broad tongue-like process called the Riedel's lobe. In my opinion, the term lobe should be abandoned. The only exception may be the caudate lobe, or the quadrate lobe when the term is used to describe segment 4b.

By definition and in common usage, a 'lobe' is a portion of an organ demarcated by fissures, sulci, connective tissues or shape. The caudate lobe falls into this category. However, the term 'caudate lobe' is still vague, and it can still have different meanings to different people (see Chap. 3). The quadrate lobe also falls into this category; however, it is better to call it segment 4b for clarity and for ease of communication.

As for the term 'segment', it has a completely different meaning in the European, American and Japanese terminologies. In the European terminology, segment means Couinaud segment. In the American terminology, a Healey's segment is equivalent to a section in the Brisbane 2000 Terminology. That explains why for the Americans, left lateral segmentec-

Fig. 2.9 The face continuous with segment 4 is the dorsal fissure

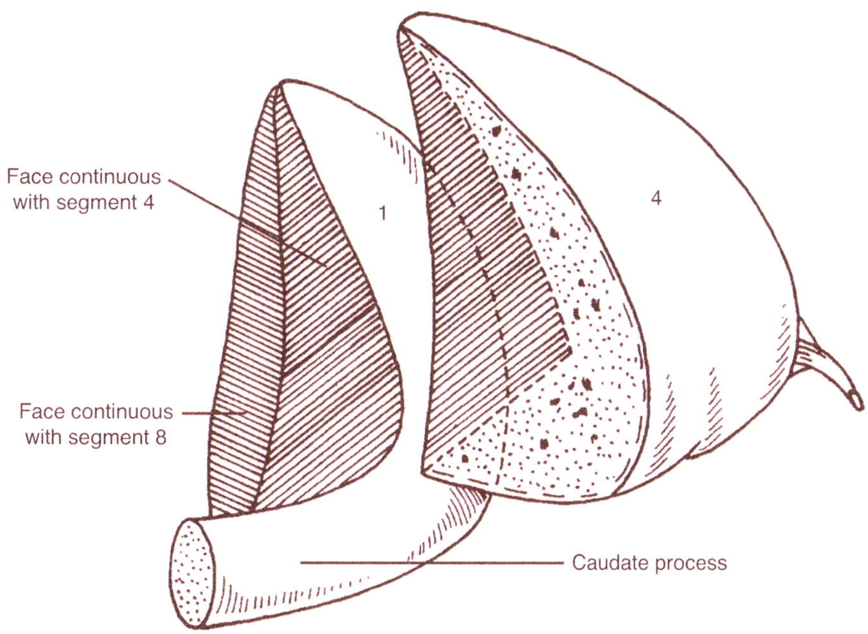

tomy is equivalent to resection of the lateral section of the liver (resection of 2 Couinaud segments, Sg 2 and 3). An American right trisegmentectomy is equivalent to the Brisbane 2000 right trisectionectomy with resection of 5 Couinaud segments, Sg 4, 5, 6, 7, 8. To add further to this chaos, Takasaki from Japan divided the liver into three almost equal parts: the right segment (equivalent to Couinaud segments 6, 7, or the right posterior sector); the middle segment (Couinaud segments 5, 8 or the right anterior sector); and the left segment (Couinaud segments 4, 3, 2 or the left medial and lateral sectors or the left hemiliver) (see Sect. 15.3.1). As Couinaud's segments have gained universal acceptance even in America and Japan, it would be better to use the term segment only to refer to the Couinaud's liver segments.

2.5 Couinaud's Portal Segmentation Versus Healey's Arteriobiliary Segmentation

There has been a lot of dispute amongst anatomists regarding the division of the portal vein on the left side. Couinaud is in disagreement with other anatomists who state that in the prevailing pattern, the left hepatic vein receives a significant tributary from segment 4 of the liver, variously called the umbilical or scissural vein (see Chap. 7). As a result, according to these anatomists, the main left hepatic vein must lie in a plane passing through the umbilical fissure and the attachment of the falciform ligament, i.e. Healey's left intersectional plane. Consequently, they argue that the position of the left hepatic vein cannot be used as the rationale for a second-order watershed plane lying between segments 2 and 3 as Couinaud advocated. In this view, the left hepatic vein is seen to have an upstream portion that lies between segments 2 and 3 and this portion is really in a third-order watershed. Couinaud argues that the branch to segment 2 and the umbilical portion of the portal vein are the two terminal branches of the transverse portion of the portal vein. He drew this conclusion from a corrosive cast in which the ligamentum venosum had been digested away. However, in the intact liver in which the ligamentum venosum is available to demarcate the transition from transverse to umbilical portions of the left portal vein, it becomes clear that there is a smooth transition from transverse to umbilical portion without branching at this point. The branch to segment 2 does not originate for about 1 cm beyond the point of attachment of the ligamentum venosum, and is, therefore, a branch of the umbilican portion, as are the branches to segments 3 and 4. As the branch to segment 2 arises downstream from the ligamentum venosum, it cannot be a terminal branch of the transverse portion as suggested by Couinaud. Couinaud's other contention is that the left hepatic vein does not run in the plane that separates segment 4 from segments 2 and 3, i.e. the plane of the umbilical fissure and the attachment of the falciform ligament or the intersectional plane. The counter-argument is although the left hepatic vein runs between segment 2 and segment 3 for most of its course, its terminal portion, which begins where it receives the umbilical vein, must be in the plane between segment 4 and segments 2/3, i.e. the intersectional plane. Therefore, the anatomical basis for dividing the left liver into the sectors is suspect (Fig. 2.10).

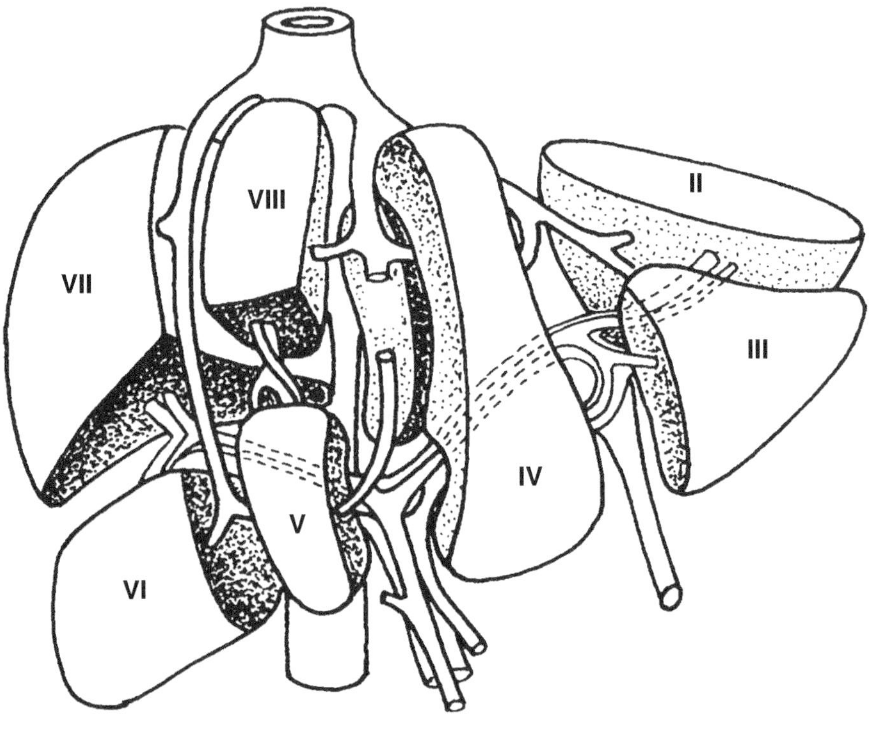

Fig. 2.10 Couinaud's portal segmentations into hemilivers, sectors and segments

Healey avoided all these arguments by treating the left portal vein as being irregular because of embryological requirements to return blood from the placenta to the heart. He did not use the hepatic and the portal vein as a basis of division of the liver but instead used the arteriobiliary branches. In the right liver, the division of the liver using Couinaud's portal segmentation or Healey's arteriobiliary segmentation is the same. In the left liver, the Couinaud's portal segmentation looks different from that of Healey's arteriobiliary segmentation. On looking at the original papers from Couinaud and Healey, these anatomists are in agreement with each other in the anatomy of the left hepatic artery and its branches, and the left hepatic duct and its branches.

In the original article by Couinaud, the left hepatic artery gave a single branch to segment 4 before branching to several branches to supply different parts of segment 4, then entered as a single artery to the left of the umbilical fissure before dividing into segment 2 and segment 3 branches. The left hepatic arterial supply to liver segments 4, 3 and 2 was exactly the same as described by Healey.

In the paper by Couinaud, there were 1–3 bile ducts from segment 4 that drained into the left hepatic duct, while the bile ducts from segments 2 and 3 joined as a single duct before it entered into segment 4. Healey described the same ductal arrangement on the left side for segments 2 and 3 ducts, but he described only a single duct in segment 4 which branched off into 2–3 branches.

The portal supply to the left liver differed slightly from that described by Couinaud and by Healey, but the difference is very small (Fig. 2.11).

Please compare this figure with Fig. 6.9 Common variations in Anatomy of Left Bile Duct. Couinaud's description is similar to type 3 and Healey's description is similar to Type 1.

Thus, taking away the difference in the segmentation in the left side of the liver by the Couinaud's portal segmentation and the Healey's arteriobiliary segmentation, the anatomical descriptions by the two anatomists are in agreement with each other. Because Healey's arteriobiliary segmentation has been commonly used in the United States, the terms left lateral segmentectomy (resection of Couinaud's segments 2, 3) and right trisegmentectomy (resection of Couinaud's segments 4, 5, 6, 7, 8) are still in common use, although in these terms, the segment refers to Healey's segment (= section in the Brisbane 2000 Terminology).

2.6 An Easy Way to Remember the Liver Segments

An easy way to remember the liver segments is to use your left hand. With the left hand open and with the left thumb adducted and flexed, the portal pedicle runs superficially to the caudate lobe to enter into the different segments of the main liver (Fig. 2.12a). The left thumb represents the caudate lobe (Fig. 2.12b) and the other four fingers represent the remaining seven liver segments when a fist is formed (Fig. 2.12c). The purple pen represents the inferior vena cava which runs posterior to the liver separating the two hemilivers (Fig. 2.12d).

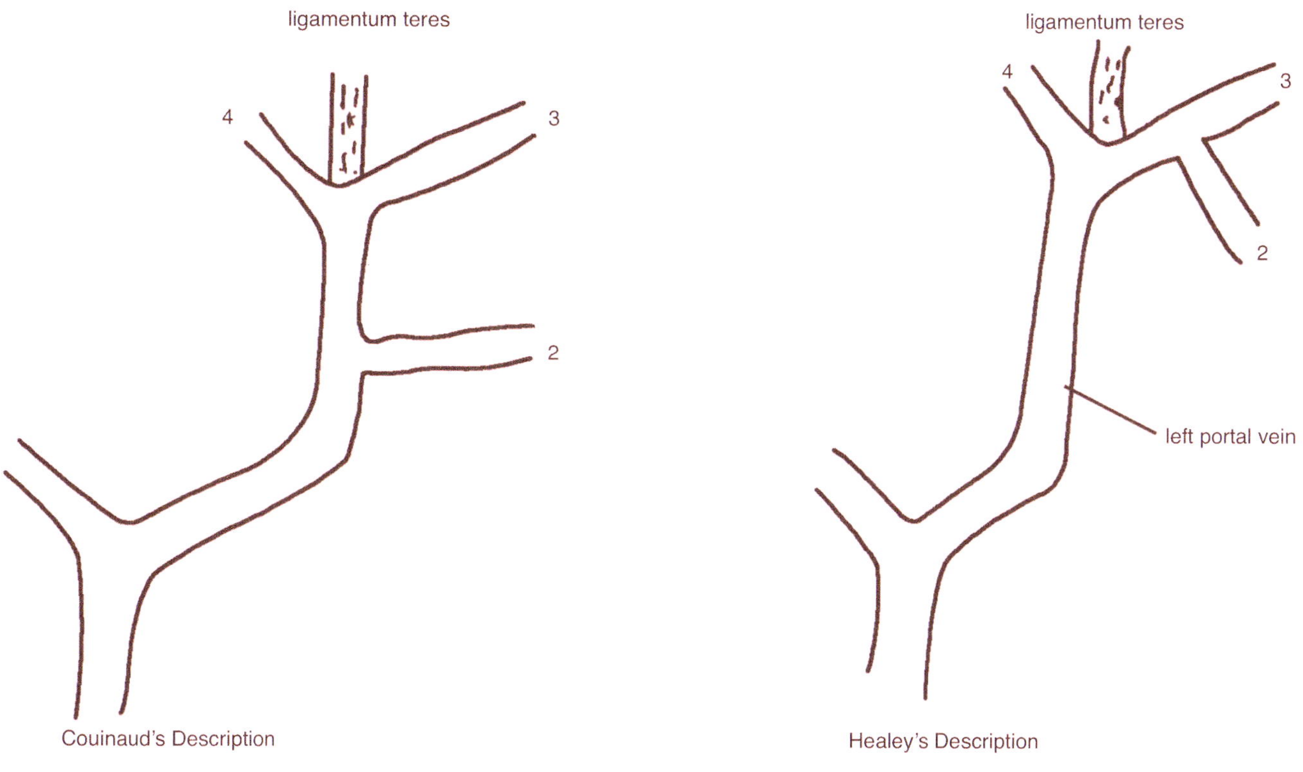

Fig. 2.11 Couinaud and Healey's description of the portal venous supply to the left liver

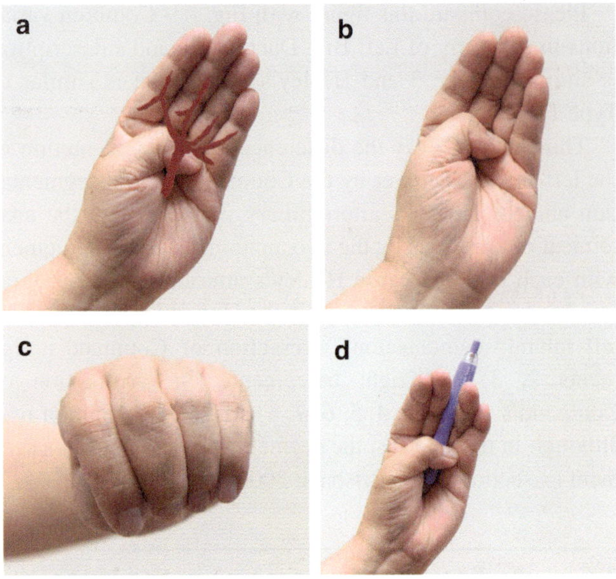

Fig. 2.12 An easy way to remember the liver segments is to use your left hand

Further Reading

Couinaud C. Lefoie. Etudes anatomiques et chirugicales. Paris: Masson & Cie; 1957.

Healey JE Jr, Schroy PC. Anatomy of the biliary ducts within the human liver: analysis of the prevailing pattern of branchings and the major variations of the biliary ducts. Arch Surg. 1953;66:599–616.

Terminology Committee of the IHPBA. The Brisbane 2000 terminology of liver anatomy and resection. HPB. 2000;2:333–9.

Caudate Lobe

3

3.1 Anatomy

The caudate lobe is the dorsal portion of the liver lying posteriorly and embracing the retrohepatic inferior vena cava in a semicircumferential fashion. The caudate lobe lies between the major vascular structures with the inferior vena cava posteriorly, the portal triad inferiorly and the hepatic venous confluence superiorly (Fig. 3.1). There is a series of short hepatic veins which drain directly from the caudate lobe into the retrohepatic inferior vena cava. Thus, the caudate lobe is surrounded by important and potentially dangerous structures deep in the centre of the liver. In view of the unique anatomical location, caudate lobe resection, especially isolated caudate lobe resection, has been considered technically challenging.

The caudate lobe can be divided into three parts: (1) the Spiegelian lobe, which is located behind the lesser omentum and extends to the left of the retrohepatic inferior vena cava (Figs. 3.2, 3.3, and 3.4).

The ligamentum venosum crosses in front of the caudate lobe as it runs from the left portal vein to the inferior vena cava at the posterior part of the trunk of the middle/left hepatic veins (Figs. 3.4, 3.5, and 3.6). On the left of the ligamentum venosum is (1) the Spiegelian lobe; (2) the paracaval portion, which lies in front of the retrohepatic inferior vena cava, just to the right of the Spiegelian lobe and is closely attached to the right and middle hepatic veins (Figs. 3.2, 3.3, and 3.4); and (3) the caudate process which is a small projection between the inferior vena cava and the adjacent portal vein anteriorly, lies just to the right of the paracaval portion (Figs. 3.2, 3.3, and 3.4).

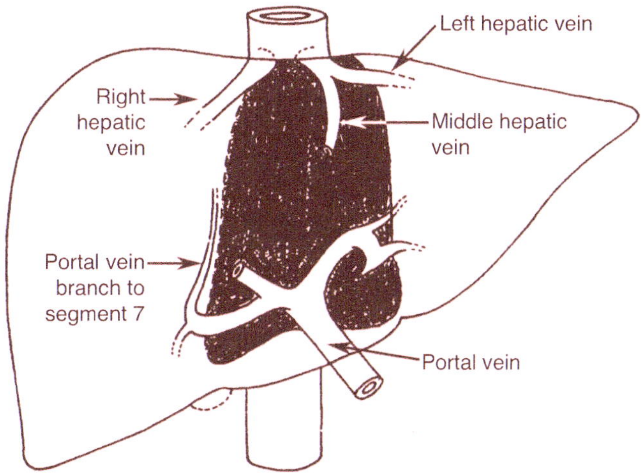

Fig. 3.1 The Caudate lobe: front view

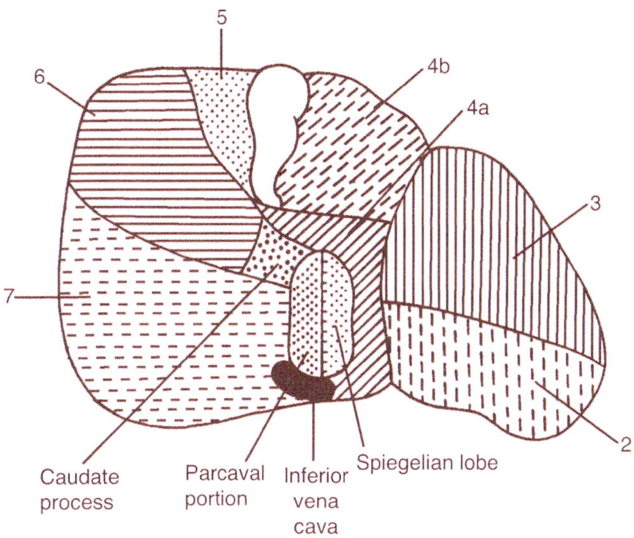

Fig. 3.2 The Caudate lobe: visceral surface

W. Y. Lau, *Applied Anatomy in Liver Resection and Liver Transplantation*, https://doi.org/10.1007/978-981-16-0800-1_3

Couinaud divides the liver into the main part containing liver segments 2–8, and the dorsal sector containing segments 1 and 9, the dividing line between these two segments is the middle hepatic vein. The dorsal sector is situated between the hepatic veins (right, middle and left) and the retrohepatic portion of the inferior vena cava. Anteriorly, it is connected (from right to left) to segments 7, 8 and 4. The dorsal fissure separates part of the posterior aspect of segment 4 from segment 1.

The dorsal sector is posterior and inferior to an axis when seen from the right side of the liver. This axis starts from the inferior vena cava at the level of emergence of the three main hepatic veins (right, middle and left), passes through the portal bifurcation to reach the inferior surface of the liver (Fig. 3.7).

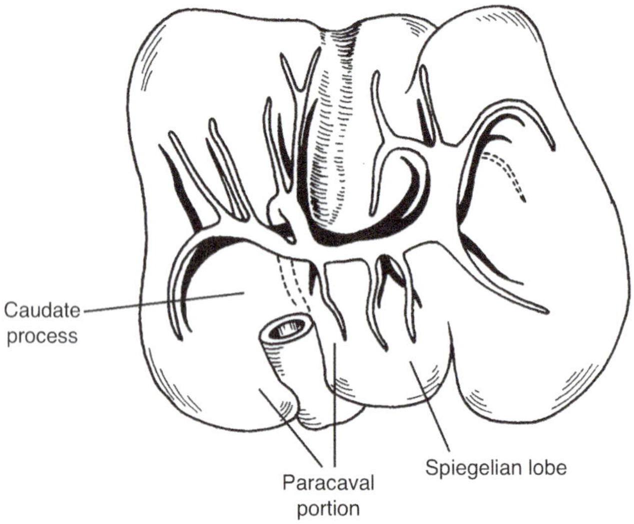

Fig. 3.3 The Caudate lobe and its three parts

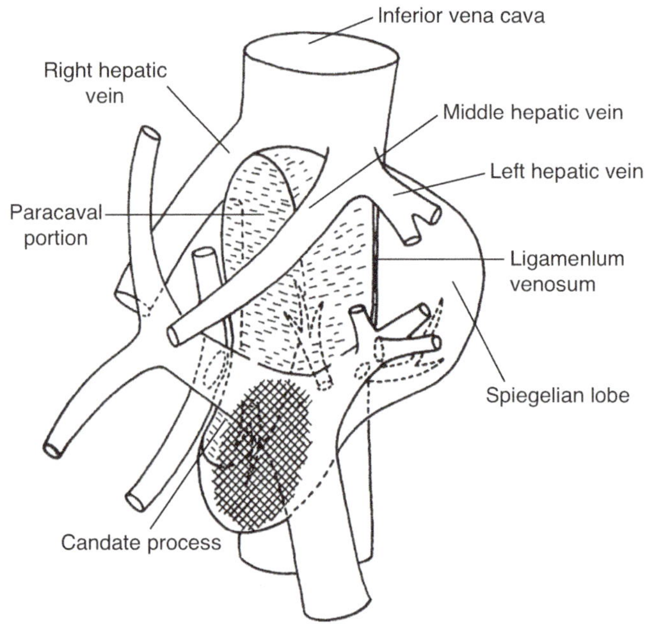

Fig. 3.4 Dorsal sector: after removal of the main part of the liver (i.e. Couinaud segments 2–8)

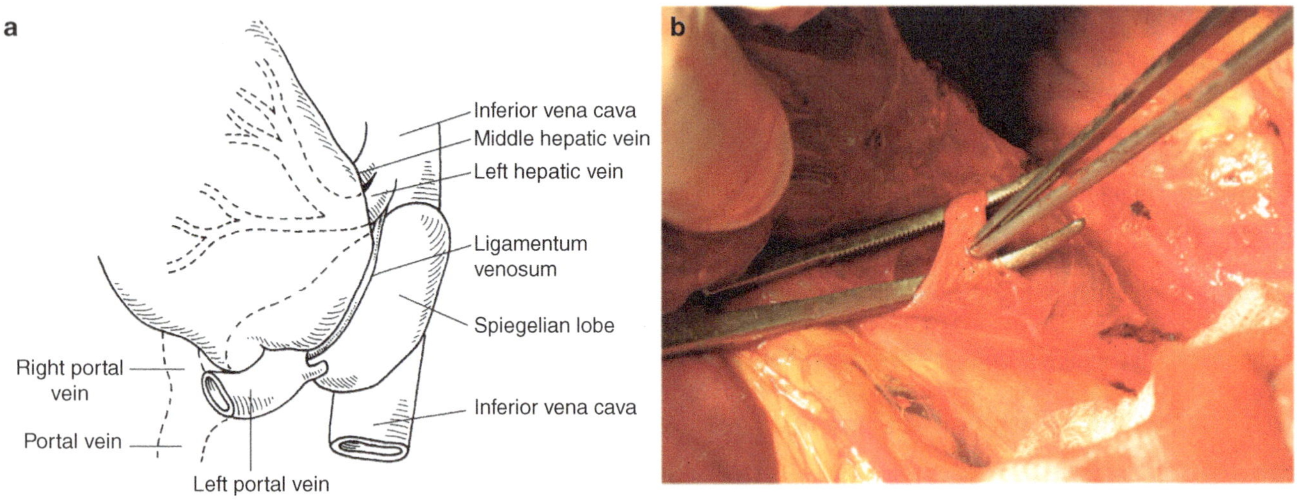

Fig. 3.5 (**a**) Relation between ligamentum venosum and the dorsal sector (viewed from the front and the left side of the abdomen and with the left liver being pushed upwards and to the right); (**b**) Operative photograph showing ligamentum venosum (pointed by the tip of forceps) running to attach to the posterior surface of the trunk of the middle/left hepatic veins

When seen from the inferior surface of the liver, segment 1 is on the left of the middle hepatic vein. It is roughly equivalent to the Spiegelian lobe, although slightly larger. Segment 9 lies on the right of the middle hepatic vein and it is divided into 9b (between the right and middle hepatic veins), 9c (under the right hepatic vein) and 9d (posterior to the right hepatic vein). The paracaval portion and the caudate process form segment 9.

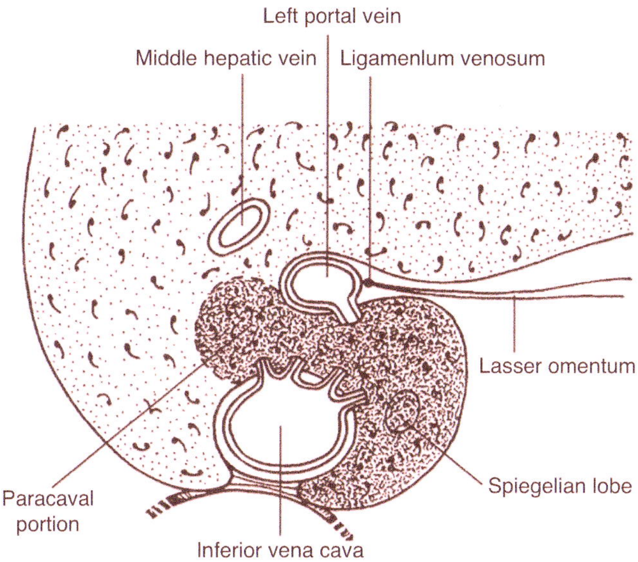

Fig. 3.6 Cross section at the dorsal sector

3.2 Blood Supply of the Caudate Lobe

The caudate lobe is supplied by blood vessels (branches from the hepatic artery and portal vein) and is drained by biliary tributaries. These vasculobiliary branches form the caudate portal triads (Fig. 3.8).

The prevailing pattern of portal vein branches of the caudate lobe is shown in Fig. 3.9. Each portal vein branch is enclosed within a Glisson's sheath almost immediately after branching. The portal vein branch is joined by the arterial and ductal elements to form a Glisson's triad although the arteries and the bile ducts of the caudate lobe ramify in the hilar plate before they go into the Glisson's sheath (see Sect. 5.3).

The Spiegelian lobe is usually supplied by two (which can join to form one) caudate portal triads most commonly originating from the left pedicle of the portal triad. The paracaval portion is usually supplied by one or two caudate portal triad(s) which originates from the right posterior sectional pedicle. The caudate process receives its blood supply originating from the right pedicle or from the bifurcation of the main portal triad.

Fig. 3.7 The dorsal sector

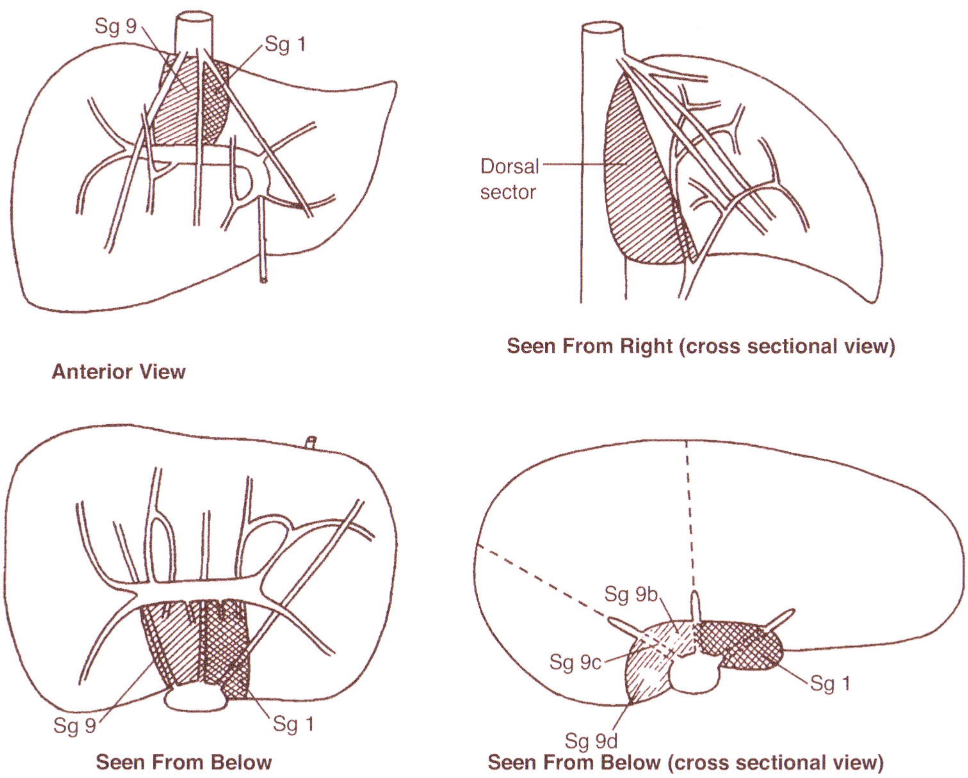

Variations in the portal vein supply which essentially represents the portal triad to the caudate lobe are common and are shown in Fig. 3.10.

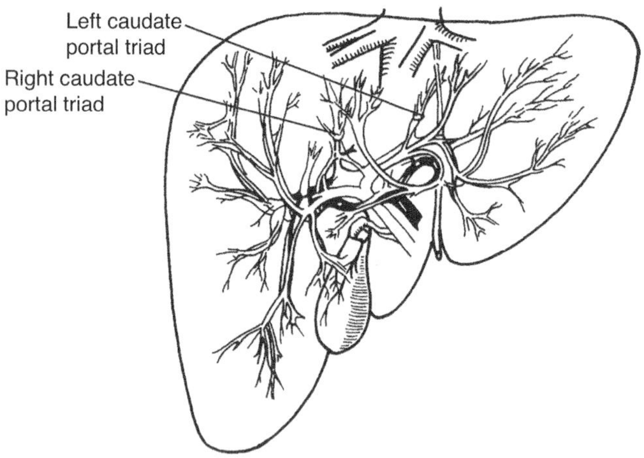

Fig. 3.8 Caudate portal triads

3.2.1 Clinical Importance

As the Spigelian lobe, paracaval portion and the caudate process receive separate portal triads, partial caudate lobectomy is technically feasible.

Variations in the patterns of the caudate portal triad commonly happen, with vessels and bile ducts crossing from one part of the caudate lobe to another part.

Transection of the caudate lobe can lead to problems of bleeding and bile leaks if not enough attention is paid to deal with the raw surfaces of the divided caudate lobe.

Ischaemic remnant of the caudate lobe can be left behind after transecting through the caudate lobe if the portal triad supplying that part of the caudate lobe is compromised.

As the bile ducts to the caudate lobe arise very near to the bifurcation of the hepatic hilus, cholangiocarcinoma of the hilus involves the caudate lobe early. Curative resection of hilar bile duct cancers should be combined with caudate lobectomy.

Fig. 3.9 Prevailing pattern of portal vein branches of the Caudate lobe. ① Spiegelian lobe; ② paracaval portion; ③ Caudate process

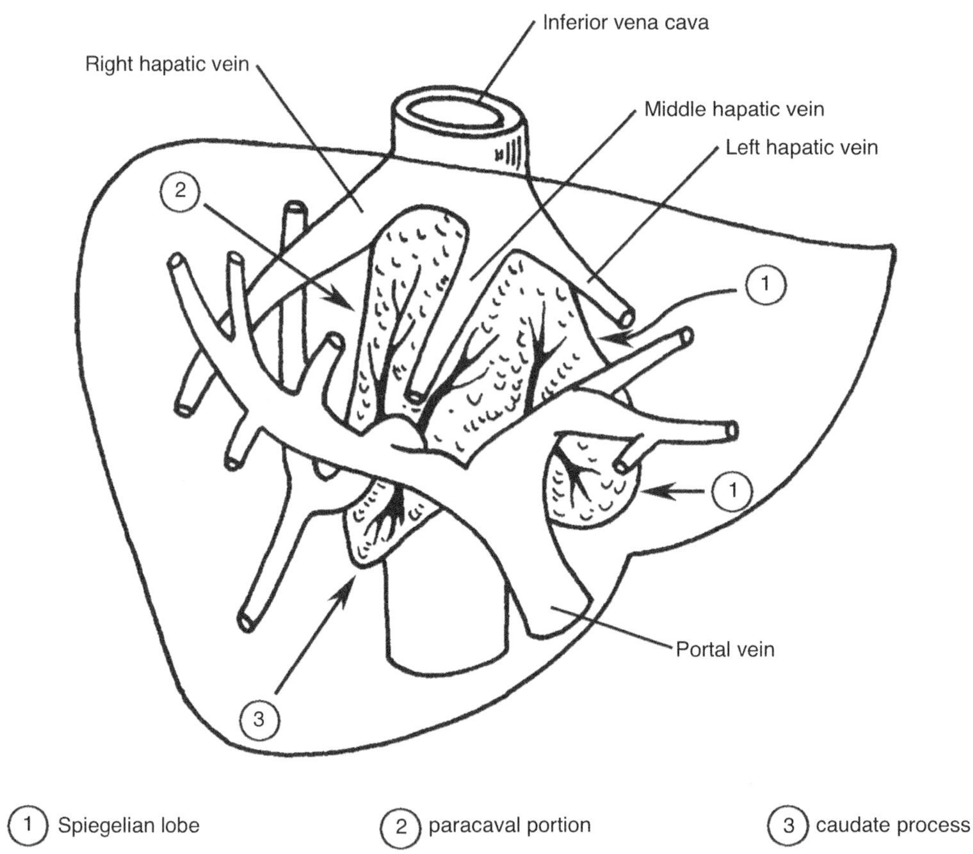

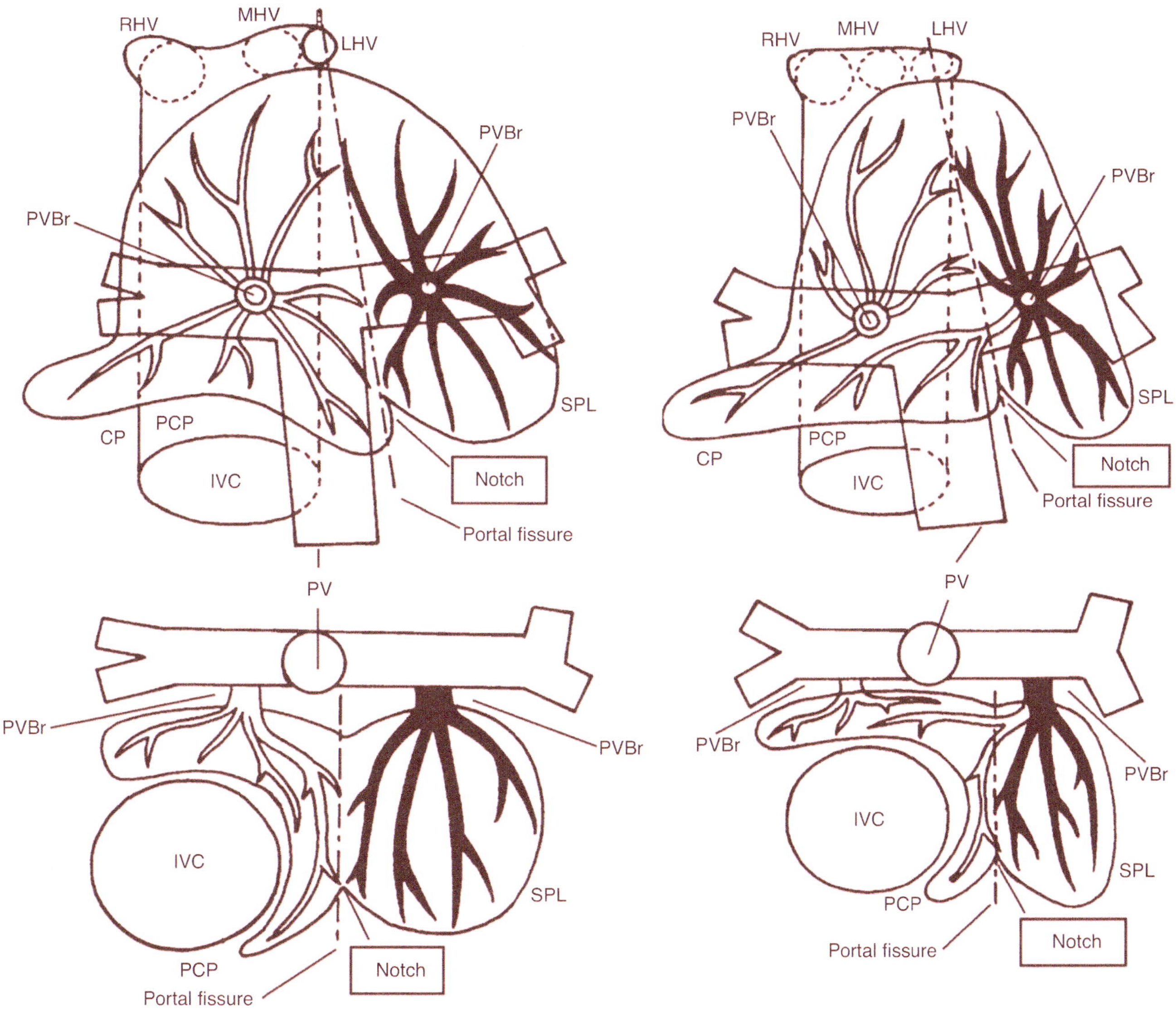

Fig. 3.10 Variations in the portal vein supply of the Caudate Lobe. *RHV* right portal vein, *MHV* middle portal vein, *LHV* left portal vein, *IVC* inferior vena cava, *SPL* Spiegelian lobe, *PCP* paracaval portion, CP Caudate process, *PVBr* portal vein branch

3.3 Venous Drainage of the Caudate Lobe

The venous drainage of the caudate lobe drains directly through the short hepatic veins into the inferior vena cava. Usually, there are two to four veins of significant size on the right side. The large short hepatic veins usually emerge from the lower or middle third of the caudate lobe but virtually never from the upper third. Very small branches from the upper third sometimes drain into the right hepatic vein or inferior vena cava, but these are nearly too small to be of surgical significance. On the left side, there are also two to four large short hepatic veins. The short hepatic veins are usually arranged on the two sides of the inferior vena cava, so that an avascular tunnel can be made safely in the majority of patients between the origins of these short hepatic veins to appear between the origins of the right and the middle hepatic vein superiorly (Fig. 3.11). This anatomy is made use of by Belghiti for his hanging technique in right/left hepatectomies (see Sect. 15.2.1.1).

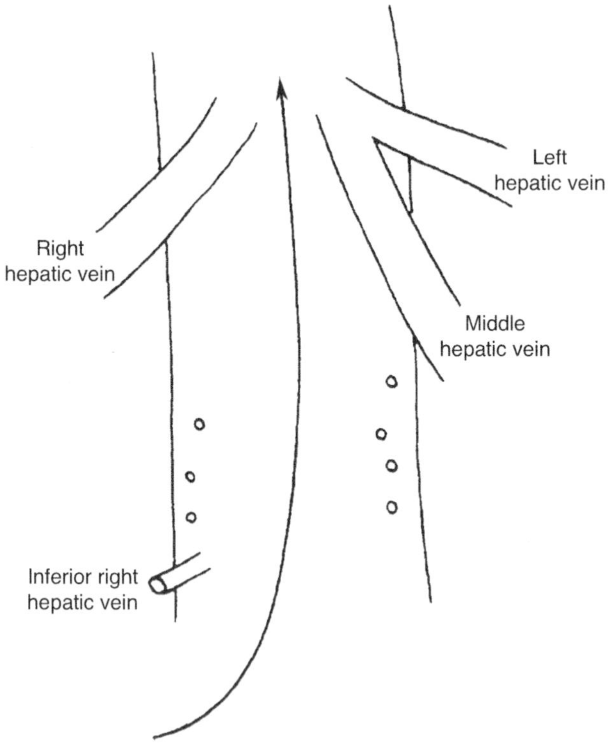

Fig. 3.11 An 'Avascular' tunnel between the short hepatic veins on the right and left sides

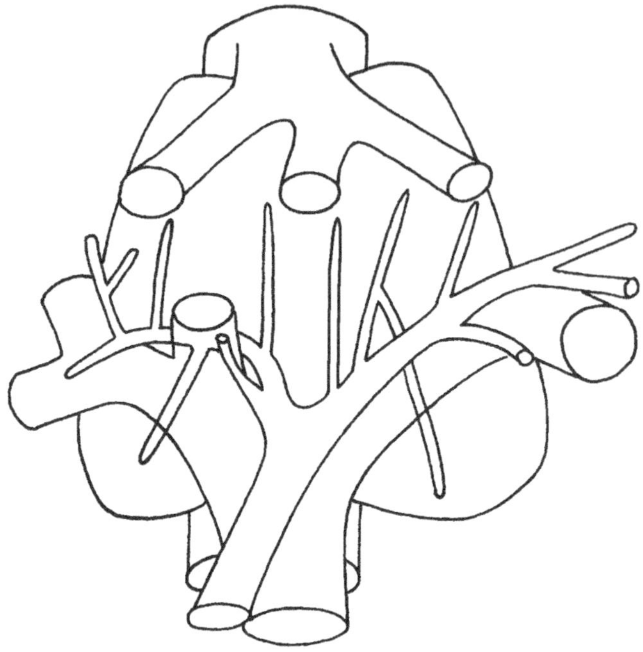

Fig. 3.12 Prevailing pattern of bile duct drainage of the Caudate Lobe

3.4 Biliary Branches from the Caudate Lobe

There are usually 2–3 biliary branches from the Spiegelian lobe to join the left bile duct. The paracaval portion is usually drained by two to three biliary branches into the right posterior sectional duct. Occasionally a biliary branch from the paracaval portion near to the middle hepatic vein area is drained into the left hepatic duct. The caudate process is usually drained into the right posterior sectional duct (Fig. 3.12).

3.5 Hepato-caval Ligament

This is a fibrous band of tissue occasionally being replaced by a bridge of liver tissue. On the right side, it arises from the right posterior liver at a level just below where the right hepatic vein branches off from the inferior vena cava. At this level, the elevation of the right hemiliver leads to a medial compression of the right lateral side of the inferior vena cava and right hepatic vein as this hepato-caval ligament is drawn tight because these three structures are in close contact (Fig. 3.13).

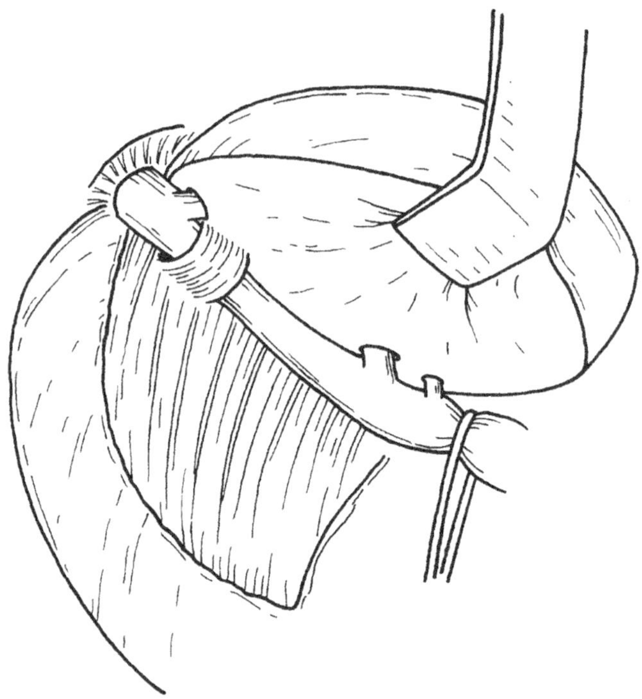

Fig. 3.13 Hepato-caval ligament after mobilisation of the right liver

To dissect the right hepatic vein, the right wall of the inferior vena cava is pushed gently to the left by blunt dissection while the traction onto the liver is released, so as to isolate the hepato-caval ligament. This ligament should be clamped, divided and ligated as it may contain a large hepatic vein. Only after this procedure has been completed should the right hepatic vein be approached from below. Blunt dissection is then done along the anterior surface of the inferior vena cava to the left of the right hepatic vein. The right hepatic vein can then be dissected and encircled with a tape (Fig. 3.14).

The hepato-caval ligament wraps around the inferior vena cava to appear on the left side of this vein. At the anterior lateral surface of the left edge of the inferior vena cava, it merges with the caudate lobe. This ligament attaches the caudate lobe to the inferior vena cava and it needs to be divided to free the caudate lobe during caudate lobectomy (Fig. 3.15).

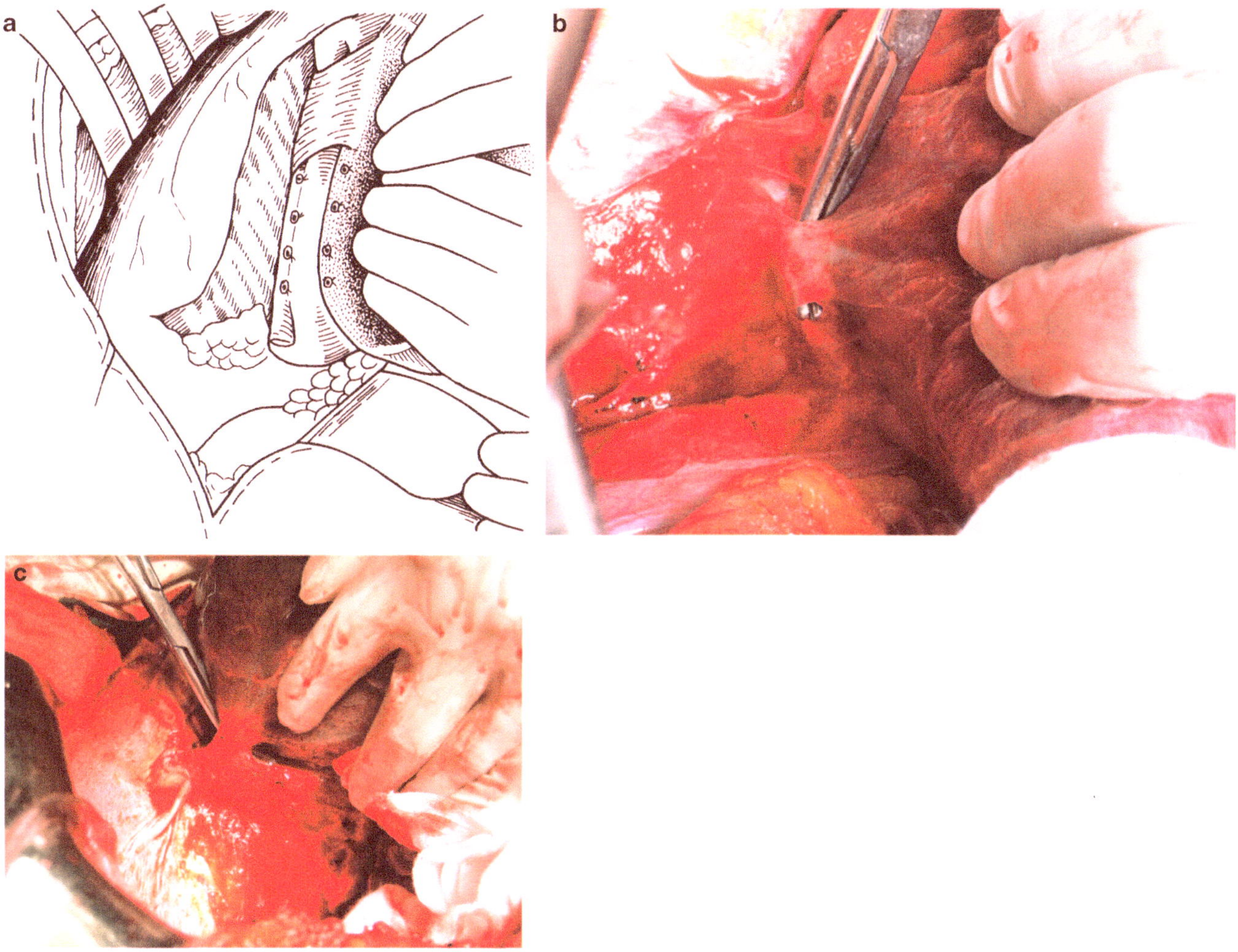

Fig. 3.14 (**a**) Division of short hepatic veins draining the right liver. (b) Isolation and division of the hepato-caval ligament. (**c**) Dissection around the right hepatic vein

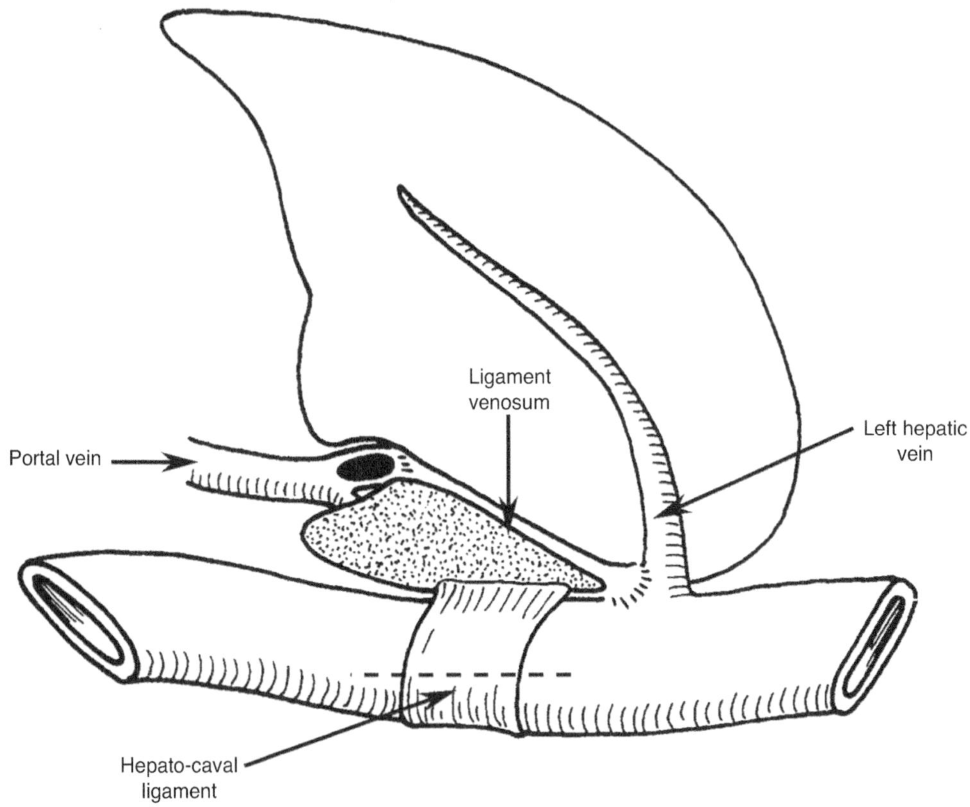

Fig. 3.15 Hepato-caval ligament on the left side of the IVC

Further Reading

Filipponi F, Romangnoli P, Mosca F, Couinaud C. The dorsal sector of human liver: embryological, anatomical and clinical relevance. Hepato-Gastroenterology. 2000;27:1726–31.

Kumon M. Anatomy of the caudate lobe with special reference to portal vein and bile duct. Acta Hepatol Jpn. 1985;26:1193–9.

Lau WY, Chi TY. Hepatic segment 9. Chin J Surg. 2000;40:342–3.

Peng SY. Chapter 26: Isolated caudate lobe resection. In: Lau WY, editor. Hepatocellular carcinoma. Singapore: World Scientific; 2008. p. 465–89.

4 Hepatic Hilar Plate System

4.1 Anatomy of Glissonian Sheath (Glisson's Sheath)

Glisson's capsule which covers the liver extends into the liver at the hilus and covers the portal triad, where it is called Glisson's sheath. Glisson's capsule also covers the Glissonian pedicles inside the liver. Couinaud called this sheath the Valoean sheath, after Valoeus, an anatomist from the Middle Ages who first described the liver capsule. The term 'Glissonian sheath' is generally used only to refer to the portion of the Glissonian pedicle inside the liver. This paragraph is reprinted from: Lunan Yan, Operative Techniques in Liver Resection. Springer, 2016. ISBN 978-94-017-7409-3.

In the extrahepatic portion of the 'Glissonian pedicle', the portal triads in the hepatoduodenal ligament are also enclosed by connective tissues and peritoneum up to the hepatic hilum. The intrahepatic and extrahepatic portions of the portal triads have the same structures anatomically. In other words, the extrahepatic and intrahepatic portal triads can be considered as part of the same Glissonian pedicle tree (Fig. 4.1). This paragraph is reprinted from: Lunan Yan, Operative Techniques in Liver Resection. Springer, 2016. ISBN 978-94-017-7409-3.

The prevailing pattern of the intrahepatic Glissonian pedicle tree has been described, and used by the Brisbane 2000 Terminology to divide the liver into hemilivers, sections (sectors) and segments (see Chap. 2). There are many variations (see Chap. 6) which make dissection of individual structures within the liver difficult and even hazardous. However, if the sheath to a particular segment is taken, it will only contain structures passing to or from that segment. Ligation of individual sheath is therefore not only simpler but safer. In Fig. 4.2, if the sheath is open at (b) and the individual structures are dissected, then the abnormally branching artery may be ligated under the supposition that is passing to the lower branch. If the lower sheath is taken en masse at (a) there is no danger to this aberrant artery.

Sometimes it is necessary to dissect structures individually within a sheath (this is particularly true for biliary-enteric anastomoses). The bile duct tends to be elliptical rather than round and the inferior aspect usually faces the corresponding artery. The relationship between the three structures within the sheaths follows two general rules of importance: first, the portal vein tends to lie posterior to the

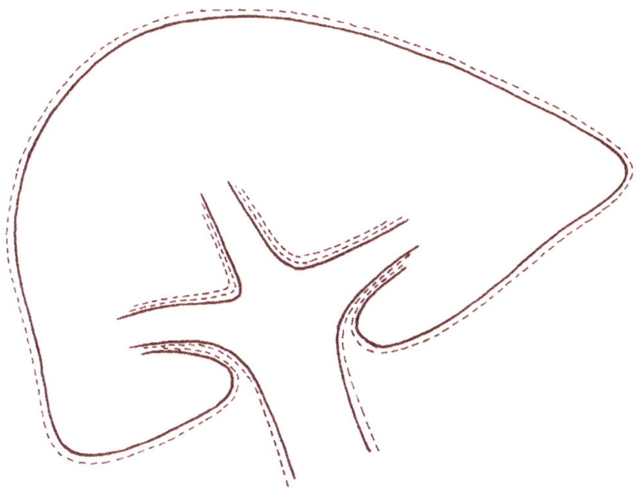

Fig. 4.1 Glisson's capsule, shown by the dotted lines

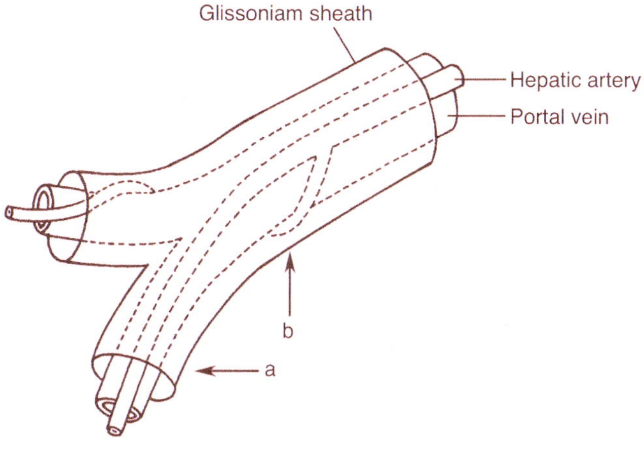

Fig. 4.2 Ligation of a Glissonian Sheath at (**a**) is safer than opening up the sheath at (**b**)

W. Y. Lau, *Applied Anatomy in Liver Resection and Liver Transplantation*, https://doi.org/10.1007/978-981-16-0800-1_4

bile duct and hepatic artery; second, the bile duct tends to lie superior to the artery and is always close to it (Fig. 4.3).

4.2 Anatomy of the Hepatic Hilar Plate System

The fusion of Glisson's capsule with the connective tissue sheaths surrounding the biliary and vascular elements at the inferior aspect of the liver constitutes the plate system. This plate system also contains a large number of lymphatics, nerves and a small vascular network. Although most workers consider the portal triad to be within the plate system, Couinaud states that the bile ducts and hepatic artery are located within the plate system, but that the portal vein is covered with a separate sheath of loose connective tissue. That is the reason why the plate containing the extrahepatic bile duct and hepatic artery can be separated easily from the portal vein. This paragraph is reprinted from: Lunan Yan, Operative Techniques in Liver Resection. Springer, 2016. ISBN 978-94-017-7409-3.

The hepatic hilar plate system includes the hilar plate above the biliary confluence, the cystic plate related to the gallbladder, the umbilical plate situated above the umbilical portion of the left portal vein and the plate of Arantius covering the ligamentum venosum (Fig. 4.4). This paragraph is reprinted from: Lunan Yan, Operative Techniques in Liver Resection. Springer, 2016. ISBN 978-94-017-7409-3.

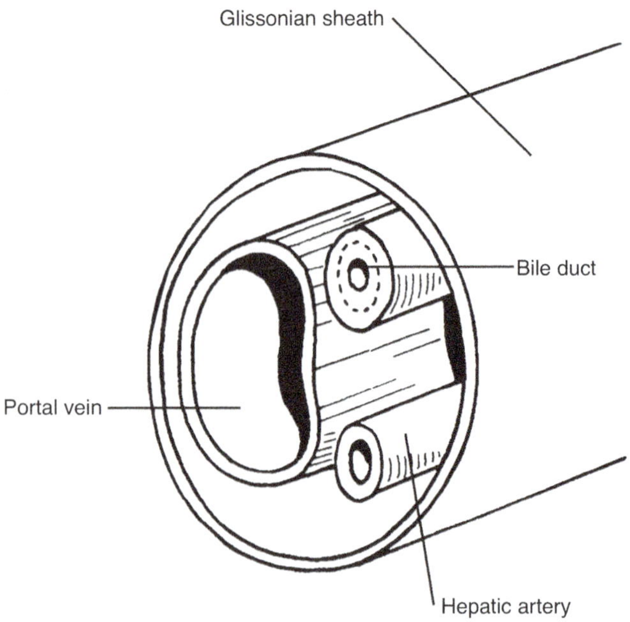

Fig. 4.3 Structures within a Glissonian Sheath

4.2.1 Hilar Plate

The hilar plate is located in the hilar area of the liver. It is bounded above by segment 4a of the liver (the posterior part of segment 4), on the right by the Rouviere sulcus (Fig. 4.5) and the cystic plate, and on the left, it is continuous with the umbilical plate anteriorly, and the plate of Arantius posteriorly. The right anterior sectional triad with the Glisson's sheath generally runs behind the junction between the cystic plate and the hilar plate to supply segments 5 and 8, and the right posterior sectional triad runs along the Rouviere sulcus into the liver to supply segments 6 and 7. As a result, the bile ducts and blood vessels of the right side can be dissected easily without widely opening the hilar plate. This paragraph is reprinted from: Lunan Yan, Operative Techniques in Liver Resection. Springer, 2016. ISBN 978-94-017-7409-3.

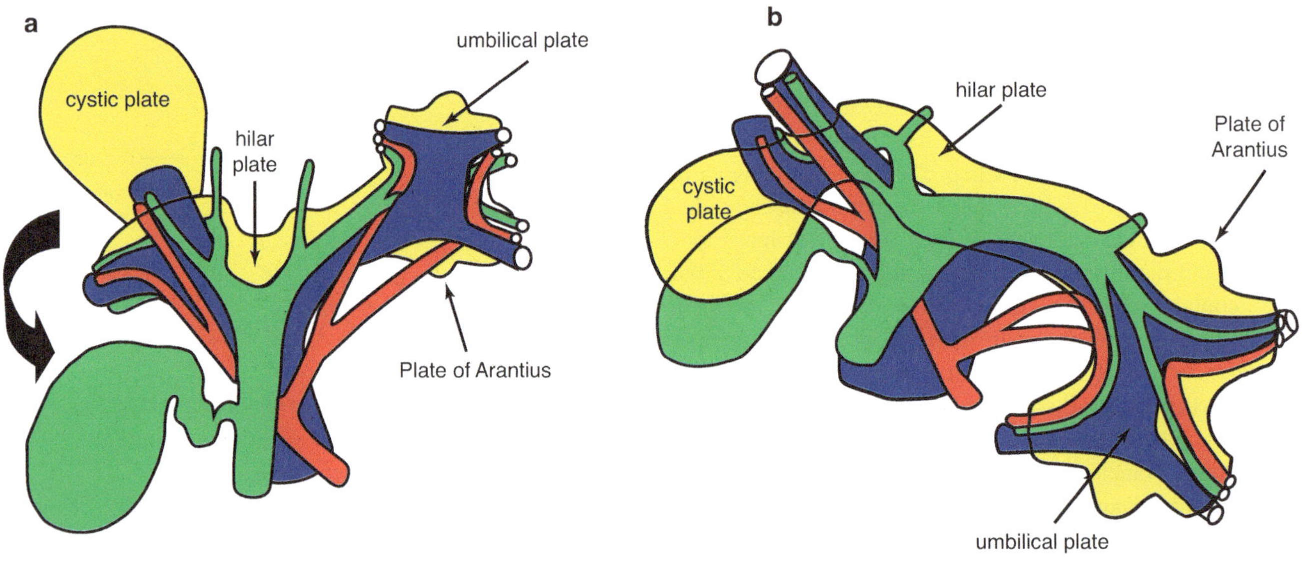

Fig. 4.4 The Hepatic Hilar plate system. (**a**) Anterior view. (**b**) Anterior-inferior view

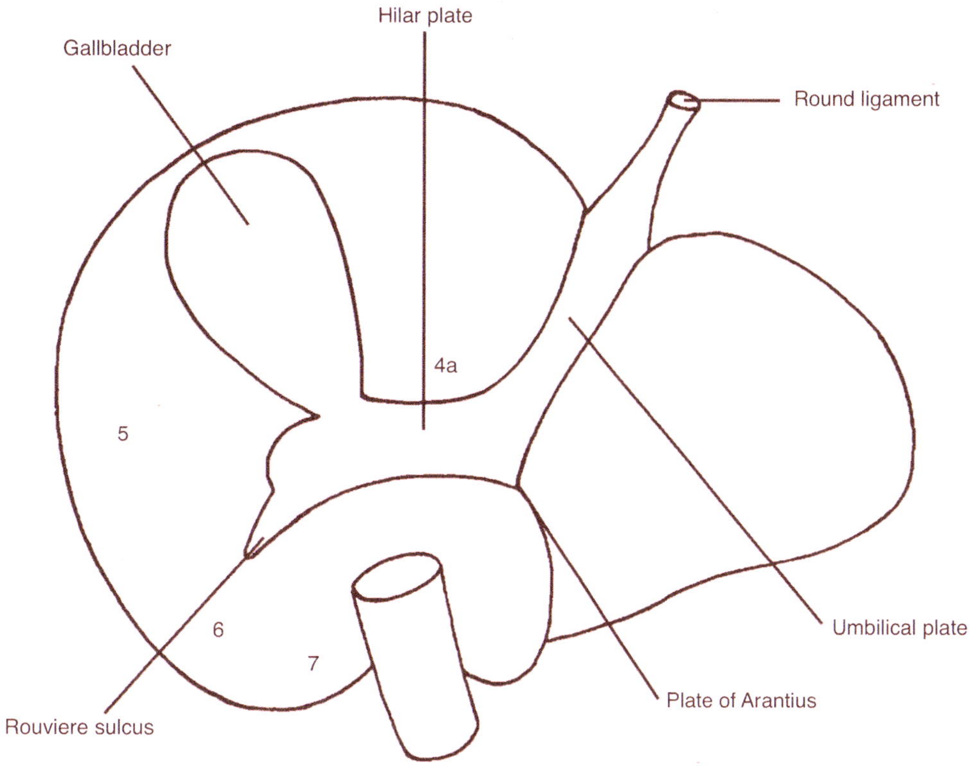

Fig. 4.5 Visceral surface of liver showing the hepatic hilar pate system. Please note that the Rouviere sulcus marks the site of entry of the right posterior sectional triad into the liver

4.2.2 Cystic Plate

The cystic plate is located in the gallbladder bed and is continuous with the capsule of segment 5, segment 4a and the Glissonian sheath of the anterior segment of the liver. The medial edge of the cystic plate lies at the midplane of the liver in the hilar area. It has also been observed by Couinaud that in most individuals (83%), the posterior edge of the cystic plate is located on the right side of the right portal vein branch. This paragraph is reprinted from: Lunan Yan, Operative Techniques in Liver Resection. Springer, 2016. ISBN 978-94-017-7409-3.

4.2.3 Umbilical Plate

The umbilical plate is located along the inferior edge of the ventral surface of the umbilical fissure. It contains the ducts and blood vessels of the segments 2, 3 and 4, and is continuous with the round ligament inferiorly. Thus, the segmental branches of the left liver divide or fuse within the umbilical plate, the upper margin of the umbilical plate can be reached by incising the superior border of the round ligament. This paragraph is reprinted from: Lunan Yan, Operative Techniques in Liver Resection. Springer, 2016. ISBN 978-94-017-7409-3.

4.2.4 Plate of Arantius

The Plate of Arantius fuses and is continuous with the ligamentum venosum posteriorly.

4.3 Surgical Approaches to Expose the Hepatic Bile Duct Confluence

Hepp and Couinaud in 1956 described a technique where, by lifting the segment 4 upwards and incising the Glisson's capsule at its base, good exposure of the hepatic hilar structures could be obtained. This technique was referred to as lowering of the hilar plate. It can be carried out with safety since there is only exceptionally (in 1% of cases) any vascular interposition between the hilar plate and the inferior aspect of the liver. The manoeuvre is of particular value when exposing the extrahepatic segment of the left hepatic duct since it has a long course beneath segment 4. It is not so effective in exposing the extrahepatic right duct or its secondary branches, which are short. The technique is of major importance for the identification of proximal biliary mucosa during bile duct repair following injury. An incision is made at the posterior edge of segment 4 where Glisson's capsule is attached to the hilar plate (Fig. 4.6). The upper surface of the hilar plate can then be separated from the hepatic paren-

Fig. 4.6 Approach of Segment 4 duct

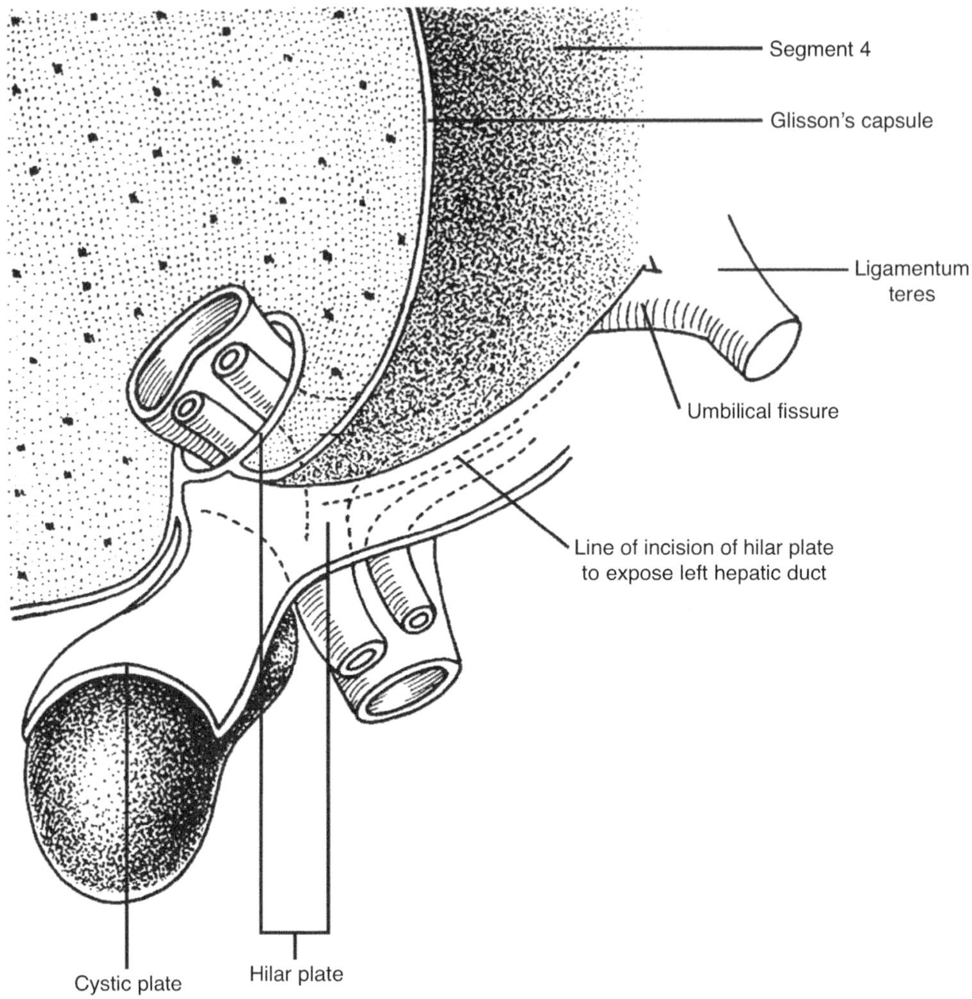

chyma and, by lifting the segment 4 upwards, display of the hepatic duct confluence, which is always extrahepatic, is effected (Fig. 4.7). In case of unresectable hilar or right ductal carcinoma, the hilar plate can be opened at the anterosuperior surface to identify the left hepatic duct. Side-to-side left duct to jejunum mucosa to mucosa anastomosis can then be made. This approach is also called the extrafascial approach to the bile duct confluence (i.e. approaching the confluence of the bile duct outside of the Glissonian sheath and the liver plate).

In the rare occasion when the extrafascial approach is hazardous, especially when anatomical deformity has been created by atrophy/hypertrophy of liver segments, and in patients where there appears to be a very deep hilus which is displaced upwards and rotated laterally, a simultaneous opening of the deepest portion of the gallbladder fossa and the umbilical fissure gives good exposure to biliary confluence and the right duct without the necessity for full hepatectomy or liver resection (Fig. 4.8). This procedure simply represents mobilisation of the inferior portion of segment 4 from the midplane (principal fissure) to the intersectional plane (umbilical fissure) to expose the left duct and the confluence of the bile duct.

In exceptionally difficult cases in bile duct strictures, part of segment 4b resection may be carried out to effect exposure of the biliary confluence.

After the confluence of the right and left hepatic pedicle is brought down, the right hepatic pedicle can be isolated by blunt dissection, and a sling be placed around it. Similar dissection can be carried out to sling the left pedicle (Fig. 4.9).

Launois devised an anterior approach to the confluence of the bile ducts called by him as the intrahepatic transfissural approach (Fig. 4.10). His steps include ligation of the lowest retrohepatic veins draining from the caudate process and the lower part of the liver to the vena cava to avoid haemorrhage by tearing these veins. After cholecystectomy, the liver plate is approached at the midplane (main fissure).

The principles of the anterior intrahepatic approach were first elaborated by Couinaud (1957) and developed by Ton That Tung (1979). The essential points are en bloc dissection of the extrahepatic pedicles, making an incision in the liver at the gallbladder bed if necessary, and blunt dissection on the

Fig. 4.7 Lowering of the Hilar Plate

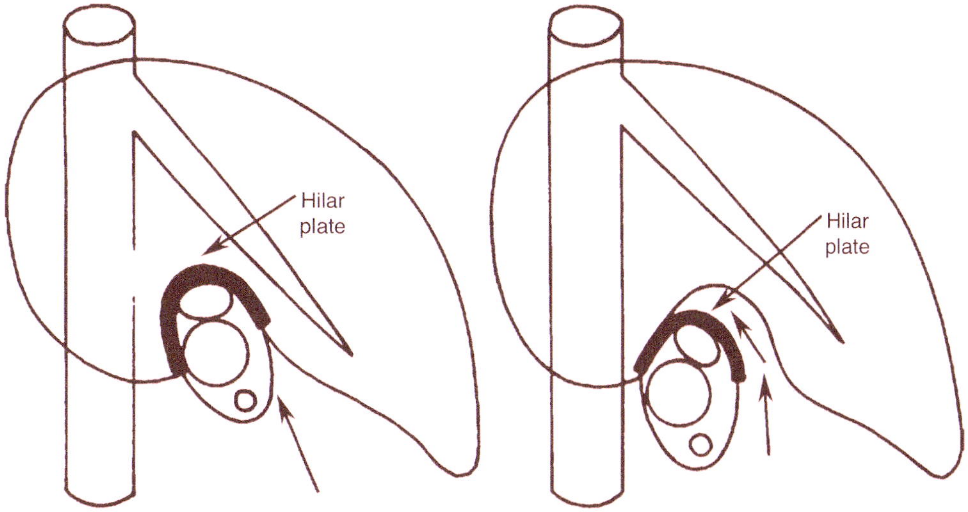

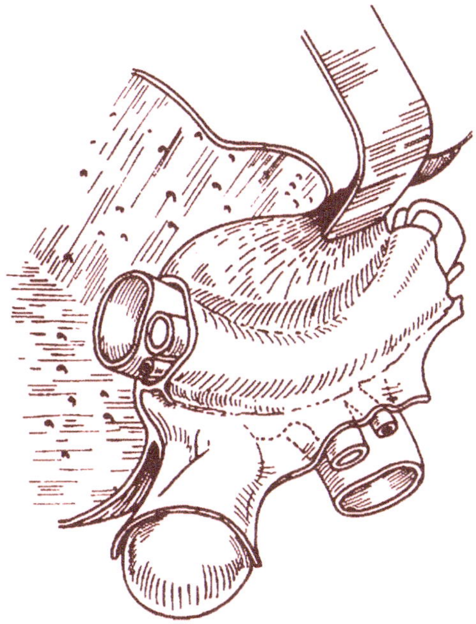

top of the pedicles at the bifurcation to come around the pedicles to exit at the caudate process to control the right pedicle. To control the left pedicle, a transverse incision is made at the liver just in front of the hilus, pass around the top and the back of the left pedicle to exit at the left side of the dorsal fissure. These approaches are very similar to the posterior intrahepatic approaches with the exception that the blunt dissection starts in the front instead of from the back of the pedicles (Fig. 4.10).

Launois devised another approach to control the right/left hepatic pedicles called the posterior intrahepatic approach to the Glissonian Sheath.

The caudate lobe immediately behind the hilus is incised for 30 mm in length. A second incision is made in front of the hilum and parallel to the first incision, extending from the gallbladder bed on the right to the umbilical fissure on the left (Fig. 4.11). The incision is deepened and the liver parenchyma is pushed upwards and away from the hilum in front in order to expose the Glissonian sheath of the confluence of the hepatic pedicle structures.

The dissection in front of the hilum corresponds to that described previously as detachment of the hilar plate. An index finger is now passed into the incision behind the hilum and the undersurface of the sheath is kept above the finger which is insinuated between the sheath anteriorly and the caudate process posteriorly until the superior part of the previously dissected sheath is reached. The surgeon's index finger and thumb of one hand are now placed in the liver substance, with the index finger in the caudate process incision and the thumb in front of the hilar plate (Fig. 4.12).

A large curved clamp is then used to pass a tape around the region of the confluence. Traction on the tape tends to exteriorize both the right and left sheaths (Fig. 4.13). By fur-

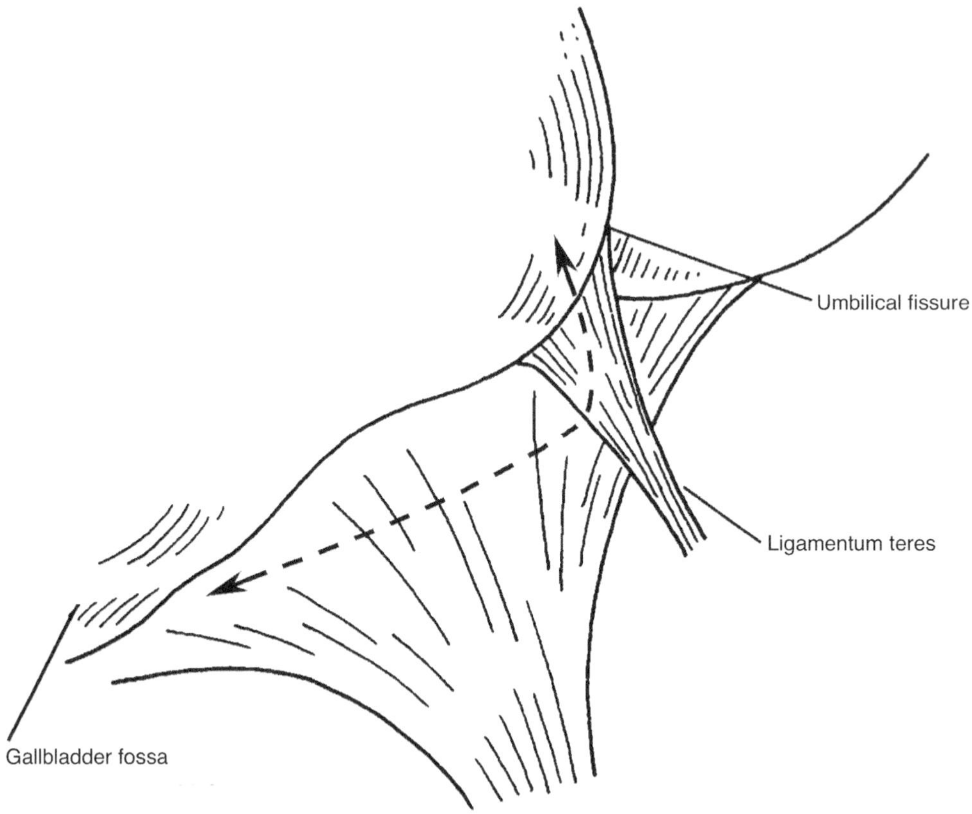

Fig. 4.8 Exposure of the confluence of the bile duct and the left Hepatic duct

ther dissection distally tapes can be passed around whichever sheath is required for further dissection.

4.4 The Umbilical Fissure and the Segment 3 (Ligamentum Teres) Approach

The round ligament, which is the remnant of the obliterated umbilical vein runs from the umbilical fissure to connect with the left branch of the portal vein within the umbilical fissure (Fig. 4.14). The ligament joins the termination of the left portal veins at which point prolongations containing channels which are elements of the left-portal system course into the liver. The bile ducts of the left liver are located above the left branch of the portal vein and lying behind these prolongations, whereas the corresponding artery is situated below the vein. Dissection of the round ligament on its left side and division of one or two vascular prolongations to segments 3 allows display of the pedicle or anterior branch of the duct of segment 3. In the event of biliary obstruction with intrahepatic ductal dilatation, the segment 3 duct is generally easily located above the left branch of the portal vein (Fig. 4.15).

It is often preferable to split the liver just to the left of the umbilical fissure to widen the fissure further. This allows access to the ductal system without the necessity to divide any elements of the portal blood supply to segment 3 (Fig. 4.16).

4.5 Surgical Approaches to the Right Hepatic Biliary Ductal System

The exposure of the right intrahepatic ductal system is much more hazardous and imprecise than the left because of the lack of precise anatomical landmarks. The use of intraoperative ultrasound helps to identify the intrahepatic structures better.

In some cases of hilar cholangiocarcinoma, the planned surgical procedure appears impossible at operation. In such a situation, intrahepatic right ductal system drainage is an option, especially in situations when the left ductal system cannot be used. Anatomically the right anterior sectional duct and its branches run on the left side of the corresponding vein. In essence, an incision on the liver is to expose the anterior sectional duct on the left aspect of the portal vein. The dilated duct is opened longitudinally and anastomosed to a Roux-en-Y loop of jejunum (Fig. 4.17).

An alternative method is open into the segment 5 duct through the gallbladder fossa by excision of part of the liver (Fig. 4.18).

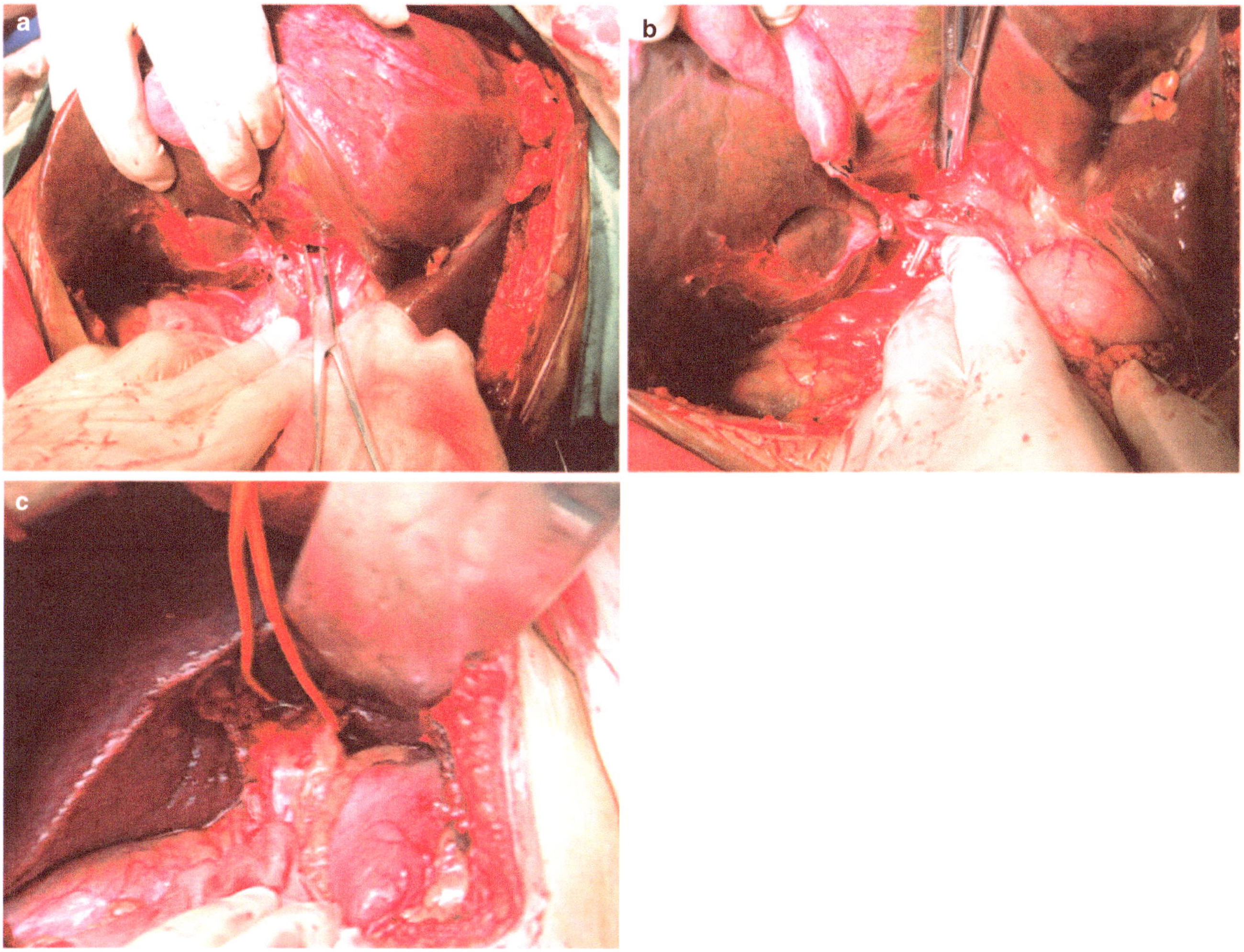

Fig. 4.9 (**a**) Blunt dissection with a forceps to lower hilar plate; (**b**) Blunt dissection around right hepatic pedicle; (**c**) Sling put around left pedicle

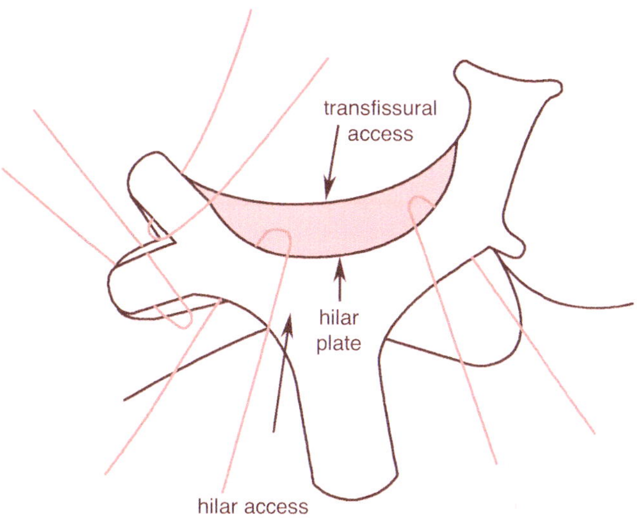

Fig. 4.10 Intrahepatic transfissural approach to the confluence of the bile duct

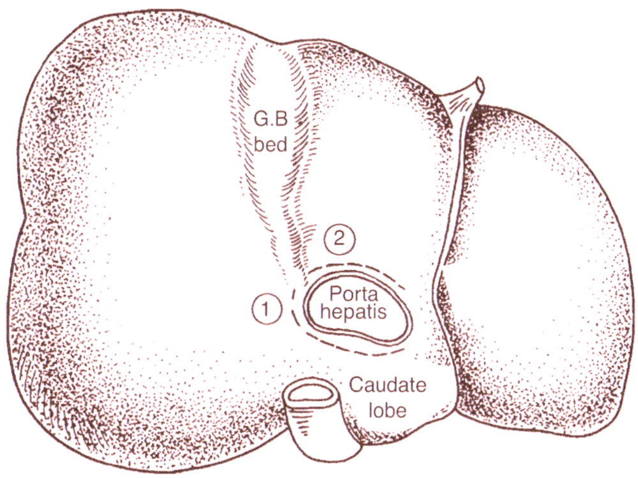

Fig. 4.11 Incisions in the liver for the posterior intrahepatic approach to the right/left hepatic pedicle confluence

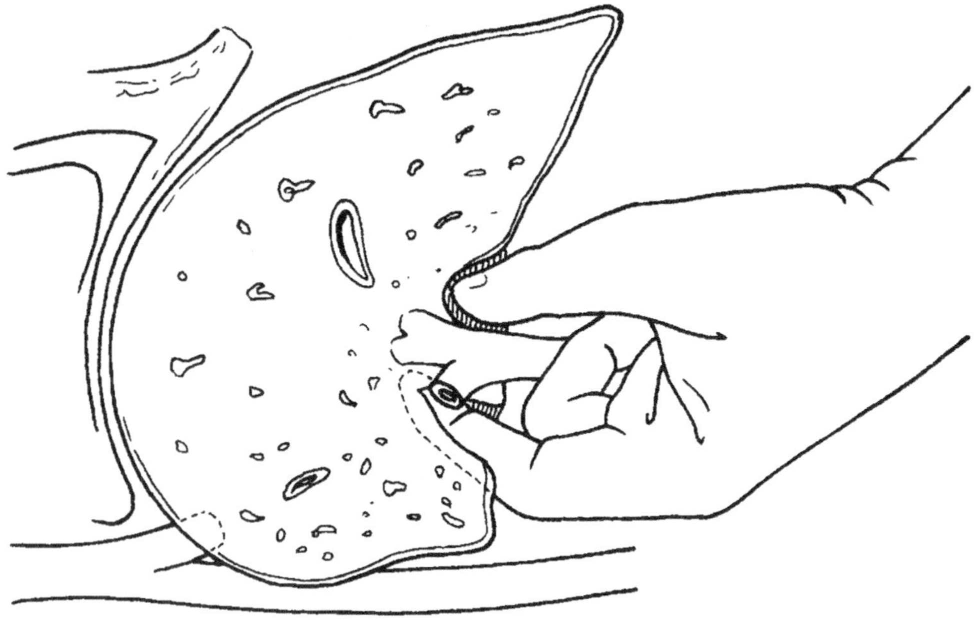

Fig. 4.12 Using the thumb and the index finger to dissect above the confluence of the Glissonian pedicles

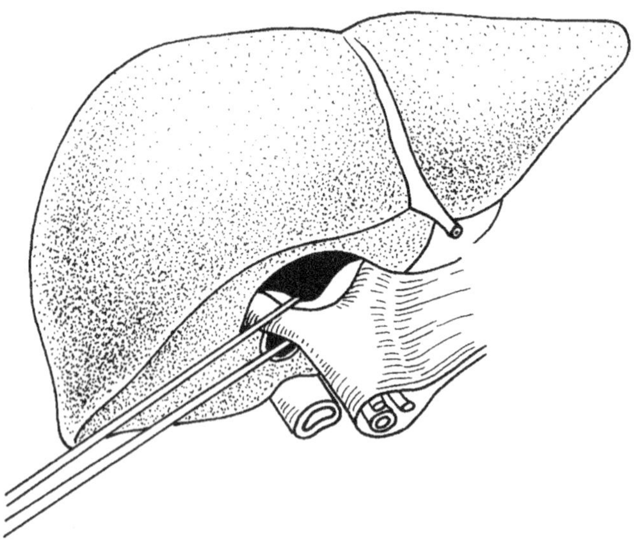

Fig. 4.13 Slinging the portal triad to the right liver

4.6 Exposure of the Bile Duct by Liver Resection

The lateral part of the left lateral section of the liver can be amputated to expose the segments 2 and 3 ducts for anastomosis to a Roux-en-Y loop of jejunum (The Longmire procedure).

As mentioned previously, removal of part of segment 4b can be carried out to effect exposure of the biliary confluence.

Although a similar procedure can be carried out to remove the inferior tip of the right liver to expose bile ducts for anastomosis, this is very rarely done because the ducts exposed are peripheral, with thin walls and small lumen, making the anastomosis technically difficult.

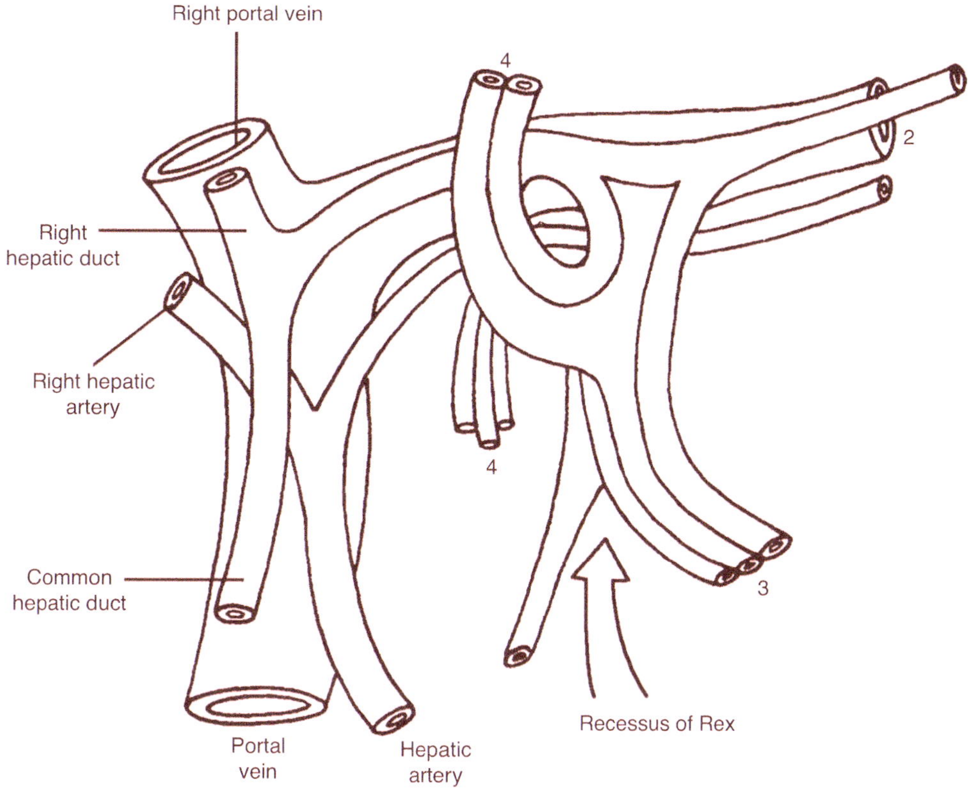

Fig. 4.14 Ligamentum teres approach through the Recessus of rex

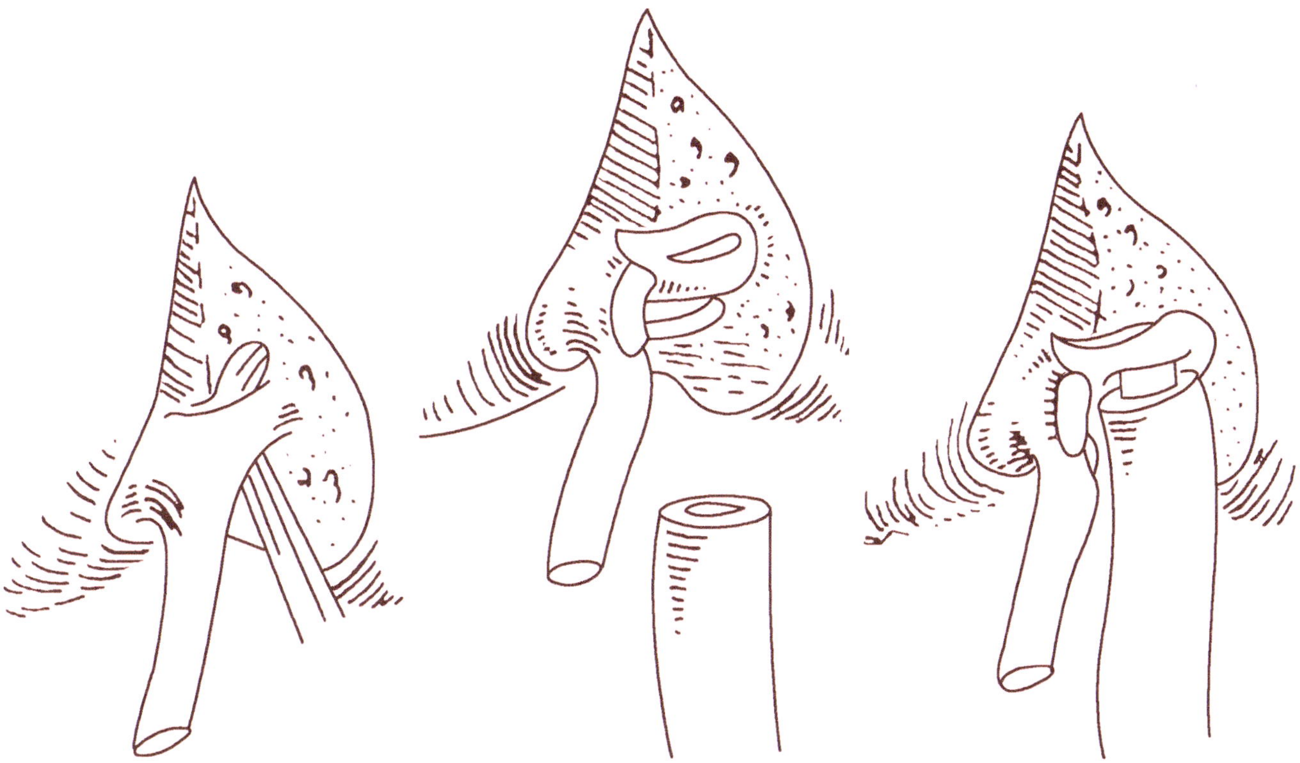

Fig. 4.15 Exposure of the segment 3 duct. (**a**) Dissection of round ligament on its left side. (**b**) Division of a branch of the portal vein to expose duct of segment. (**c**) Anastomosis to a Roux-en-Y loop of jejunum

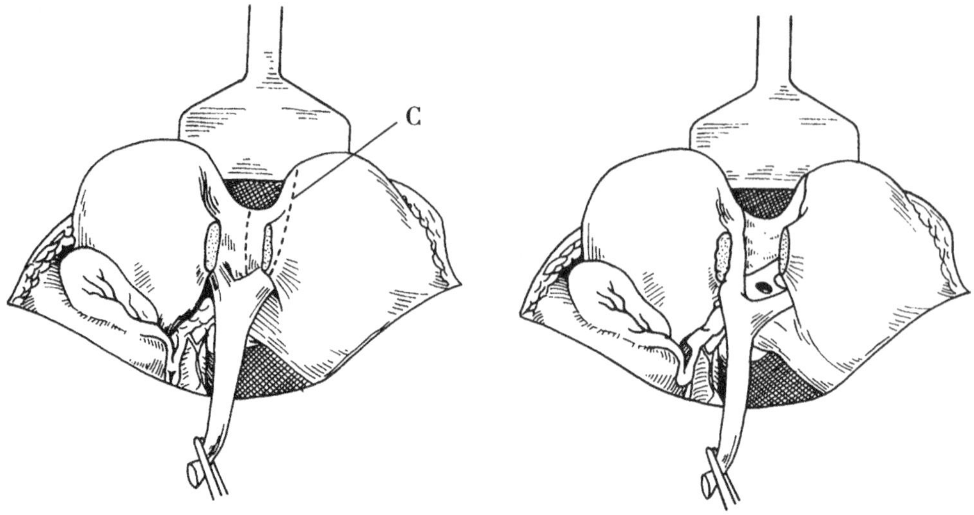

Fig. 4.16 (**a**) Splitting liver to left of umbilical fissure. (**b**) Access to bile duct system without need to divide any portal vein branch

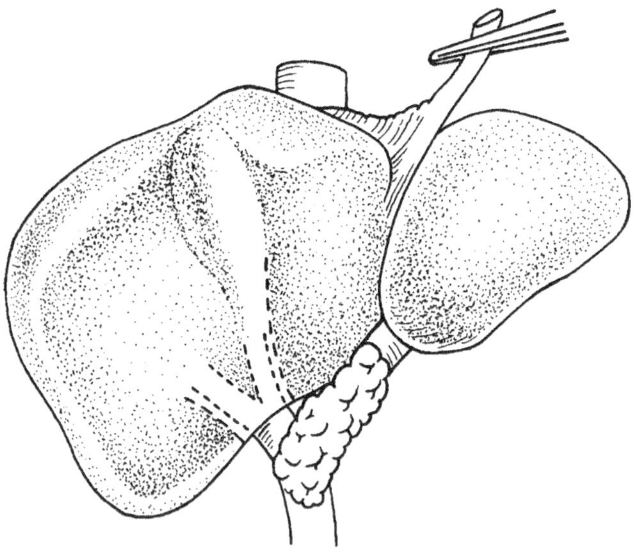

Fig. 4.17 Incisions on right liver to expose the right sectional duct

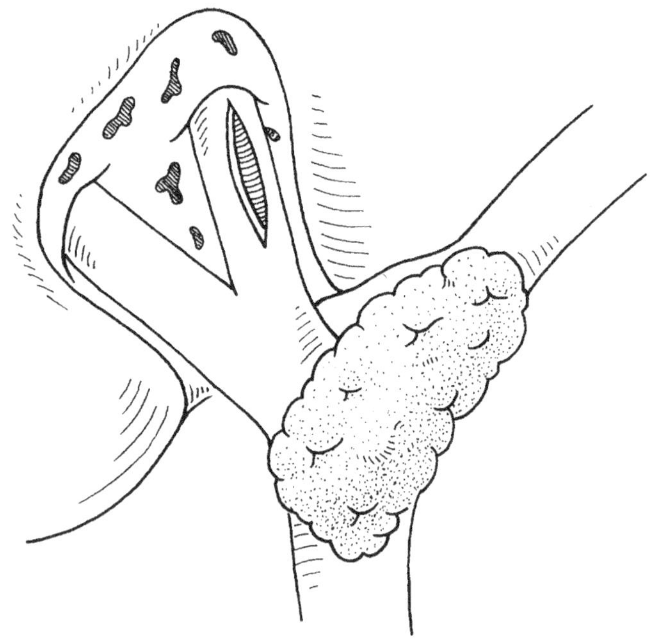

Fig. 4.18 Excision of part of liver exposes the right sectional ducts better

Further Reading

Blumgart LH, Hann LE. Chapter 1: Surgical and radiologic anatomy of the liver and biliary tract. In: Blumgart LH, Fong Y, editors. Surgery of the liver and biliary tract, vol. 1. 3rd ed. London: W.B. Saunders; 2000. p. 3–33.

Couinaud C. Controlled hepatectomies and exposure of the intrahepatic bile ducts. Anatomical and technical study. Paris: C. Couinaud; 1981.

Kawarada Y, Das BC, Taoka H. Anatomy of the hepatic hilar area: the plate system. Journal of HBP Surgery. 2000;7:580–6.

Launois B, Tay KH. Chapter 19: Intrahepatic Glissonian approach. In: Lau WY, editor. Hepatocellular carcinoma. Singapore: World Scientific; 2008. p. 429–46.

5 Anatomy of the Hepatic Hilar Region

To understand the hilar region, we must first of all understand that although the extrahepatic portal triad consisting of the hepatic artery, the portal vein and the bile duct is enclosed by connective tissues and peritoneum up to the hepatic hilum, the term Glisson's sheath is reserved for the Glisson's capsule which extends into the intrahepatic portion of the liver beyond the hilum. Also, because the hepatic hilar plate system is formed by the fusion of Glisson's capsule with the connective tissue sheaths surrounding the biliary and vascular elements at the inferior aspect of the liver, branches of the biliary and vascular elements can ramify in the hepatic hilar plate system before they go into the Glisson's sheath of a particular sector or segment of the liver. Alternatively, the elements can go straight into a particular Glisson's sheath after branching. Sometimes, for a particular liver sector or segment, branches of the artery, and bile duct can act differently from branches of the portal vein before they go into the Glisson's sheath.

5.1 Lymphatics in the Plate System

The lymphatics of the hilar area are distributed in a very complex manner, consisting of superficial and deep lymphatic systems, coursing through the plate system of the hilar area. The deep system is also called the lymphatics of Glisson's sheath which drains into this system. The superficial lymph vessels travel through the junction between the upper edge of the hilar plate and the capsule of segment 4a. Histology showed a thick layer of connective tissues with numerous lymph vessels between segment 4a and the ventral aspect of the portal vein. In contrast, the connective tissue between the dorsal aspect of the portal vein and the caudate lobe is thin and contains very few lymph vessels.

5.2 Anatomy of the Hilar Region

The confluence of the right and left hepatic ducts is located in the hilar plate. In the treatment of carcinoma of the hilar bile ducts, it is important to resect the hilar duct confluence en bloc with the hilar plate because tumour cells can easily invade into the adjacent plate tissues.

On the right side of the hilar plate, the right hepatic artery courses between the cystic plate and the hilar plate, and the branches of the right portal vein run dorsally to the artery before they enter into the Glisson's sheaths.

On the left side, the middle hepatic artery, after branching off from the left hepatic artery, courses through the umbilical plate, and the left branch of the portal vein lies dorsal to the artery. At these sites, the hilar ducts can be separated easily from the portal vein by dividing the common bile duct caudally and lifting it cranially (Fig. 5.1).

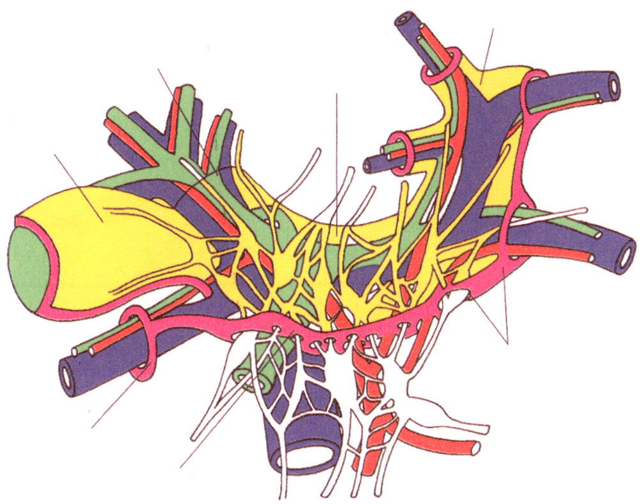

Fig. 5.1 Components of the plate system. *GB* Gallbladder, *G* Glisson's Sheath

W. Y. Lau, *Applied Anatomy in Liver Resection and Liver Transplantation*, https://doi.org/10.1007/978-981-16-0800-1_5

5.3 Vascular and Biliary Branches to the Caudate Lobe and to Segment 4

The hepatic plate system is continuous with the Glisson's sheath. The bile ducts and blood vessels usually course through the plate system to enter into the Glisson's sheath intrahepatically. The arrangement of the bile ducts and blood vessels going into the Glisson's sheaths is different in the caudate lobe and segment 4 of the liver.

For the caudate lobe, the bile ducts and arteries of the caudate lobe ramify in the hilar plate, but the branches of the portal vein run directly to the caudate lobe. Thus, while the caudate lobe artery and bile duct branches do not form a Glisson's triad immediately after branching within the liver, the portal vein branches do. Each portal vein branch is enclosed within a Glisson's sheath together with the arterial and ductal elements immediately after branching to form a Glisson's triad (Fig. 5.2).

The intrahepatic bile duct branches of segment 4 of the liver make a J-shape turn. Each branch then runs out of the Glisson's sheath, pieces the umbilical plate and ultimately joins the left hepatic duct. Branches of the middle hepatic artery enter into the Glisson's sheaths of segment 4 soon after branching. By contrast, the portal vein branches to segment 4 arises from the right side of the umbilical portion of the portal vein in a straight line, and each branch enters into its corresponding Glisson's sheath to form a Glisson's triad together with its hepatic arterial and bile ductal elements after travelling for a distance intrahepatically. The vessels and ducts of segment 4 do not form a Glisson's sheath immediately to the right of the umbilical portion of the portal vein, but at some distance away from it (Fig. 5.3). These anatomical differences in that ducts and blood vessels of segment 4 are attributable to the fact that this segment develops later than the other segments of the liver, and that the portal vein branches of segment 4 develop from the tertiary level of the portal system in the embryo. Such a view, however, is still controversial and not universally accepted (see Chap. 2).

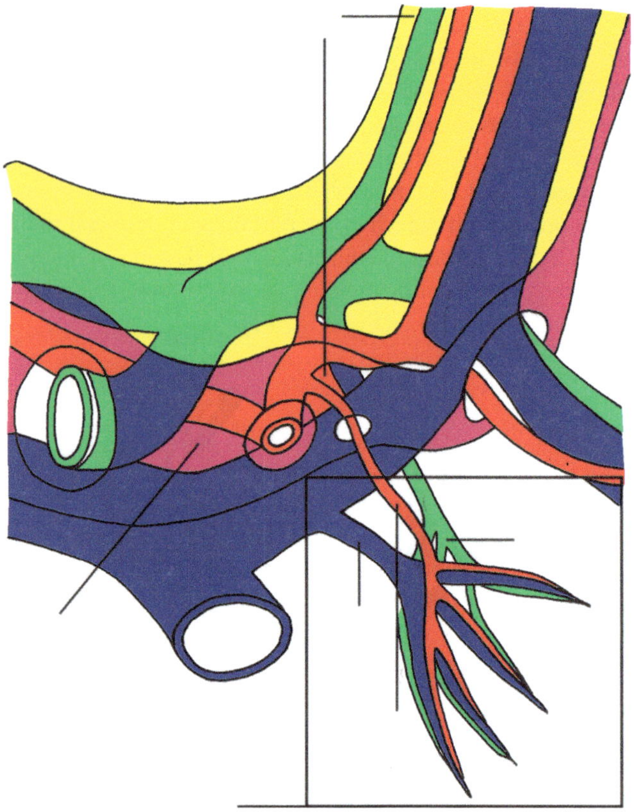

Fig. 5.2 Vascular and biliary branches to the Caudate Lobe. *LHD* left hepatic duct, *PV* portal vein, *HA* hepatic artery, *P* portal vein branch of S1 (left side), *A* hepatic artery branch S1 (left side), *B* bile duct branch of S1 (left side), *G* Glisson's sheath of S1 (left side)

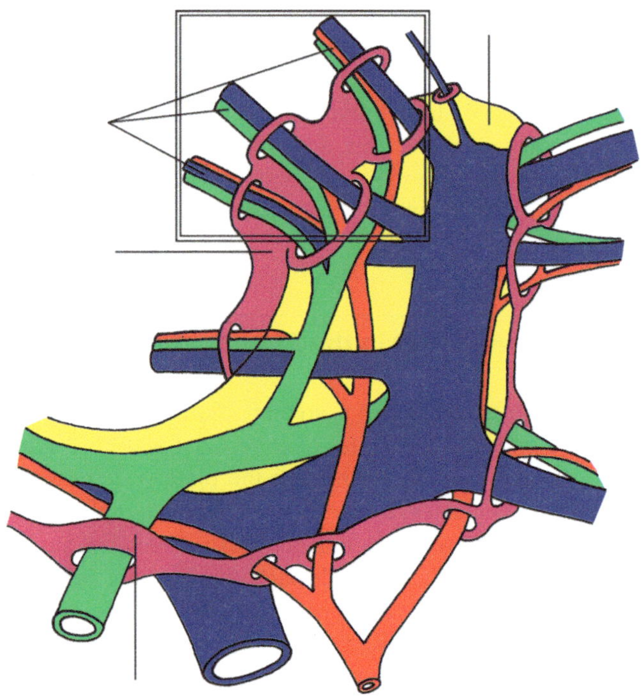

Fig. 5.3 Vascular and biliary branches to segment 4. *PV* portal vein, *LPV* left portal vein, *UP* umbilical portion of the left portal vein, *LHD* left hepatic duct, *G2* Glisson's sheath to segment 2, *G3* Glisson's sheath to segment 3, *G4b* Glisson's sheath to segment 4b

Further Reading

Couinaud C. Controlled hepatectomies and exposure of the intrahepatic bile ducts. Anatomical and technical study. Paris: C. Couinaud; 1981.

Healey JE, Schroy PC. Anatomy of the biliary ducts within the human liver. Arch Surg. 1985;68:599–616.

Ishiyama S, Yamada Y, Narishima Y, Yamaki T, Kuaii Y, Yamauchi H. Surgical anatomy of the hilar bile duct carcinoma. J Biliary Tract Pancreas. 1999;20:811–29.

Kawarada Y, Das BC, Taoka H. Anatomy of the hepatic hilar area: the plate system. J Hepato-Biliary-Pancreat Surg. 2000;7:580–6.

Kida H, Uchimura H, Okamoto K. Intrahepatic architecture of bile and portal vein. J Biliary Tract Pancreas. 1987;8:1–7.

Kumon M. Portal vein and bile duct branches of the caudate lobe: analysis of liver casts and clinical cases. Acta Hepatol Jpn. 1985;55:1193–9.

Michels NA. New anatomy of the liver and its variation in blood supply and collateral circulation. Am J Surg. 1966;112:337–46.

Suzuki H. Correlation and anomalies of the vascular structure in Glisson's area around the hepatic hilum, from the standpoint of hepatobiliary surgery. Arch Jpn Chir. 1982;51:713–31.

6 Intrahepatic Glissonian Triad: Anatomy Relevant to Liver Resection and Liver Transplantation

After entering into the liver, each bile duct, hepatic artery and portal vein unit is surrounded by a fibrous sheath called the Glisson's sheath (or Glissonian sheath, Valoean sheath). Each liver segment is supplied individually by one or more of these Glisson's triad (or Glissonian triad, portal pedicle, portal trinity) contained within the Glisson's sheath, thus making each liver segment an independent unit that can be resected individually or in combination. As some liver segments are supplied by more than one Glisson's triad for a single liver segment, subsegmentectomy can be done for these liver segments. However, before these bile duct or vascular elements enter into their respective Glissonian sheaths, they can ramify independently of each other outside of the liver, or even in the liver hilar plate system. This explains why sectoral bile duct anomalies can be different from sectoral portal vein anomalies as described in this chapter.

6.1 Number of Glissonian Sheaths to Each Liver Segment

If we look at the number of Glissonian sheaths going into each individual liver segment, the number varies tremendously (Table 6.1).

Table 6.1 Number of Glissonian Sheaths to each liver segment

	Usually	Occasionally	
Segment 1	Spiegelian lobe	1	2–3
	Paracaval	1	2–3
	Caudate process	1	2
Segment 2	1	2	
Segment 3	1–2	3	
Segment 4	3–10	Can be more	
Segment 5	Several	1	
Segment 6	1 (<50%)	2–3	
Segment 7	1	2	
Segment 8	3–4	Can be more	

Scheele in 1989, in studying corrosive casts of the human liver after filling of the portal structures and subtotal removal of small tributaries, concluded that the portal pedicles leading to the peripheral segments 2, 3, 6 and 7 are characterised by a large main trunk, and in segments 6 and 7 by a treetop-like peripheral arborization. In contrast, structures to the central segments 4, 5 and 8 show an early ramification, sometimes bush-like, often fan-shaped, aligned on the body longitudinal axis. Consequently, as a rule, in a monosegmentectomy for these central segments, several Glissonian sheaths at various depths in the parenchyma must be dealt with. Later studies by Lau showed that segment 1, although not mentioned by Scheele, like the other central segments, also has a bush-like or fan-like distribution of the Glissonian sheaths (Fig. 6.1).

As a consequence of these arrangements of the Glissonian sheaths, resection of a peripheral liver segment is technically easier than a central liver segment. Also, because of the multiple Glissonian sheaths to a central segment, subsegmental resection is technically easier for a central liver segment (1, 4, 5, 8) than a peripheral segment (2, 3, 6, 7). This is fortunate because most of the extended liver resections involve resection of part of a central liver segment/subsegment, e.g. extended right hepatectomy involving resection of part of segment 4, extended left hepatectomy involving part of segments 8 and/or 5. Also isolated partial resection of the caudate lobe and subsegmentectomy of segments 4, 5, 8 is possible, thus opening the door to the further development of subsegment-based liver resection (see Chap. 15).

W. Y. Lau, *Applied Anatomy in Liver Resection and Liver Transplantation*, https://doi.org/10.1007/978-981-16-0800-1_6

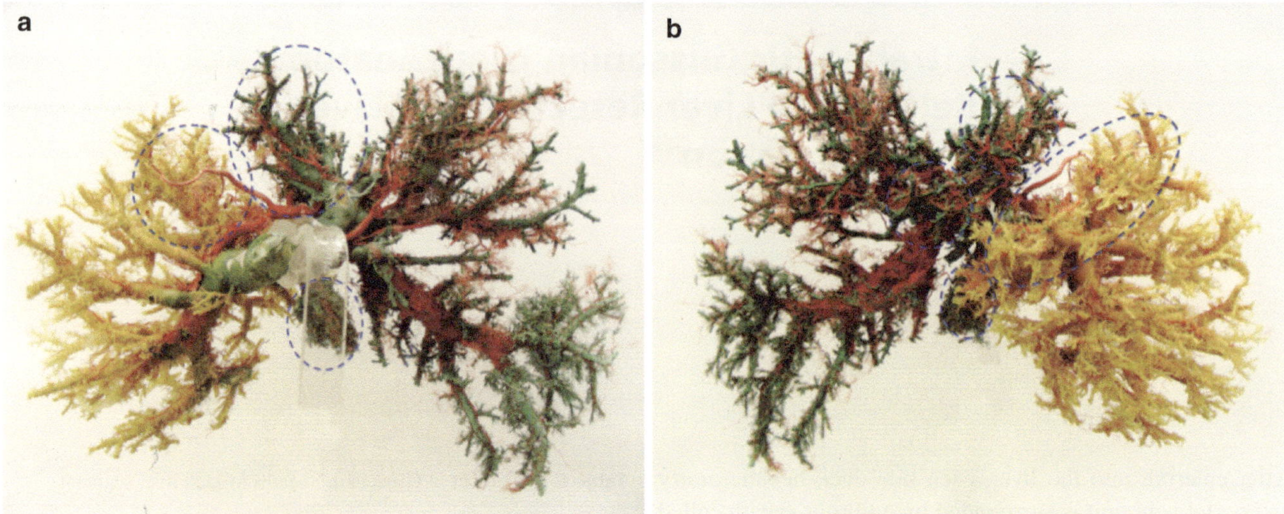

Fig. 6.1 Corrosion cost of the human liver after filling of the portal structures and subtotal removal of small tributaries. Numbers indicate segments. (**a**) Anterior aspect demonstrating the difference in branching characteristic of peripheral versus central segments. (**b**) Posterior aspect showing the distribution pattern of the portal triad

6.2 Glissonian Sheaths to Segment 4

The pedicles to segment 4 have more variations than any other segments in the liver. The portal pedicles commonly are between three and ten in number, and there may be many more. The arterial and biliary pedicles are even more variable. When these vasculo-biliary structures are dissected individually for isolated resection of segment 4, the surgeon is putting the blood supply and/or biliary drainage of segments 2 and 3 at considerable risk (see Sect. 2.3).

In isolated segment 4 liver resection, in split and reduced liver graft techniques in liver transplantation and in ALPPS, it is important to preserve intact blood and biliary supply to segments 2 and 3 by dissecting outside the Glissonian sheath. The surgeon should avoid entering into the Glissonian sheath by dissecting outside the hilar plate in the porta hepatis and incising 5 mm to the right of the umbilical fissure within the liver parenchyma. The major sheaths to segment 4 come from the right side of the sheath lying in the umbilical fissure. There are usually two or three such sheaths. Additional sheaths arise from the main left sheath as it transverses the base of segment 4. These sheaths supply the posterior part of segment 4, which is the reason why the posterior part (segment 4a) can be safely left undisturbed if necessary, even though the main sheaths to segment 4b have been divided. This anatomy also allows segment 2, 3 resection to be combined with 4b resection.

The major venous drainage of segment 4 is via the middle hepatic vein through the segment 4 vein. Occasionally the vein of the umbilical fissure provides enough drainage to allow segment 4 to survive even if the middle hepatic vein is resected together with the segment 4 vein (see Chap. 7).

In resection of segments 2 and 3, the surgeon should dissect the liver parenchyma 5 mm to the left of the umbilical fissure. Segments 2 and 3 are drained by the left hepatic vein.

6.3 Anomalies of Bile Ducts Draining the Right Hemiliver

6.3.1 Right Hepatic Duct

Studies of casts of normal liver revealed that the most common anatomy is the presence of a right hepatic duct that joins the left hepatic duct (53–72%). In most individuals (83–89%), the posterior sectoral bile duct runs superiorly, dorsally and then inferiorly (Hjortsjo crook) to the right branch of the portal vein (the 'north-turning bile duct branch') (Fig. 6.2). In the minority (11–17%), the posterior sectoral bile duct courses ventrally and inferiorly to the right branch of the portal vein (the 'south-turning' bile duct branch).

The Hjortsjo crook occurs in the majority of the population. As the right posterior sectoral bile duct courses superiorly, dorsally and inferiorly to the right branch of the portal vein and hooks over the origin of the right anterior sectoral portal vein, resection of the anterior right sector of the liver (segments 5 and 8) can damage the right posterior sectoral duct if the resection is done too close to the bifurcation of the right portal vein into the anterior and posterior sectoral branches. The correct procedure is to stay away from the bifurcation of the right portal vein (Fig. 6.3).

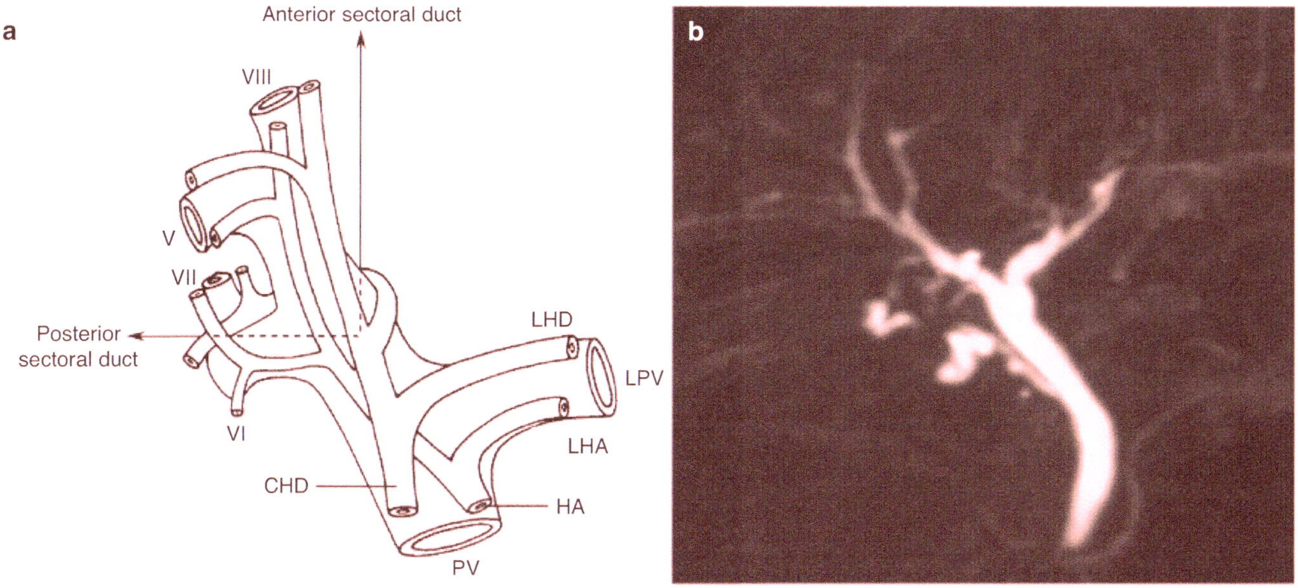

Fig. 6.2 (**a**) Hjortsjo Crook. (**b**) Magnetic resonance cholangiopancreatogram showing Hjortsjo Crook

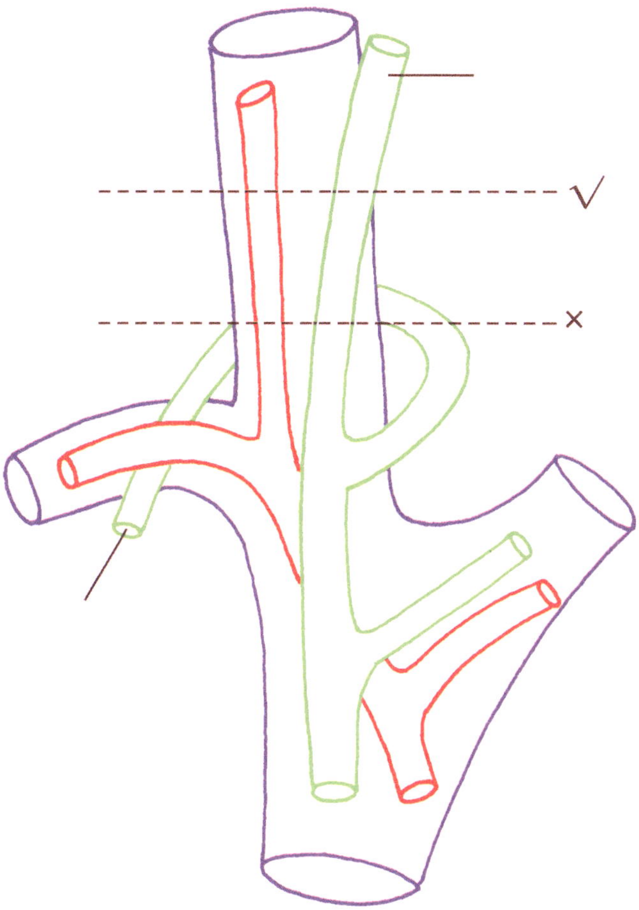

Fig. 6.3 Hjortsjo Crook and its clinical significance

6.3.2 Absence of Right Hepatic Duct

The absence of the right hepatic duct is an anatomical variation that results during development.

There are three anatomical variations in which the anterior and posterior sectoral ducts do not form the right hepatic duct, thus resulting in absence of the right hepatic duct (Fig. 6.4 shows the incidences of these anomalies): This paragraph is reprinted from: W.Y. Lau, Hilar Cholangiocarcinoma. Springer, 2013. ISBN 978-94-007-6472-9.

The anomalies which can result in absence of the right hepatic ducts are:

6.3.2.1 Shifting of the Entry of the Right Bile Duct Inferiorly

This set of anomalies involves the insertion of the right bile duct, or one of its branches, inferiorly into the biliary tree at a lower point than the prevailing site of confluence (Fig. 6.5). This paragraph is reprinted from: W.Y. Lau, Hilar Cholangiocarcinoma. Springer, 2013. ISBN 978-94-007-6472-9.

Low union may affect the main right bile duct, a sectoral right duct (usually the anterior one and this anomaly results in absence of right hepatic duct), a segmental duct, or a subsegmental duct. The duct unites with the common hepatic duct below the prevailing site of confluence, or in about 2% of patients, unites first with the cystic duct and then with the common hepatic duct. These anomalies place a greater risk of

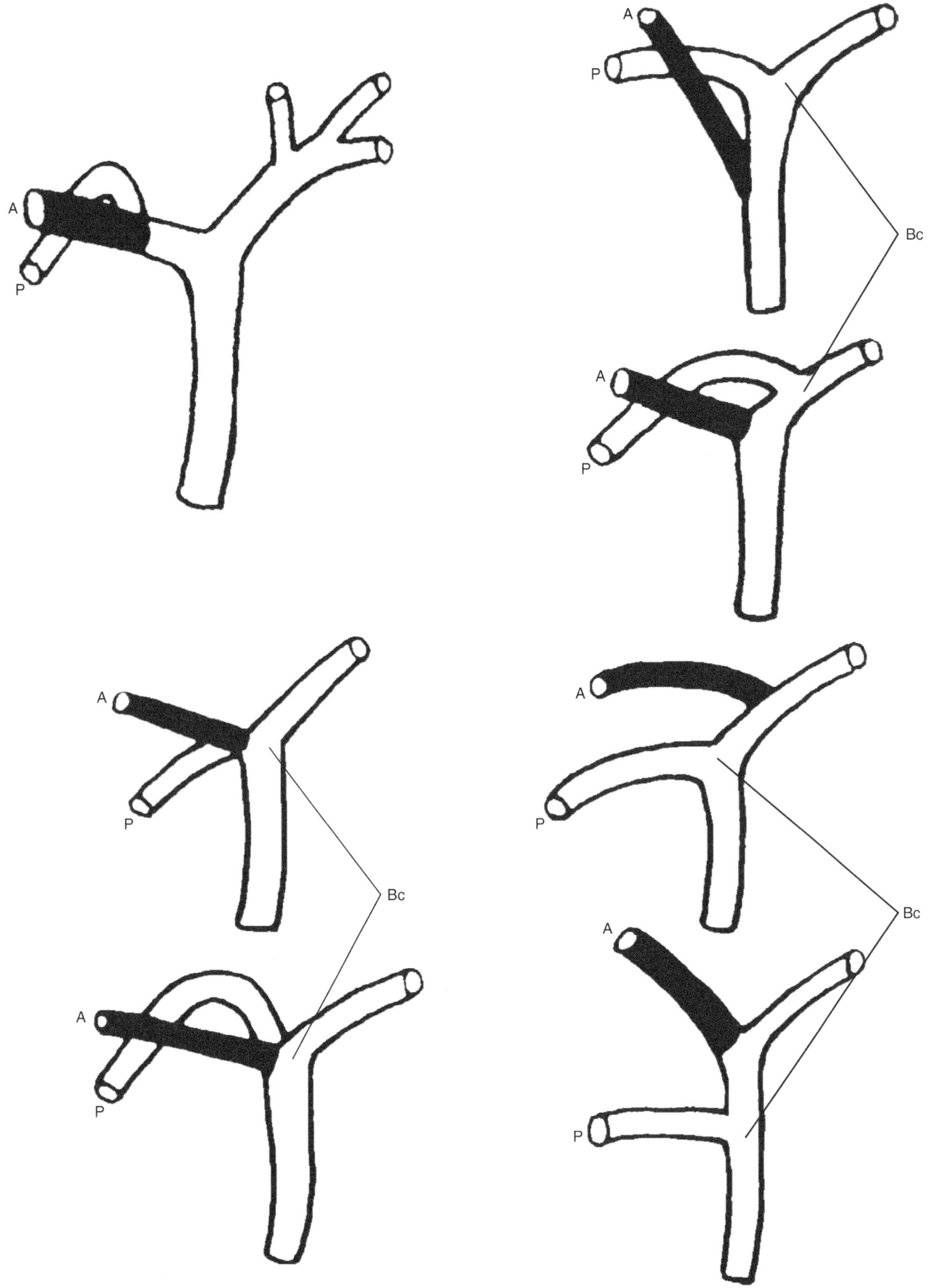

Fig. 6.4 Variations in the anatomy of the right hepatic duct and their incidences, based on analysis of liver casts. *A* anterior sectoral branch, *P* posterior sectoral branch, *Bc* bile duct confluence

ductal injury during laparoscopic cholecystectomy. This paragraph is reprinted from: W.Y. Lau, Hilar Cholangiocarcinoma. Springer, 2013. ISBN 978-94-007-6472-9.

6.3.2.2 Trifurcation of Bile Duct

When performing a right hemihepatectomy, a left hemihepatectomy, an anterior right sectionectomy, a right posterior sectionectomy, a right trisectionectomy or a left trisectionectomy, a stricture is likely to develop at the biliary trifurcation site if no normal biliary safety margin is left at the site of the transection. It is always safer to divide the biliary tree with a safety margin of at least 1 cm from the site of the biliary confluence (Fig. 6.6). This paragraph is reprinted from: W.Y. Lau, Hilar Cholangiocarcinoma. Springer, 2013. ISBN 978-94-007-6472-9.

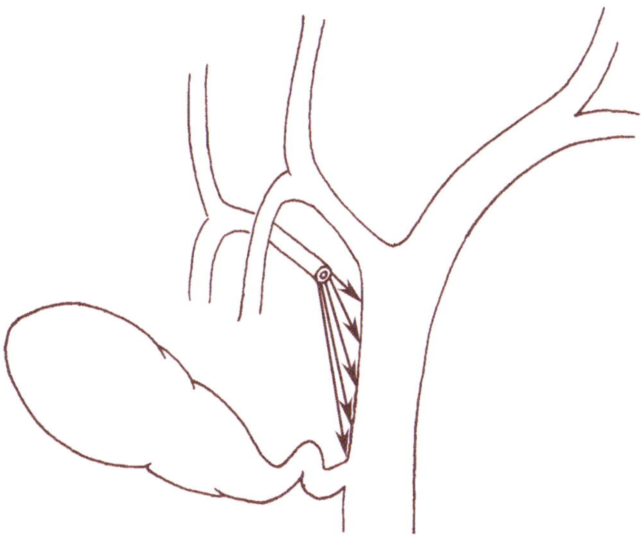

Fig. 6.5 Shifting of entry of right bile duct inferiorly

6.3.2.3 Anterior or Posterior Sectoral Branch Joining the Left Hepatic Duct

The right posterior sectoral duct inserts with the left bile duct in 20% of patients and the right anterior sectoral bile duct does so in 6%. In both cases, there is no right hepatic duct as both join the left duct, one to the left of the midline and the other in the midplane. A right sectoral bile duct inserting into the left bile duct to the left of the midplane is in danger of injury during left hepatectomy. Therefore, in the left hepatectomy, the left bile duct should be divided close to the umbilical fissure so as to avoid injury to a right sectoral duct. If the left duct is divided at the normal site of confluence of the right and left hepatic ducts, the right sectoral duct can be injured (Fig. 6.7). It is good practice to obtain an intraoperative cystic duct cholangiogram when performing a left hepatectomy to detect this anomaly. Please take note that even with these anomalies, a right hepatectomy is safe. This paragraph is reprinted from: W.Y. Lau, Hilar Cholangiocarcinoma. Springer, 2013. ISBN 978-94-007-6472-9.

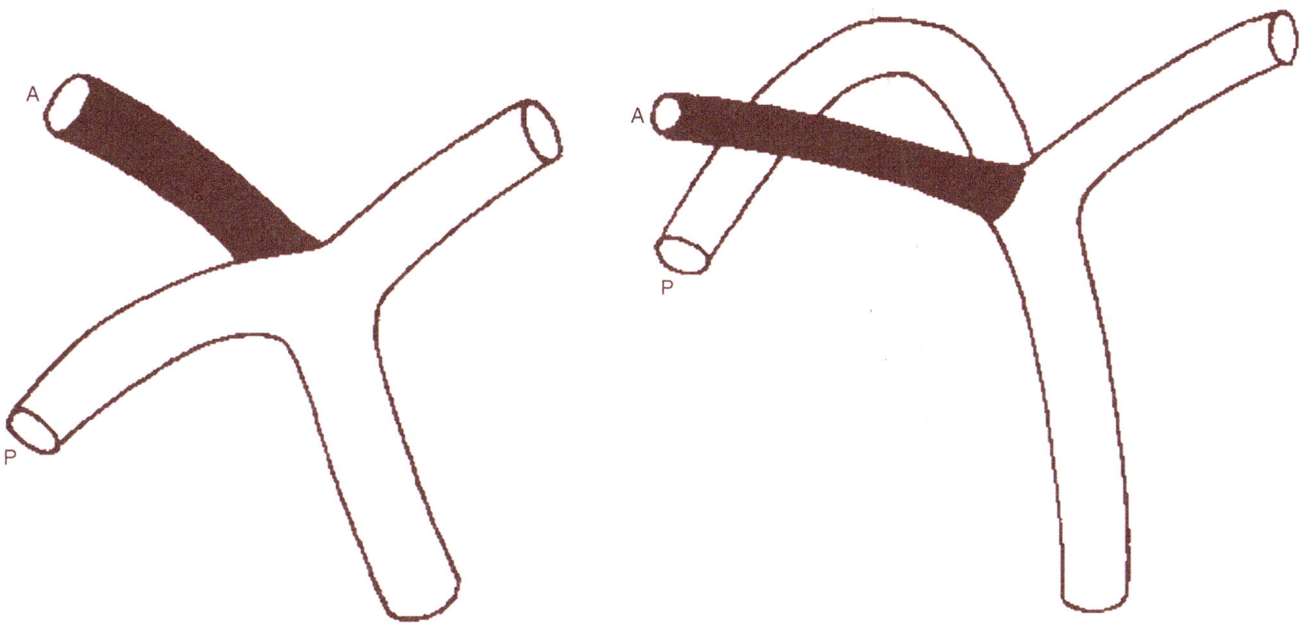

Fig. 6.6 Trifurcation of bile duct. *A* right anterior sectoral branch, *B* right posterior sectoral branch

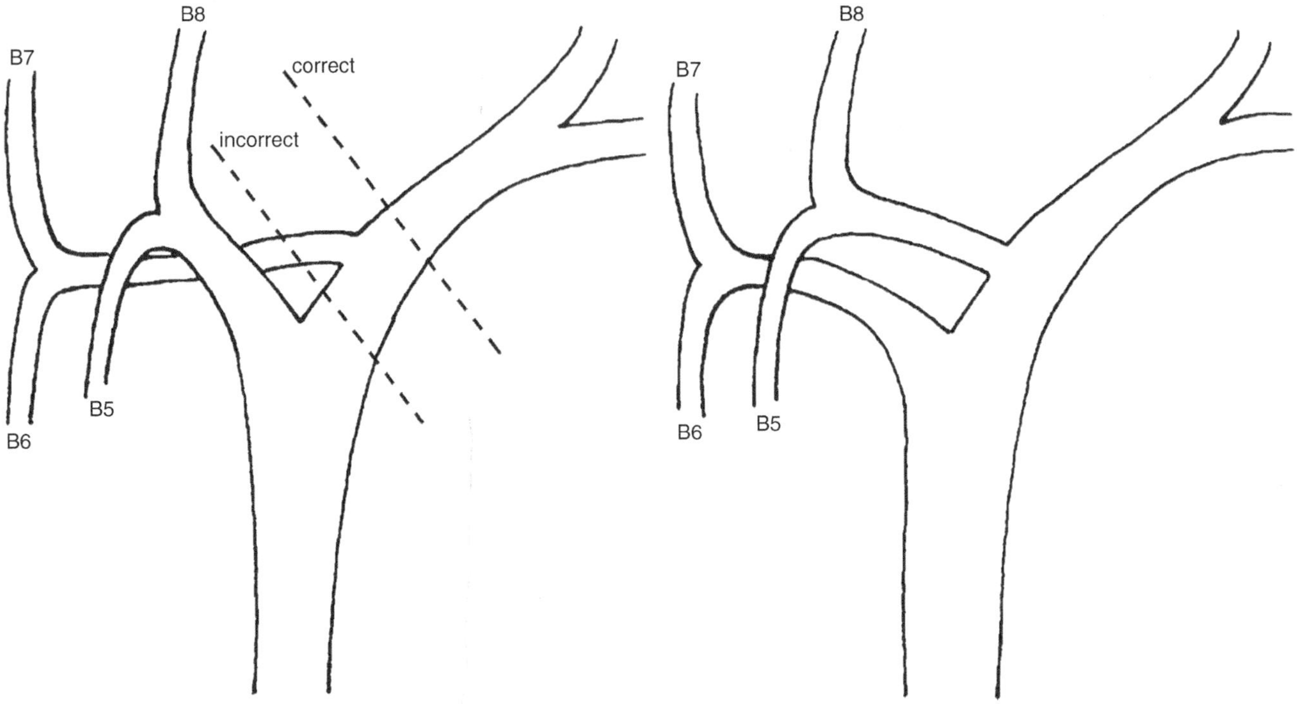

Fig. 6.7 Separate entry of right anterior and right posterior sectoral ducts (no right hepatic duct)

6.4 Anomalies of Bile Ducts Draining the Left Hemiliver

The left hepatic duct has a much longer extrahepatic course than the right bile duct. Normally, a 2–3 cm length of the left hepatic bile duct is in an extrahepatic position.

The prevailing pattern of bile duct drainage from the left liver is shown in Fig. 6.8a and is present only in 30% of individuals. Thus, variations are present in the majority of individuals.

The segmental ducts from segments 2 and 3 (B2 and B3, respectively) unite to form the left lateral sectional duct. This duct passes behind the umbilical portion of the portal vein and unites with the duct from segment 4 (B4), also called the left medial sectional duct. The union of these ducts to form the left hepatic duct occurs about one-third of the distance between the umbilical fissure and the confluence of the left and right bile ducts.

The surgically important anomalies of the left ductal system involve variations in the site of insertion of B4 (Fig. 6.8b), multiple ducts coming from B4 (Fig. 6.8c), and the primary union of B3 and B4 with subsequent union of B2 (Fig. 6.8d). B4 may join the left lateral sectional duct to the left or right of its point of union in the prevailing pattern (Fig. 6.8b). In the former case, the insertion may occur at any place to the right of the prevailing location up to the point where the left lateral sectional duct unites with the right bile duct. In the latter instance, which according to Couinaud, is present in 8% of individuals, there is no left hepatic duct, instead the common hepatic duct is formed by the confluence of three ducts—the right hepatic duct and two left hepatic ducts (B4 and the left lateral sectional duct).

The commonest variations in the left bile duct are: Type 1, common joining of B2 with B3, B4 then joins near to the left hepatic duct; Type 2, common joining of B4, B3 and B2 near to the same point; and Type 3, common joining of B4 with B3 (Fig. 6.9).

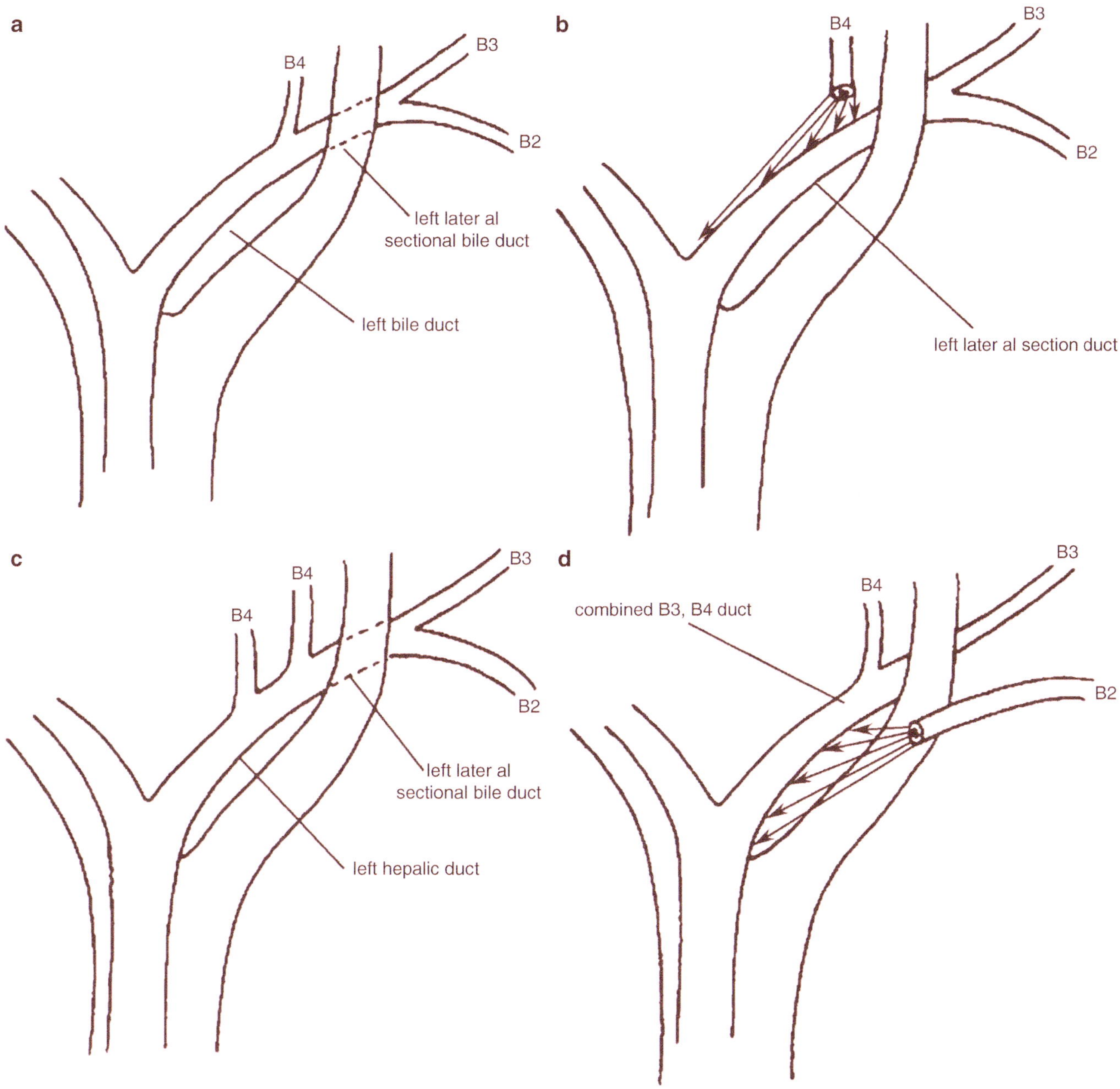

Fig. 6.8 Variations in formation of left hepatic ducts. (**a**) Prevailing pattern of left bile duct. (**b**) Insertion of B4 shifted to right or left. (**c**) Multiple ducts draining B4. (**d**) B3, B4 form common channel before insertion of B2

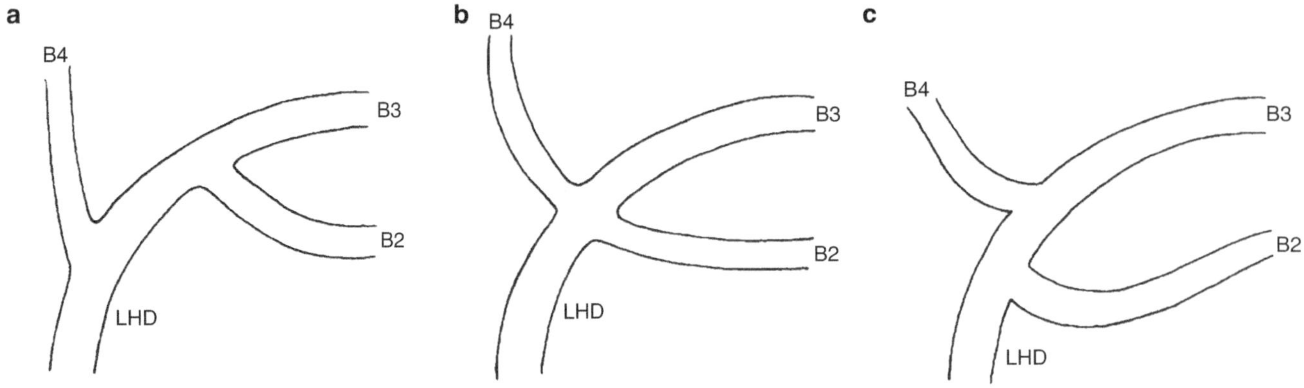

Fig. 6.9 Common variations in the anatomy of the left bile duct. (**a**) Common Joining of B2, 3 Type 1. (**b**) Common joining of B2, 3, 4, Type 2. (**c**) Common joining of B3, 4 Type 3

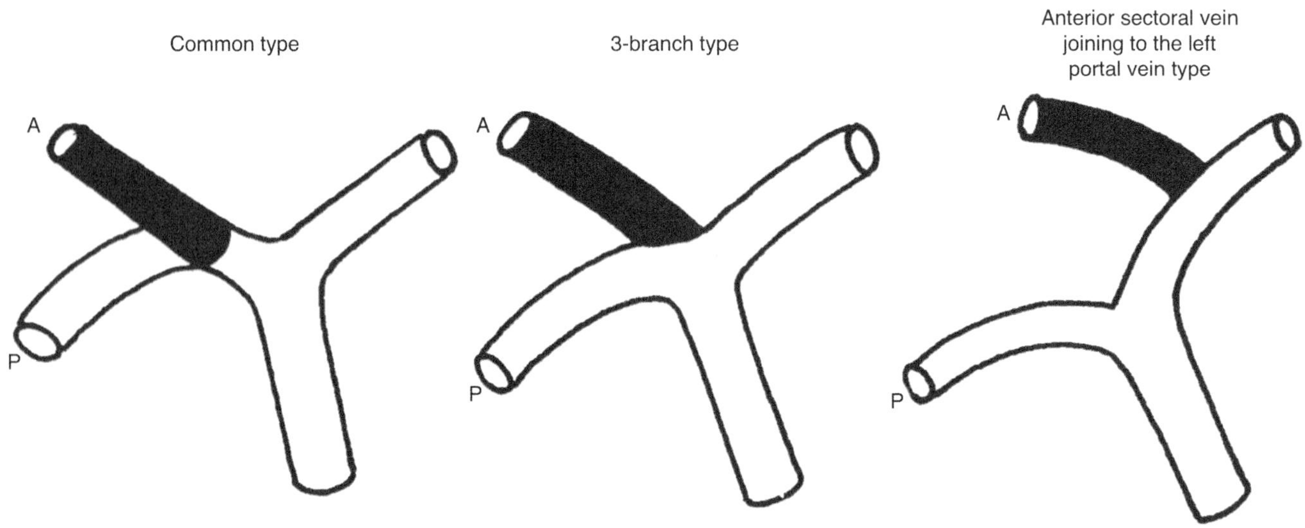

Fig. 6.10 Variations in the anatomy of the right portal vein (analysis of liver cast). *A* Anterior sectoral branch. *P* Postterior sectoral branch

The left hepatic duct runs at a variable angle. In some individuals, it is almost horizontal, but in others, it runs sharply upward. It is much easier to expose a long length of duct in the former type.

The prevailing pattern of bile ducts draining the caudate lobe has been described in detail in Chap. 3.

6.5 Variations in Anatomy of the Portal Vein Branches in the Hilar Area

Few variations are found in the major portal vein branches because the portal vein develops during the very earliest part of the gestational period. Reports by three investigators revealed three principal portal vein branching patterns in the hilar area (Fig. 6.10):

Kida et al. reported that variations in the anatomy of the biliary tract are mostly (81%) associated with variations in the anatomy of the portal vein.

6.6 Portal Vein and Liver Resections

How the portal vein and the hepatic veins divide the liver into hemilivers, sections/sectors and segments has been described in detail in Chap. 2. On the right side of the liver, the portal vein division corresponds exactly to those of the hepatic artery and bile duct. On the left side of the liver, arteriobiliary segmentation differs from portal vein segmentation. The left portal vein consists of a transverse and an umbilical portion. The transverse portion sends only a few small branches to segment 4 and one or two branches to segment 1. All larger branches from the portal vein to the left liver arise exclusively beyond the attachment of the ligamentum venosum, i.e. from the umbilical portion of the vein (Fig. 6.11).

The umbilical portion of the portal vein has a unique pattern of ramification. The pattern is similar to an air-conditioning duct that sends branches at the right angle from both of its sides to supply rooms (liver segments). On the right side usually more than one branch pass to segment 4. On the left side usually, one branch goes to segment 2, but more than one branch to segment 3. The left portal vein terminates where it joins the ligamentum teres at the free edge of the liver (Fig. 6.11). The explanation for this unique structure of the umbilical portion of the left portal vein is that it is adapted to a dual function: in utero, it acts as a conduit between the umbilical vein and the ductus venosus (Fig. 6.12), a conduit in which blood flows towards the duc-

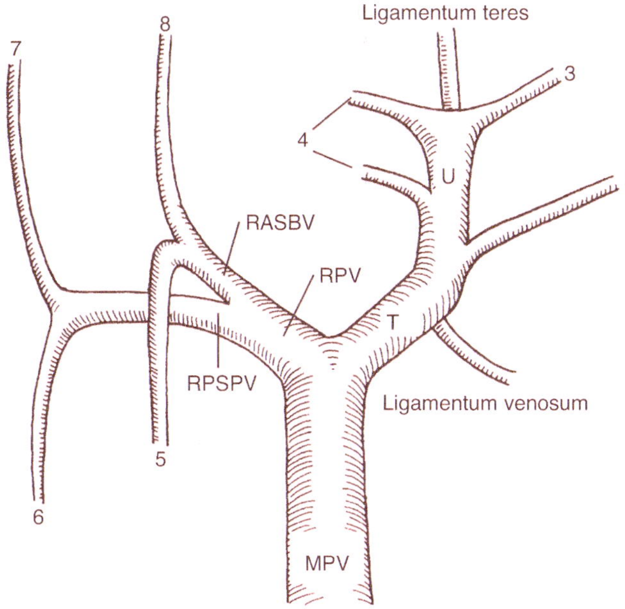

Fig. 6.11 The portal vein and its intrahepatic branches. *MPV* main portal vein, *RPV* right portal vein, *RASBV* right anterior sectoral portal vein, *RPSPV* right posterior sectoral portal vein, *U* umbilical portion of left portal vein, *T* transverse portion of left portal vein

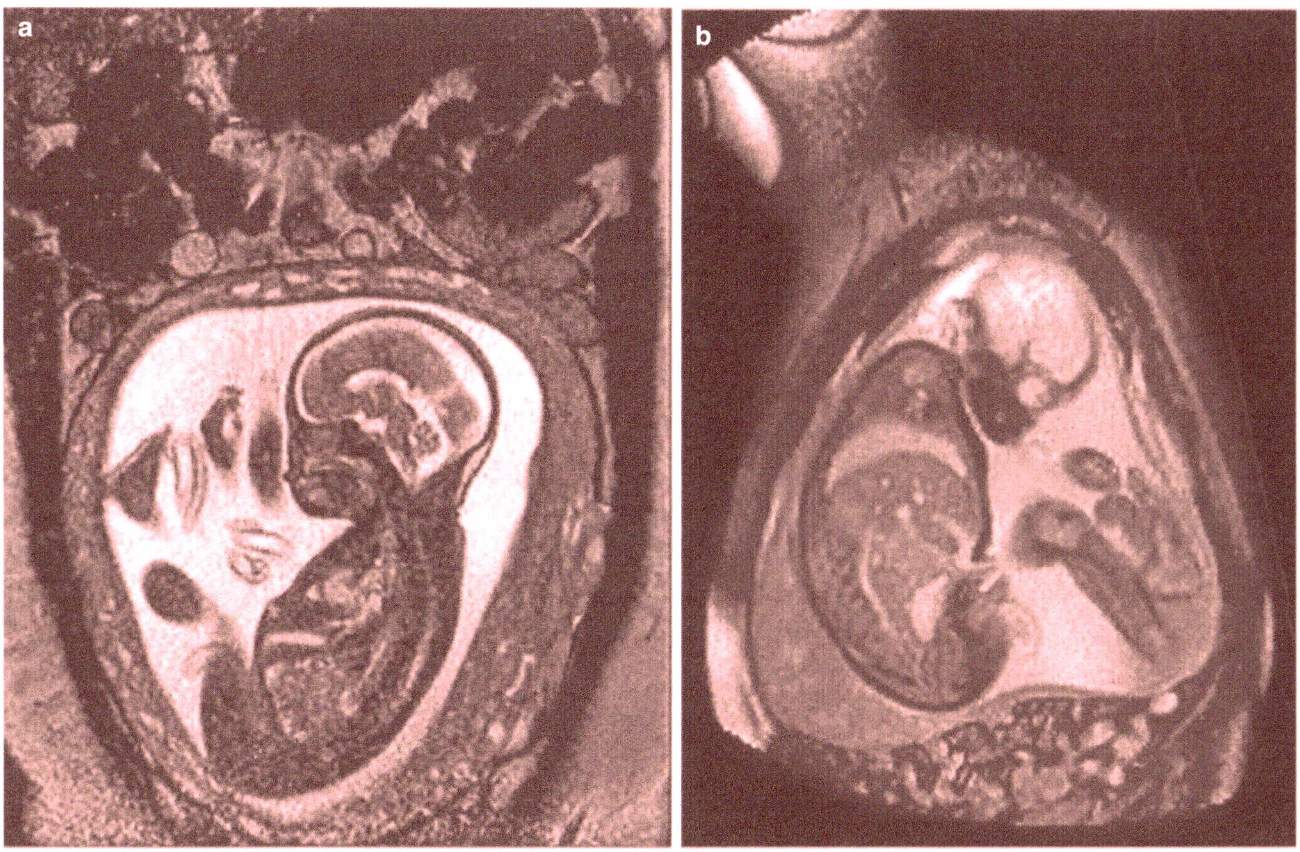

Fig. 6.12 (**a**) Magnetic resonance imaging showing ductus venosus (D) in a 25 weeks foetus. (**b**) Magnetic resonance imaging showing ductus venosus (D) and umbilical vein (U) in a 35-week foetus

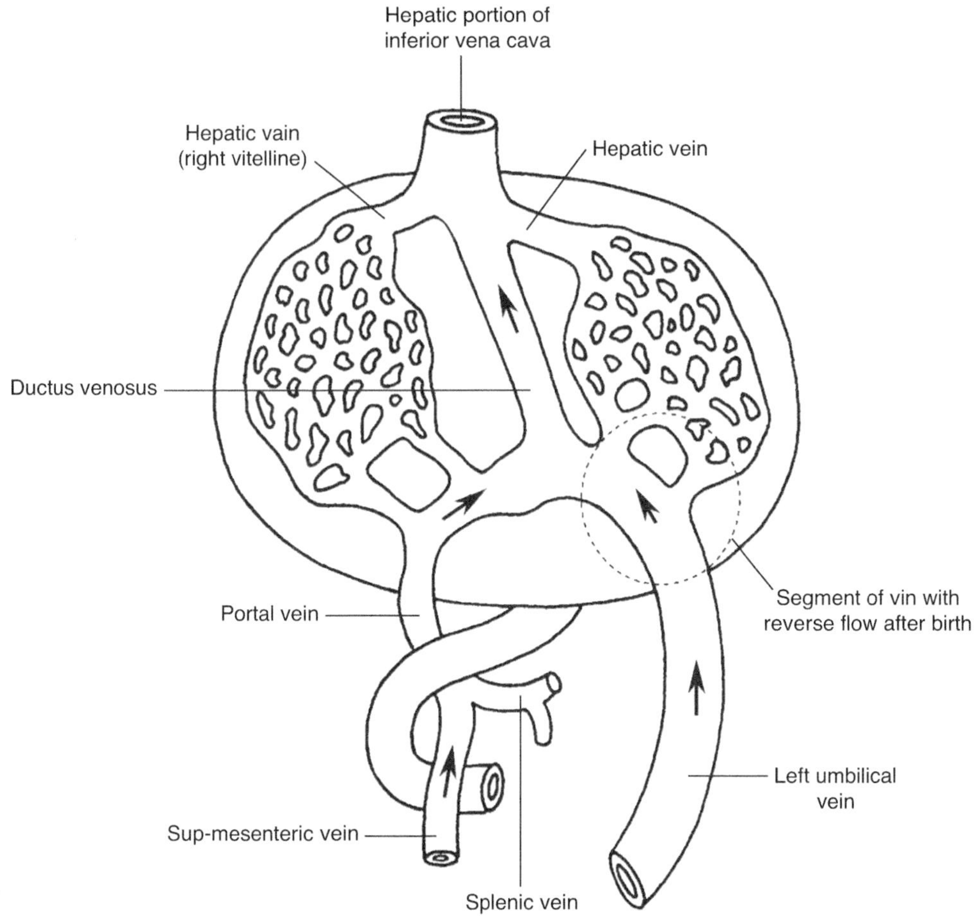

Fig. 6.13 Portal venous blood flow in utero

tus venosus (Fig. 6.13); in adult life, it acts as a conduit for portal vein supply to the left liver, and blood flows in the reverse direction from the ligamentum venosum towards the ligamentum teres (Figs. 6.14 and 6.15). It is for this reason that the umbilical portion of the portal vein is like an air-conditioning duct. It is also because its branching pattern on the left side of the liver is so structurally different from the branching pattern on the right side so that we have the different terms of sector and section on the left side while the term sector is equivalent to section on the right side of the liver.

It is uncommon to have variations of the portal vein division. Probably the most common is the right anterior sectoral vein joins to the left portal vein (see Sect. 6.5) while the right posterior sectoral vein originates independently from the main portal vein. When this occurs, the anterior sectoral vein is usually quite high in the porta hepatis and may not be obvious. An unsuspecting surgeon may divide the posterior sectoral vein, thinking that it is the right portal vein, and will consequently be confused when the anterior sectoral vein is come upon during hepatic transection carried out for right hepatectomy. In the left hepatectomy, the anterior sectoral portal vein may inadvertently be damaged, resulting in ischaemia to liver segments 5 and 8.

Another less common portal vein anomaly is trifurcation of the right anterior sectoral vein, right posterior sectoral vein and the left portal vein at the hilum. Again to avoid subsequent stricture developing in the portal vein that is left after liver transection, a safety margin of at least 1 cm should be left at the transection plane. A rare but potentially devastating anomaly is the absent extrahepatic left portal

Fig. 6.14 Blood circulation in utero

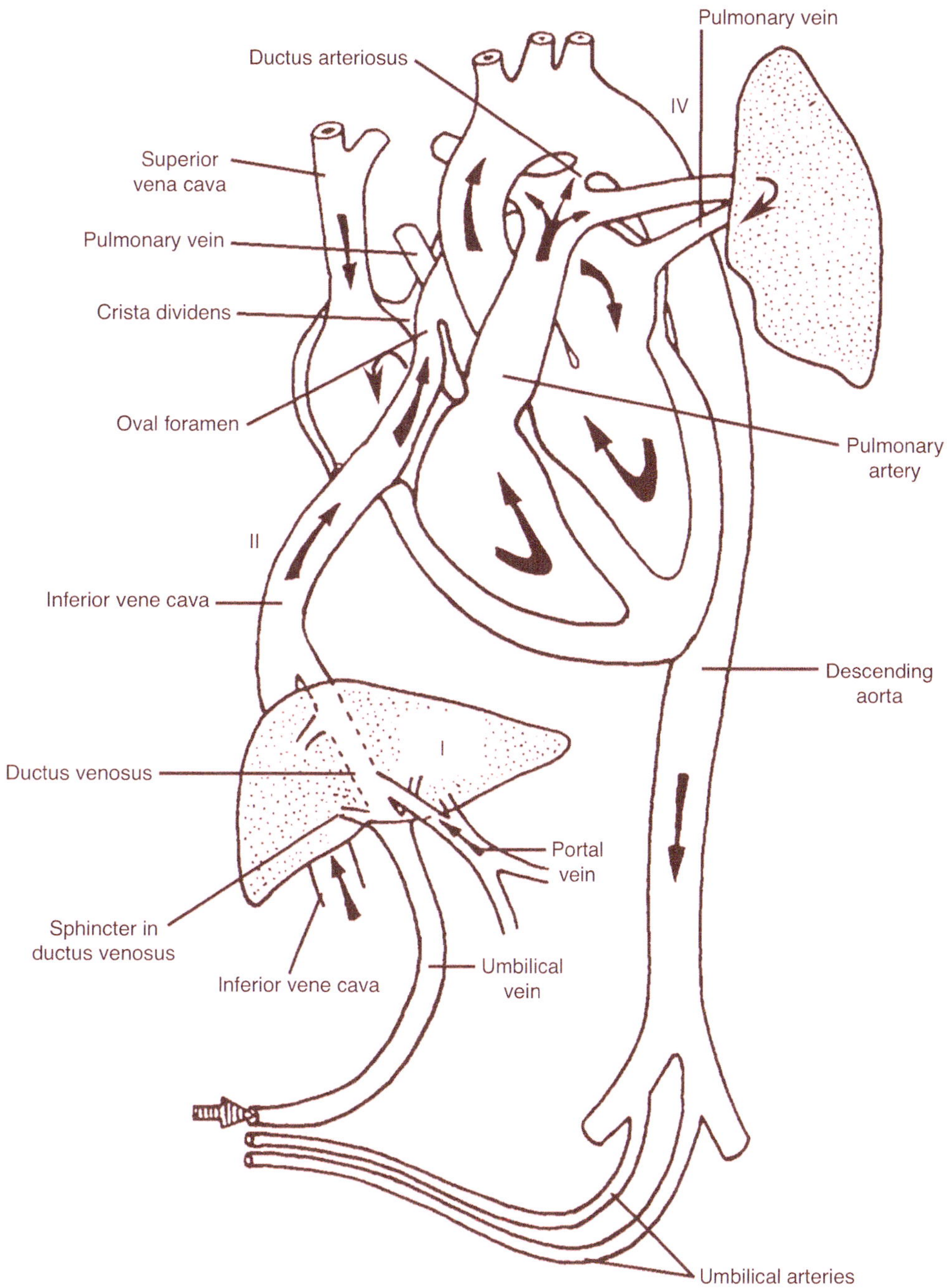

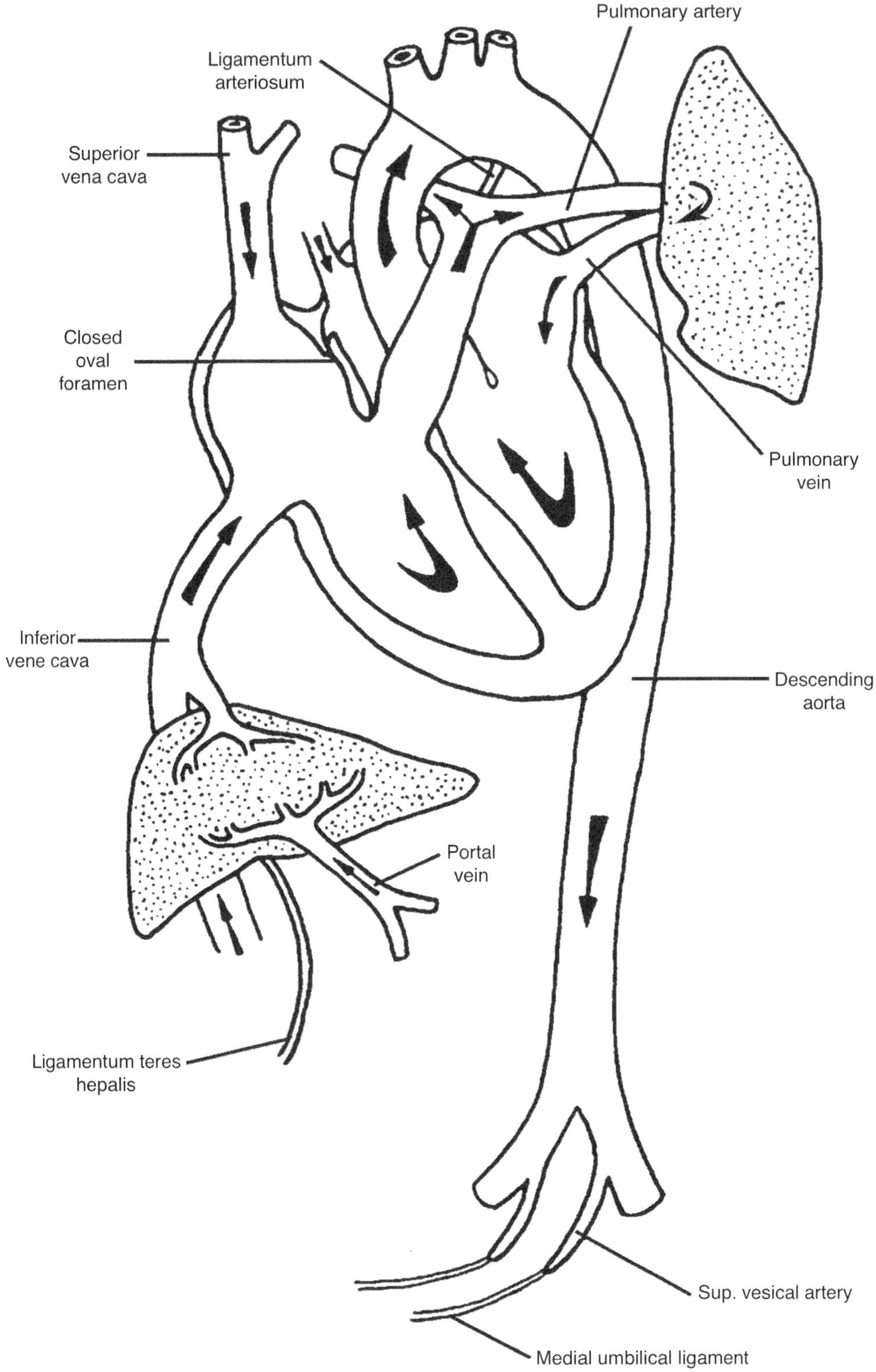

Fig. 6.15 Blood circulation after birth

vein (Fig. 6.16). In this case, the apparent right vein is actually the main portal vein, a structure that enters the liver, gives off the right vein, and then loops back within the liver substance to supply the left side. The vein looks like a right vein in terms of position, but it is larger. Transection results in total portal vein disconnection from the liver. This anomaly should always be searched for in computed tomography (CT), as right hepatectomy is not usually possible when it is present. Identification of the umbilical portion of the left vein in the umbilical fissure on CT can preclude the presence of this problem. Please note that a left hepatectomy is possible with this anomaly. This paragraph is reprinted from: W.Y. Lau, Hilar Cholangiocarcinoma. Springer, 2013. ISBN 978-94-007-6472-9.

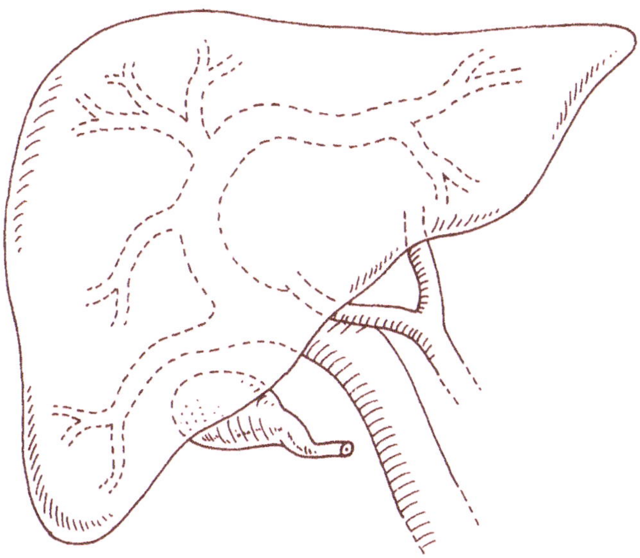

Fig. 6.16 Absent extrahepatic left portal vein

6.7 Variations in Anatomy of the Hepatic Arterial Branches in the Hilar Area

Embryologically, the hepatic artery develops late in the gestational period, and thus, variations are found in a large proportion of the populations (33–45%). More than ten variations in the anatomy of the hepatic artery, including an accessory or replaced artery, have been identified. In addition to the presence of an accessory hepatic artery, the pattern of the hepatic artery in the hilar area has been divided into the following types (Fig. 6.17).

Analysis of the anatomical relationship between the hepatic artery and the portal vein and hepatic duct branches revealed that the hepatic artery usually courses dorsal to the hepatic duct in 76% of the population, and sometimes ventral to it in 24% of the population. In addition, in 9% of the population, the right hepatic artery runs dorsal to the portal vein, making it necessary to pay special attention to the anatomy of the vessels and ducts of the hilar area during surgical dissection of this area.

6.8 Umbilical Fissure and Liver Resections

The umbilical portion of the left portal vein runs in the umbilical fissure. For resection of liver segments 2 and 3 (left lateral sectionectomy), the liver should be transected 5 mm away from the umbilical fissure on the left side. Care should be taken to avoid injury of the umbilical portion of the left portal vein, or else segments 2, 3 and 4 would be deprived of portal venous supply. For a tumour that impinges on the umbilical fissure, an extra resection margin can be obtained by dissecting into the umbilical fissure to isolate and protect

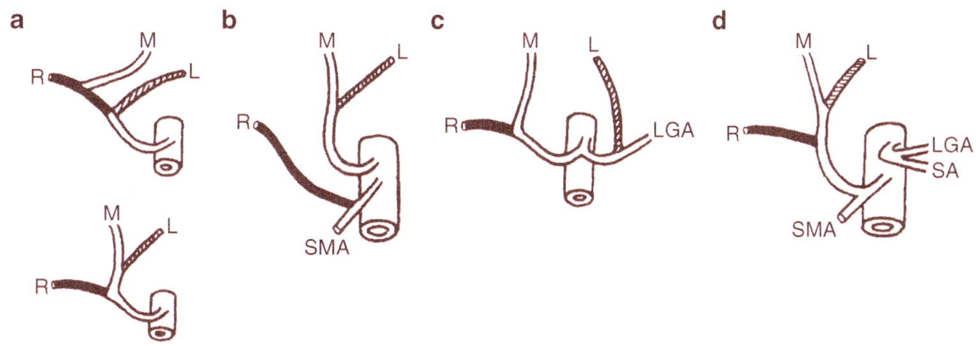

Fig. 6.17 Variations in the anatomy of the main hepatic arteries exclusive of accessory hepatic arteries. *R* right hepatic artery, *M* middle hepatic artery, *L* left hepatic artery, *SMA* superior mesenteric artery, *LGA* left gastric artery, *SA* splenic artery

the umbilical portion of the left portal vein and the portal branches to segment 4.

The approach to the umbilical fissure in isolated resection of segment 4b, and combined segments 2, 3 and 4b resection (with preservation of 4a), has been discussed previously (see Sect. 6.2).

6.9 Ligamentum Teres and Liver Surgery

After birth, the left umbilical vein becomes reduced to a fibrous cord, the ligamentum teres. The ligamentum teres runs upwards from the umbilicus to the inferior surface of the liver, commencing in the median plane at the umbilicus, then inclines slightly to the right and recedes slightly from the anterior abdominal wall. It raises a triangular fold of peritoneum from the anterior abdominal wall and inferior surface of the diaphragm in this upward course, forming a fold called the falciform ligament of the liver. After reaching the inferior surface of the liver, the ligamentum teres runs in a fissure called either the umbilical fissure or the fissure for the ligamentum teres. It is bridged over partially or completely by a band of fibrous or liver tissue. The ligamentum teres end at the termination of the left portal vein at the point where segment 3 and segment 4 branch off from each other.

Surgical importance of the ligamentum teres are:

1. Embryonic origin
2. Ligamentum teres approach to segment 3 bile duct (see Sect. 4.4)
3. Together with ligamentum venosum it forms the boundary between the left lateral and the medial section of the liver (see Chap. 2)
4. In the umbilical fissure runs the umbilical portion of the left portal vein
5. In cirrhosis of the liver and in obstruction or thrombosis of the portal vein, enlargement of the connexions between the vein running within the ligamentum teres anastomoses with the epigastric veins resulting in caput medusa, a varicose condition of veins radiating from the umbilicus.

6.10 Ligamentum Venosum and Liver Surgery

The ligamentum venosum is the fibrous remnant of the ductus venosum. It runs from the junction between the transverse and the umbilical portions of the left portal vein to the junction where the common trunk of the middle and left hepatic veins enters into the inferior vena cava. Along its course, the ligamentum venosum crosses anterior to the caudate lobe.

Surgical importance of the ligamentum venosum are:

1. Embryonic origin
2. Together with ligamentum teres, it forms the boundary between the left lateral and the medial sections of the liver (see Chap. 2).
3. It divides the caudate lobe into the Spiegelian lobe and the paracaval portion/caudate process (see Chap. 3).
4. It marks the transverse portion of the left portal vein from the umbilical portion. The transverse portion only gives small branches to segments 4 and 1 while the umbilical portion gives large branches to segments 2, 3 and 4.
5. By tracing and dividing the ligamentum venosum near to its termination at the common trunk of the middle and left hepatic veins, and after dissecting the trunk in the groove between the right and the left hepatic veins (see Chap. 8), the trunk can be encircled with a tape easily (Fig. 6.18).

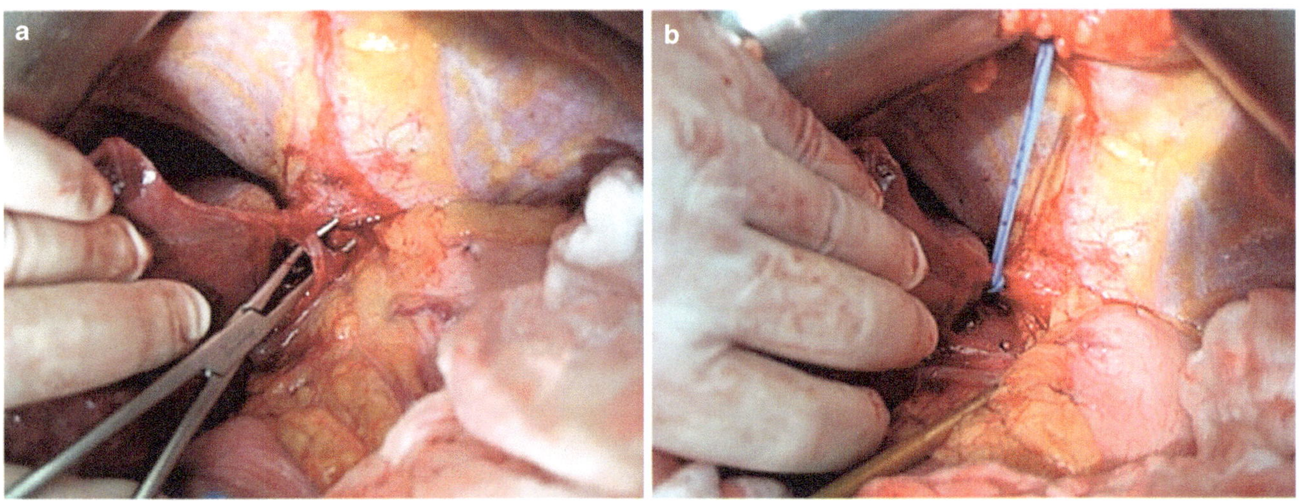

Fig. 6.18 (**a**) Dissecting and dividing the ligamentum venosum near to its termination at the common trunk of the middle and left hepatic veins. Note the dissection freed the left and the posterior parts of the common trunk. (**b**) After blunt dissection posteriorly and on the right side of the common trunk, the common trunk of the middle and left hepatic veins can be encircled with a tape

6. The beginning of left hepatic vein runs in the groove of the ligamentum venosum before the left hepatic vein turns to run between segments 2 and 3.

Further Reading

Jamieson G, Launois B. Chapter 2: Liver resections and liver transplantation: the anatomy of the liver and associated structures. In: Jamieson GG, editor. The anatomy of general surgical operations. Edinburgh: Elsevier; 2006. p. 8–23.

Scheele J, Stangl R. Chapter 86: Segment-orientated anatomical live resections. In: Blumgart L, Fong Y, editors. Surgery of the liver and biliary tract. 3rd ed. London: WB Saunders; 2000. p. 1743–63.

Strasberg SM. Chapter 2: Liver terminology and anatomy. In: Lau WY, editor. Hepatocellular carcinoma. Singapore: World Scientific; 2008. p. 25–50.

7 Anatomy of the Abdominal Inferior Vena Cava and Its Suprarenal Branches

7.1 Surface Markings

The inferior vena cava is formed by the junction of the right and left common iliac veins behind the right common iliac artery. This point corresponds to the surface of the abdomen at the transtubercular plane 2.5 cm from the right side of the median plane. The inferior vena cava is about 2.5 cm wide, runs cranially to end at a point 2.5 cm to the right of the xiphisternal joint. This point corresponds to the point where the inferior vena cava perforates the tendinous portion of the diaphragm and the pericardium to enter the inferioposterior part of the right atrium.

7.2 The Abdominal Inferior Vena Cava and Its Suprarenal Branches

There are no valves in the inferior vena cava except for a somewhat vestigial valve at its entry to the right atrium.

The relevant part of the abdominal inferior vena cava in relation to liver surgery is the suprarenal part where the bilateral renal veins join the inferior vena cava (Fig. 7.1). For ease of description, the venous branches that come off from this part of the inferior vena cava are divided into four groups:

1. The suprahepatic group
2. The hepatic veins
3. The retrohepatic group. These are the veins that drain from the caudate lobe.
4. The infrahepatic group

7.2.1 The Suprahepatic Group

The right inferior phrenic vein drains the undersurface of the right diaphragm. It drains into the inferior vena cava usually above the entrance of the right hepatic vein. The left inferior phrenic vein is often double, with a posterior branch draining into the left adrenal or left renal vein. The anterior branch normally passes in front of the oesophageal hiatus to drain into the inferior vena cava or the left hepatic vein.

In suprahepatic control of the inferior vena cava using a clamp or a sling, care should be taken to avoid damaging these branches during suprahepatic dissection of the inferior vena cava. These vessels can be plicated with sutures should they be too close to the plane of dissection.

7.2.2 The Hepatic Veins

There are three hepatic veins, the right, the middle and the left hepatic veins. The prevailing pattern and the anomalies of these veins will be discussed in detail in the latter part of this chapter (Sects. 7.3, 7.4 and 7.5).

7.2.3 The Retrohepatic Group

These are the venous drainage of the caudate lobe which have been discussed in detail in Sect. 3.3.

7.2.4 The Infrahepatic Group

The right and the left adrenal veins both exit from the adrenal hilum. The right is very short and drains into the posterior aspect of the inferior vena cava. The left adrenal vein passes down the medial border of the gland to join the left renal vein.

In hepatic surgery, even if the inferior vena cava is cross clamped suprahepatically and infrahepatically above the renal veins, there can still be blood loss when the inferior vena cava is opened if the right adrenal vein is not ligated. The blood comes from the right adrenal artery, through the right adrenal gland, then drains through the right adrenal vein to go into the inferior vena cava. As the left adrenal vein drains into the left renal vein, there is no need to ligate the left adrenal vein. For the same reason, if the suprahepatic

W. Y. Lau, *Applied Anatomy in Liver Resection and Liver Transplantation*, https://doi.org/10.1007/978-981-16-0800-1_7

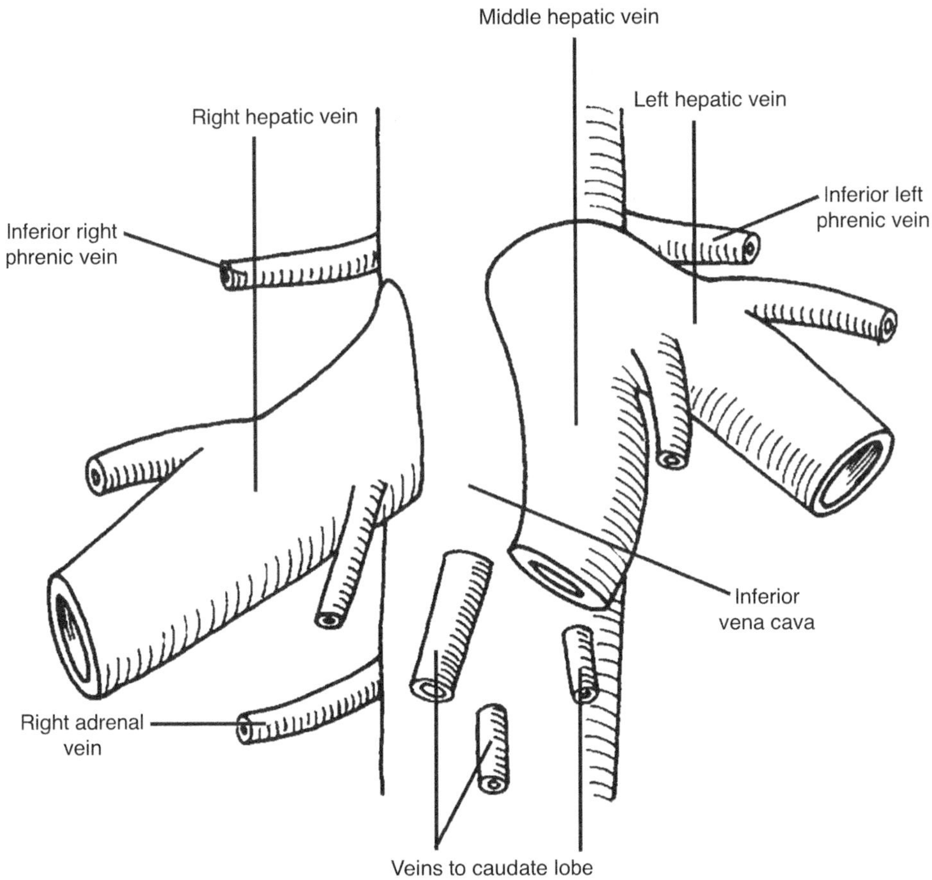

Fig. 7.1 The abdominal inferior vena cava and its suprarenal branches

clamp is placed in the inferior vena cava above the origins of the inferior right and left phrenic veins, these inferior phrenic veins should be plicated before opening the inferior vena cava.

7.3 The Prevailing Pattern of the Hepatic Veins

The right hepatic vein commences near the anterioinferior angle of the liver on the right and it has a long course, largely in the coronal plane in the liver (Fig. 7.2). It runs in the intersectional plane between the right anterior and posterior sections of the liver, receiving venous drainage from usually all of segments 6 and 7, and some of segments 5 and 8. Near to its termination, it lies almost horizontally. It enters the inferior vena cava at about the same level as the upper pole of the caudate lobe, and this level is a few millilitres lower than the entry of the trunk of the middle and left hepatic veins into the inferior vena cava. It may receive very small branches from the upper part of the caudate lobe. In extrahepatic mobilisation of the right hepatic vein, the hepatocaval ligament should first be divided (see Sect. 3.5).

The middle hepatic vein arises from the confluence of two veins. The segment 4b branch of the middle hepatic vein is long, tenuous, sagittal and enters the middle vein on its left side. It is joined by the vein from the right side draining segment 5. In 25% of cases, a substantial amount of venous drainage from segment 6 drains into the middle hepatic vein. The middle hepatic vein runs in the midplane of the liver receiving venous drainage from the parts of the right and left livers. The branch from segment 8 is large, and it usually runs transversely into the right side of the middle hepatic vein.

The middle hepatic vein ends as a single trunk in the inferior vena cava in only 3–15% of cases. In the great majority of cases, it forms a common trunk with the left hepatic vein. This trunk is usually 5 mm or less in length but there can be a common wall between the middle and the left hepatic veins. Therefore, it should be a surgical maxim that there are only two major hepatic veins entering the inferior vena cava—the right and the common trunk of the middle and left hepatic veins. Any attempt trying to dissect the middle hepatic vein from the left hepatic vein extrahepatically is dangerous as a hole made in the trunk or the common wall can result in torrential bleeding.

The left hepatic vein drains segments 2 and 3. It runs in the intersegmental plane between segments 3 and 2. It then runs in the posterior part of the fissure for the ligamentum venosum which forms part of the intersectional

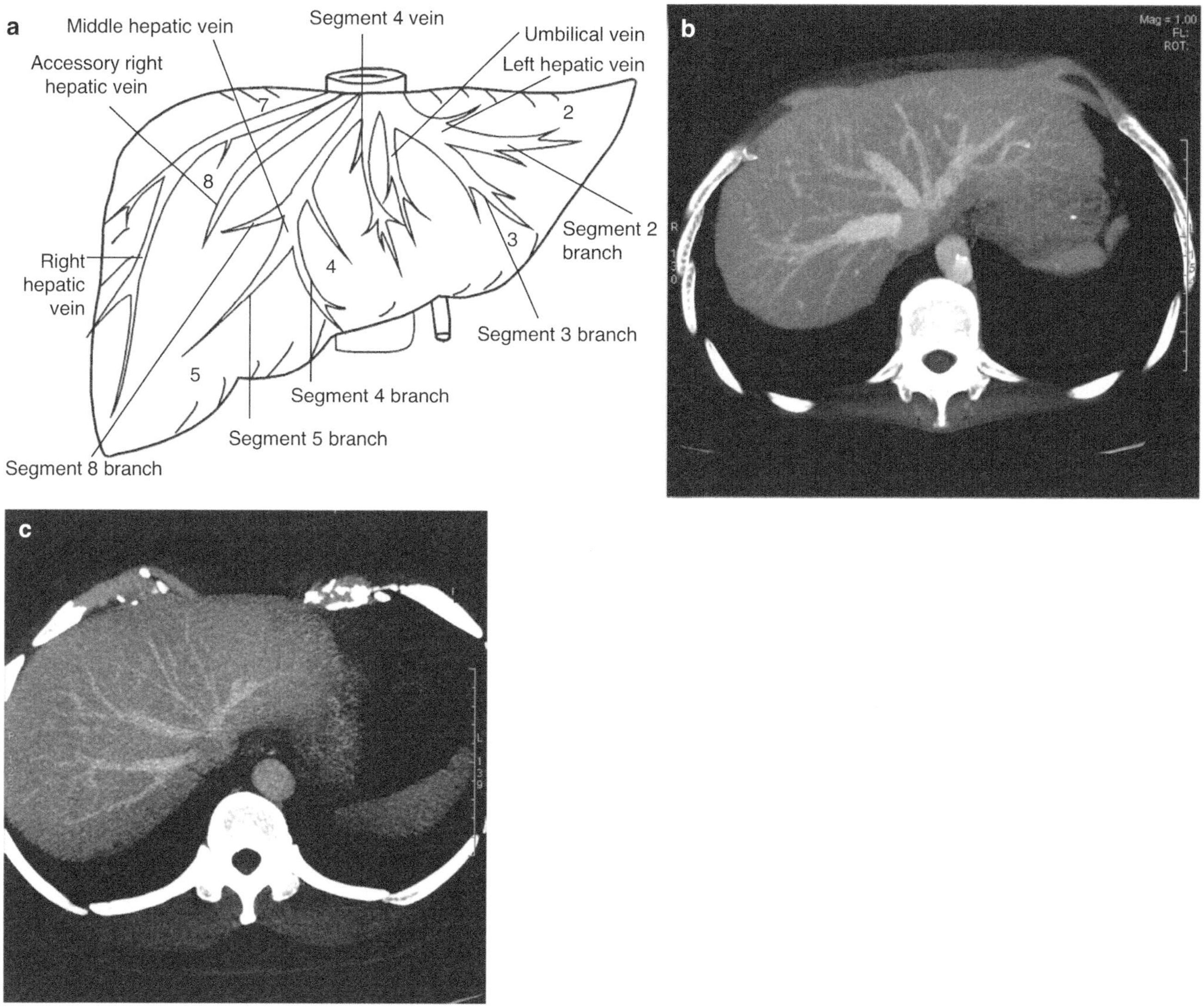

Fig. 7.2 (**a**) The hepatic veins (some veins may not be present in a proportion of patients); (**b**, **c**) The hepatic veins shown on computed tomography (intravenous contrast). *RHV* right hepatic vein, *MHV* middle hepatic vein, *LHV* left hepatic vein, *Sg 4V* segment 4 vein, *UV* umbilical vein

plane between segment 4 and segments 2 and 3. The left hepatic vein is situated in the cranial 2 cm of this fissure which divides segment 4 from segment 2, and it makes up part of the posterior edge of the liver. At this level, the vein is covered only by the connective tissues of the left triangular ligament. The vein then travels transversely and posteriorly to the right in the direction of the vena cava, following the superior edge of segment 1. It terminates in the inferior vena cava, usually receiving the middle hepatic vein to form a common trunk before it does so. The ligamentum venosum usually attaches to the common trunk. Division of ligamentum venosum at this point frees the posterior attachment of the common trunk making isolation of the trunk extrahepatically technically easier (see Sect. 6.10).

The left hepatic vein receives two main branches within the liver, an umbilical vein which runs in the umbilical fissure draining parts of segments 4 and 3. This vein is inconstant, happening in less than 60% of cases. Another vein, the accessory segment 4 vein, or the segment 4 vein by some authors, drains into the left hepatic vein in 57.5% of cases. Please note that the segment 4 vein is not the same as the segment 4 branch of the middle hepatic vein (Fig. 7.2a). It is important not to confuse the umbilical portion of the left portal vein with the umbilical vein; the latter is a tributary of the left hepatic vein that normally drains the most leftward and part of segment 4. It is also important not to confuse the umbilical vein which exists in utero but becomes obliterated after birth to form the ligamentum teres with the umbilical vein of the left hepatic vein.

7.4 The Right Hepatic Vein and Its Anomalies

The prevailing pattern of the right hepatic vein is a long trunk, with only small branches draining all of segments 6 and 7, and some of segments 5 and 8 (Fig. 7.3).

On rare occasions, the right hepatic vein has only a short trunk, and branches off a posterior branch which drains all of segments 6 and 7, and an anterior branch which drains some of segments 5 and 8 (Fig. 7.4).

The right hepatic vein may be small and drains only all of segments 7, and parts of segments 6 and 8 under the following three anomalies:

1. A small right hepatic vein, being compensated by a well-developed middle hepatic vein (Fig. 7.5a)
2. A small right hepatic vein; and a large accessory inferior right hepatic vein which arises from the inferior vena cava; this happens in about 15% of cases (Fig. 7.5b)
3. An accessory right hepatic vein (also called dorsal hepatic vein) which drains directly into the inferior vena cava (Fig. 7.5c)

In these three anomalies, division of the right hepatic vein will not result in venous congestion of segment 6. Thus, an isolated segment 7 resection can be done together with resection of the right hepatic vein when these anomalies are present.

7.5 The Trunk of the Middle/Left Hepatic Veins and Its Anomalies

The prevailing pattern of the common trunk of the middle and left hepatic veins is that the trunk is directed to the right (Fig. 7.6).

On rare occasions, the common trunk is directed to the left (Fig. 7.7), or the trunk can be completely absent (Fig. 7.8). In the latter situation, the middle and the left hepatic veins branch from the inferior vena cava in a V pattern.

Fig. 7.3 The prevailing pattern of the right hepatic vein

Fig. 7.4 Anomaly of the right hepatic vein with early branches

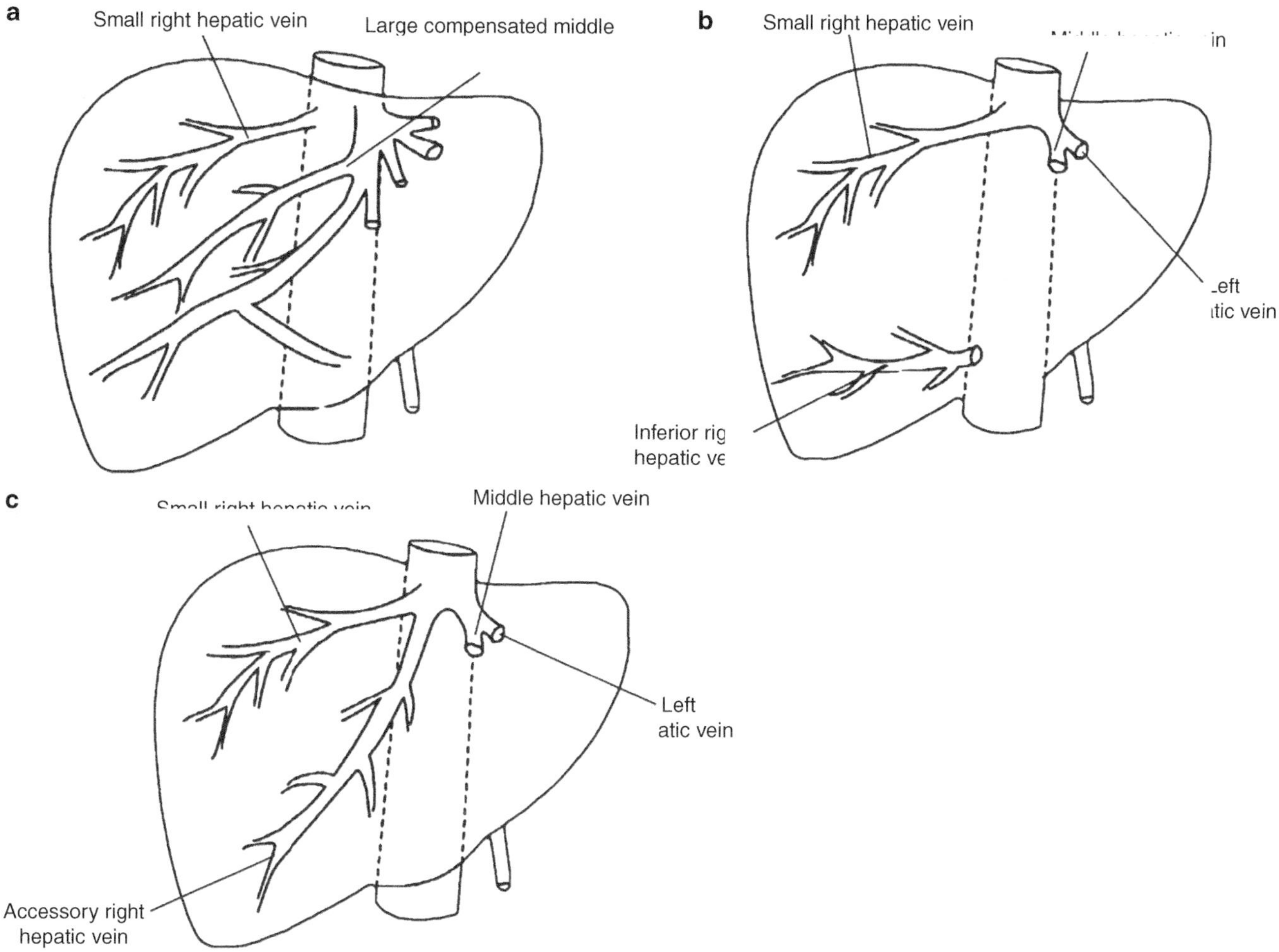

Fig. 7.5 Other anomalies of the right hepatic vein. (**a**) Small right hepatic vein compensated by a well-developed middle hepatic vein. (**b**) Small right hepatic vein with a large inferior right hepatic vein. (**c**) Small right hepatic vein with an accessory right hepatic vein

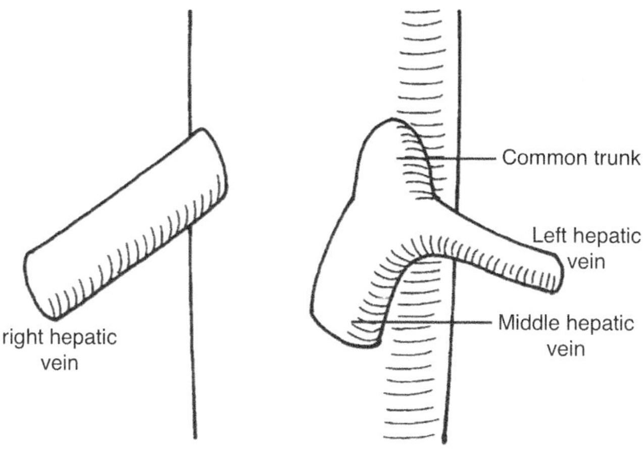

Fig. 7.6 Prevailing pattern of the common Trunk of the middle and left hepatic veins

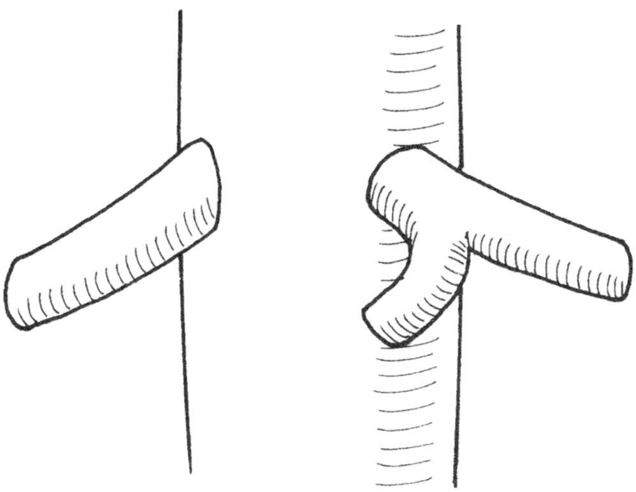

Fig. 7.7 Anomaly of the common Trunk of the middle and left hepatic veins pointing to the left

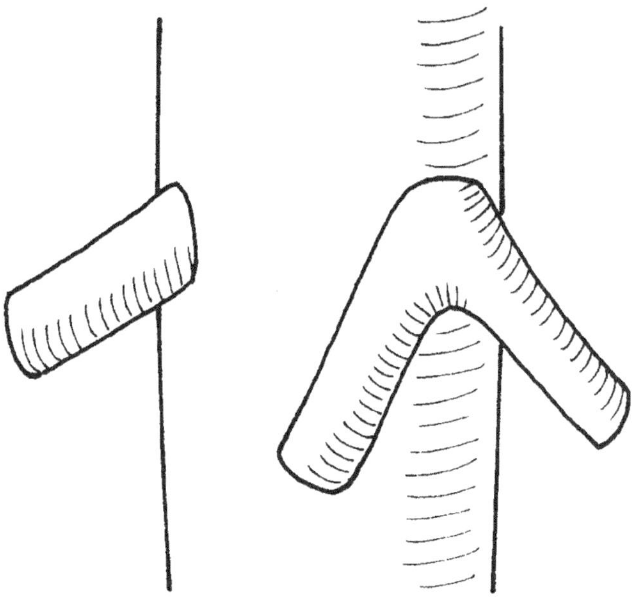

Fig. 7.8 Anomaly of the Trunk of the middle and left hepatic veins in a V pattern

7.6 Venous Drainage of Segment 4 and Its Anomalies

The venous drainage of the cranial (dorsal or posterior) portion of segment 4 (called segment 4a) is mainly by a short hepatic vein or veins that drain into the middle and/or the left hepatic vein. As previously discussed in Chap. 2, segment 4a is small and it represents only 20% of segment 4 in the studies by Couinaud in 1957. The quadrate lobe is considered by some authors to be segment 4b and it is drained by a long, tenuous and sagittal vein that enters the middle vein on the left side in the prevailing pattern. This vein is called segment 4 vein or accessory segment 4 vein by some authors. This segment 4 vein can drain into the middle hepatic vein (commonest, or prevailing pattern), into the trunk of the middle/left hepatic veins, into the left hepatic vein or even directly into the inferior vena cava (Fig. 7.9).

7.7 Renal and Infrarenal Branches of the Inferior Vena Cava

For hepatic surgery, cross-clamping of the infrahepatic inferior vena cava is done above the renal veins. There is usually no major branching from the inferior vena cava below the right adrenal vein and cross-clamping can be done safely at this site.

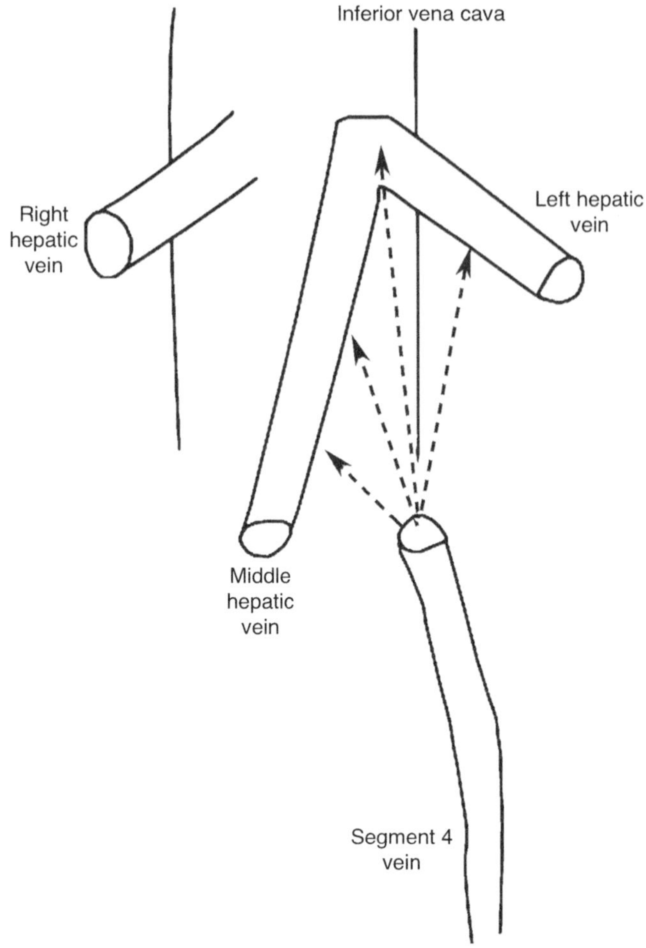

Fig. 7.9 Segment 4 vein and its anomalies

7.7.1 The Renal Veins

The renal veins enter the inferior vena cava at an angle of 45° to the vertical, and the left vein is usually higher than the right by 1–2 cm. The right vein is about 2 cm in length which is much shorter than the left renal vein which crosses in front of the aorta to reach the left kidney. The left renal vein receives a left adrenal vein from above and a left gonadal vein from below. The left renal vein is inferior and several centimetres posterior to the splenic vein and there is usually a thickened layer of fibrovascular tissue between the two veins in patients with cirrhosis and portal hypertension who requires a shunting operation.

7.7.2 The Right Gonadal Vein

On the right side, the gonadal vein drains into the inferior vena cava below the right renal vein.

7.7.3 Lumbar Veins

There are usually four pairs of lumbar veins, but it is only the caudal third and fourth pairs which regularly drain into the posterior aspect of the inferior vena cava. The surgical significance is as the two left veins pass behind the aorta, they are at risk during mobilisation of that structure. The right veins are easily damaged when the inferior vena cava is retracted anteriorly during lumbar sympathectomy. The second pair of lumbar veins may drain into the inferior vena cava at the level of the renal vein, but like the first pair of lumbar veins, they often drain into ascending lumbar or lumbar azygos veins.

Further Reading

Bismuth H. Surgical anatomy and anatomical surgery of the liver. In: Blumgart LH (Hrsg). Surgery of the liver and biliary tract, vol. 1. Edinburgh: Churchill Livingstone; 1988. p. 1ff.

Broelsch CE, Kremer K, Liidinghausen M, Leber V. Chirurgische Operationslehre, Bd. 5 (Hrsg: K Kremer, W Lierse, W Platzer, HW Schreiber, S Weller). New York: Thieme, Stuttgart; 1993. p. 76ff.

Scheuerlein H, Kőckerling F. The anatomy of the liver. In: Liver surgery, operative techniques and avoidance of complications. Heidelberg: J.A. Barth; 2001. p. 9–38.

8 Five Important Anatomical Structures in Right or Left Hepatectomy

8.1 Anatomy and Surgery

Surgery should be carried out based on a solid foundation of anatomy, for without anatomy, there is no surgery. Liver surgery is no exception. There are several anatomical structures that are especially relevant to liver surgery. The correct application of knowledge on these anatomical structures not only leads surgeons to safer and easier liver surgery but can also lead surgeons to design new approaches and new operations.

8.2 Important Anatomical Structures in Right or Left Hepatectomy

There are five important anatomical structures in right or left hepatectomy:

I. Hepatic intervenous fossa
II. Hepato-caval ligament
III. Ligamentum venosum
IV. Hepatic hilar plate
V. Short hepatic veins

Right or left hepatectomy is the most basic type of anatomical liver resection. With the exception of the left lateral sectionectomy, these two anatomical liver resections are the easiest and the most commonly performed anatomical liver resections. To carry out either right or left hepatectomy, or left sectionectomy, there is no need to have any knowledge on the Couinaud's liver segments in the main liver.

8.2.1 (I) Hepatic Intervenous Fossa

The hepatic intervenous fossa is found at a point where the two leaves of the falciform ligament reflect to join the diaphragm. It is covered by the coronary ligament anteriorly. The fossa is situated between the right hepatic vein and the common trunk of the middle/left hepatic veins. At its back is the inferior vena cava (Fig. 8.1).

As the hepatic veins join the inferior vena cava at a slight angle that runs anterior to posterior, these veins form a protuberance on the two sides of an area of depression. We, therefore, call this area of depression a fossa. Before dissection, this fossa is filled with connective tissues (Fig. 8.2).

Dissection of the connective tissues in the fossa leads to the anatomical structure called the hepatic intervenous fissure. This fissure measures 1.5–2 cm long and 1–1.5 cm wide. It is situated in the gap between the right hepatic vein and the common trunk of the middle/left hepatic veins. Anterior is segment 8 of the liver. Posterior is the anterior wall of the inferior vena cava. Inside it contains moderately dense connective tissues which lead inferiorly to the loose connective tissues between the posterior wall of the liver and the anterior wall of the inferior vena cava (Fig. 8.3).

8.2.2 Dissection of Hepatic Intervenous Fossa and Fissure

After dividing the coronary ligament, the fossa is identified. The moderately dense connective tissues in the fossa are dissected using a combination of sharp and blunt dissection. This is followed by blunt dissection using the index finger in the fissure and pushing inferiorly. This is technically possible because there is no vessel in the connective tissues in the fossa and the fissure. The push using the index finger should be gentle and as the finger moves inferiorly, the tip of the finger should be closely applied anteriorly to the posterior wall of the liver. This method is safe and reliable and it does not lead to a tear of the inferior vena cava if carried out properly. When the tip of the finger reaches the loose connective tissues, there is a sudden 'breaking

W. Y. Lau, *Applied Anatomy in Liver Resection and Liver Transplantation*, https://doi.org/10.1007/978-981-16-0800-1_8

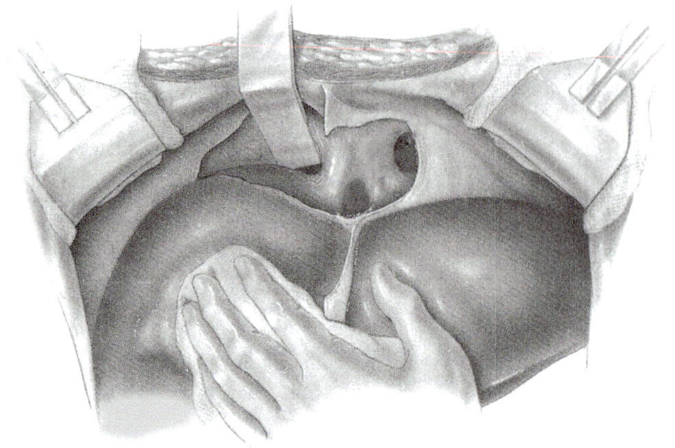

Fig. 8.1 Hepatic intervenous Fossa

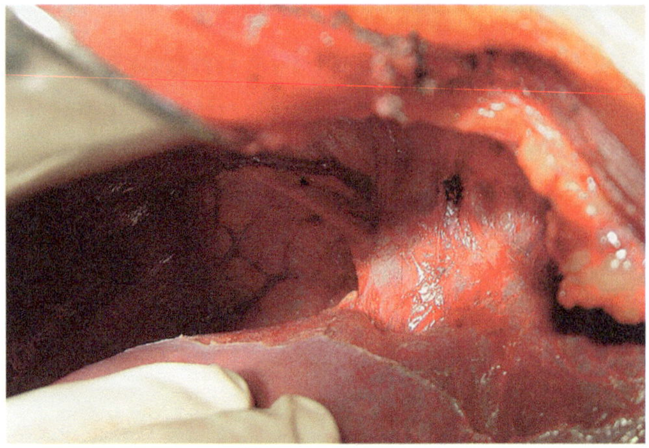

Fig. 8.2 Hepatic intervenous Fossa before dissection

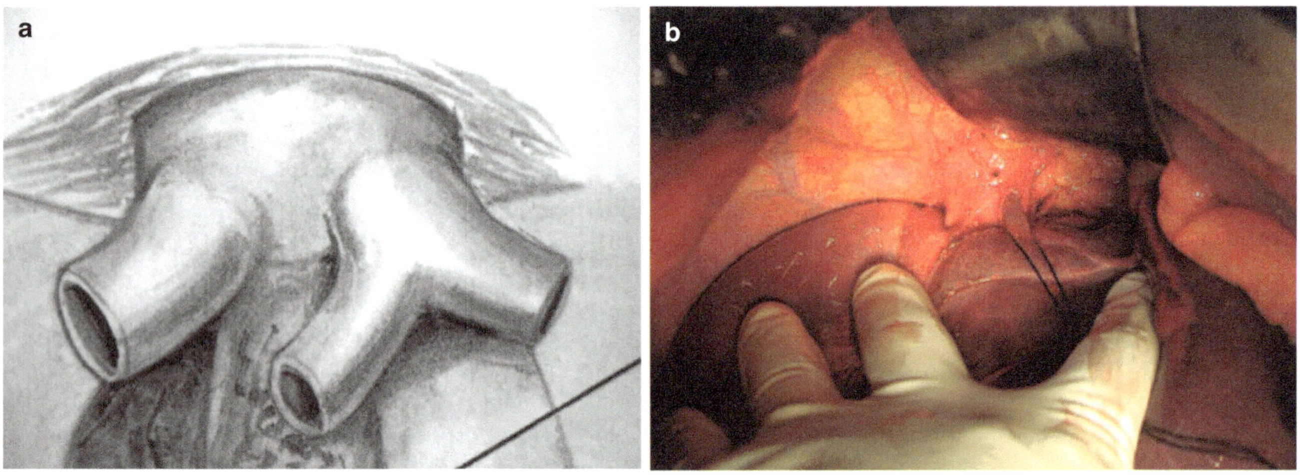

Fig. 8.3 (**a**) Hepatic intervenous Fossa and Fissure. (**b**) Fossa/Fissure after dissection (with sling around common Trunk of middle/left hepatic veins)

through' sensation and the dissection should end at this point (Fig. 8.4).

8.2.2.1 Clinical Applications

1. Easy separation of the right hepatic vein from the common trunk of the middle/left hepatic veins in piggy-back liver transplantation in the recipient operation (Fig. 8.5)
2. Used in selective hepatic vascular outflow occlusion to either occlude the right hepatic vein or the common trunk of the middle/left hepatic veins by:
   (a) Vascular clamps (Fig. 8.6a)
   (b) Slings (Fig. 8.6b)
3. To use Belghiti's hanging liver technique in liver resection (Fig. 8.7)

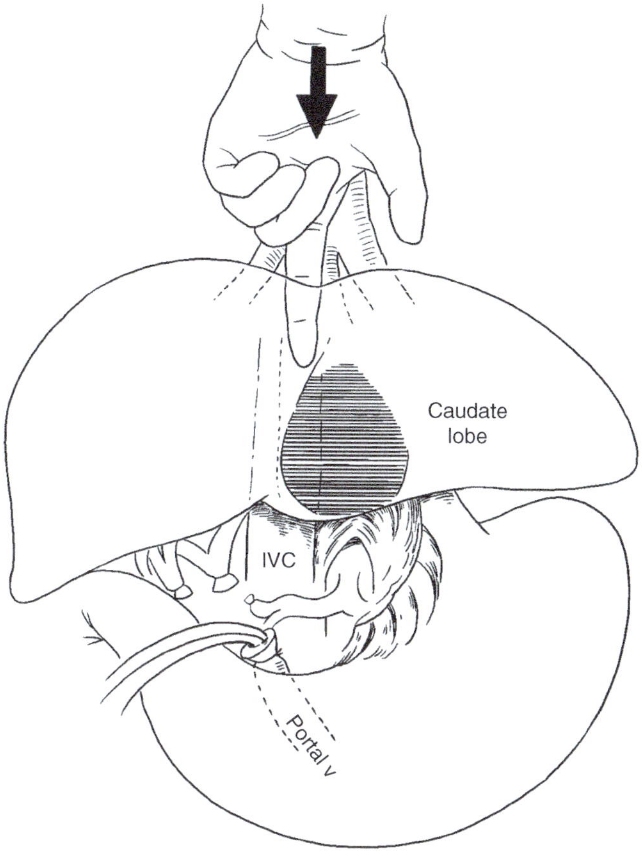

Fig. 8.4 Blunt dissection of Fossa/Fissure

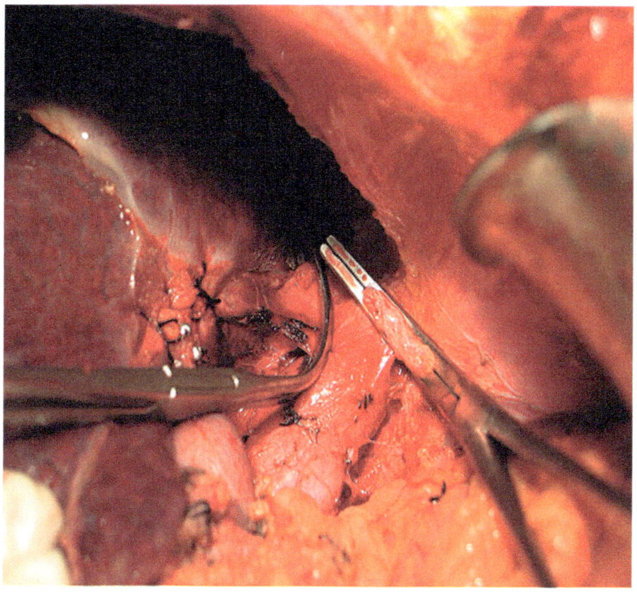

Fig. 8.5 Separation of right hepatic vein from common trunk of middle/left hepatic vein in piggy-back liver recipient operations

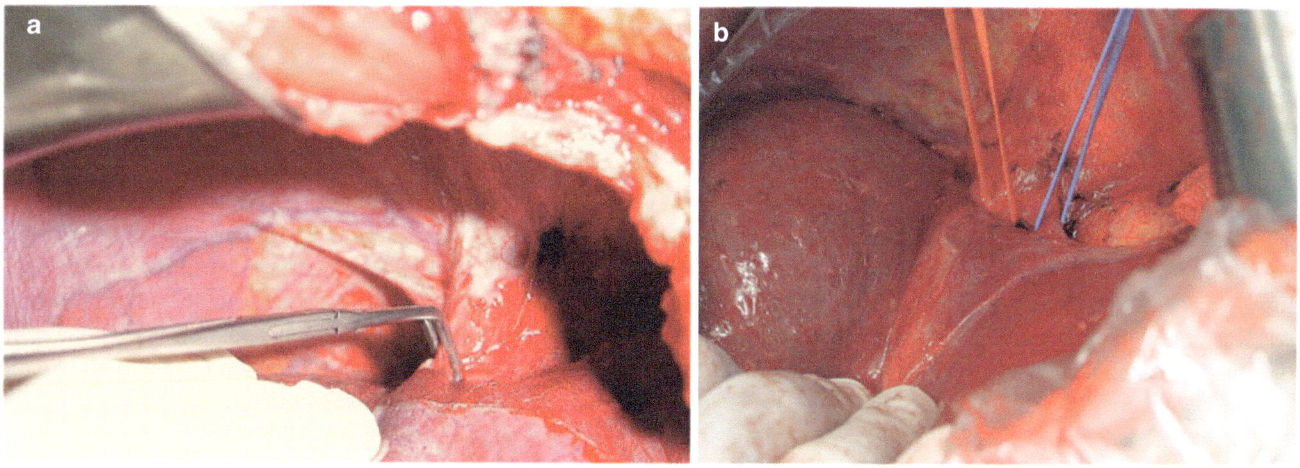

Fig. 8.6 (a) Vascular clamp on right hepatic vein. (b) Slings around right hepatic vein, and around common trunk of middle/left hepatic veins

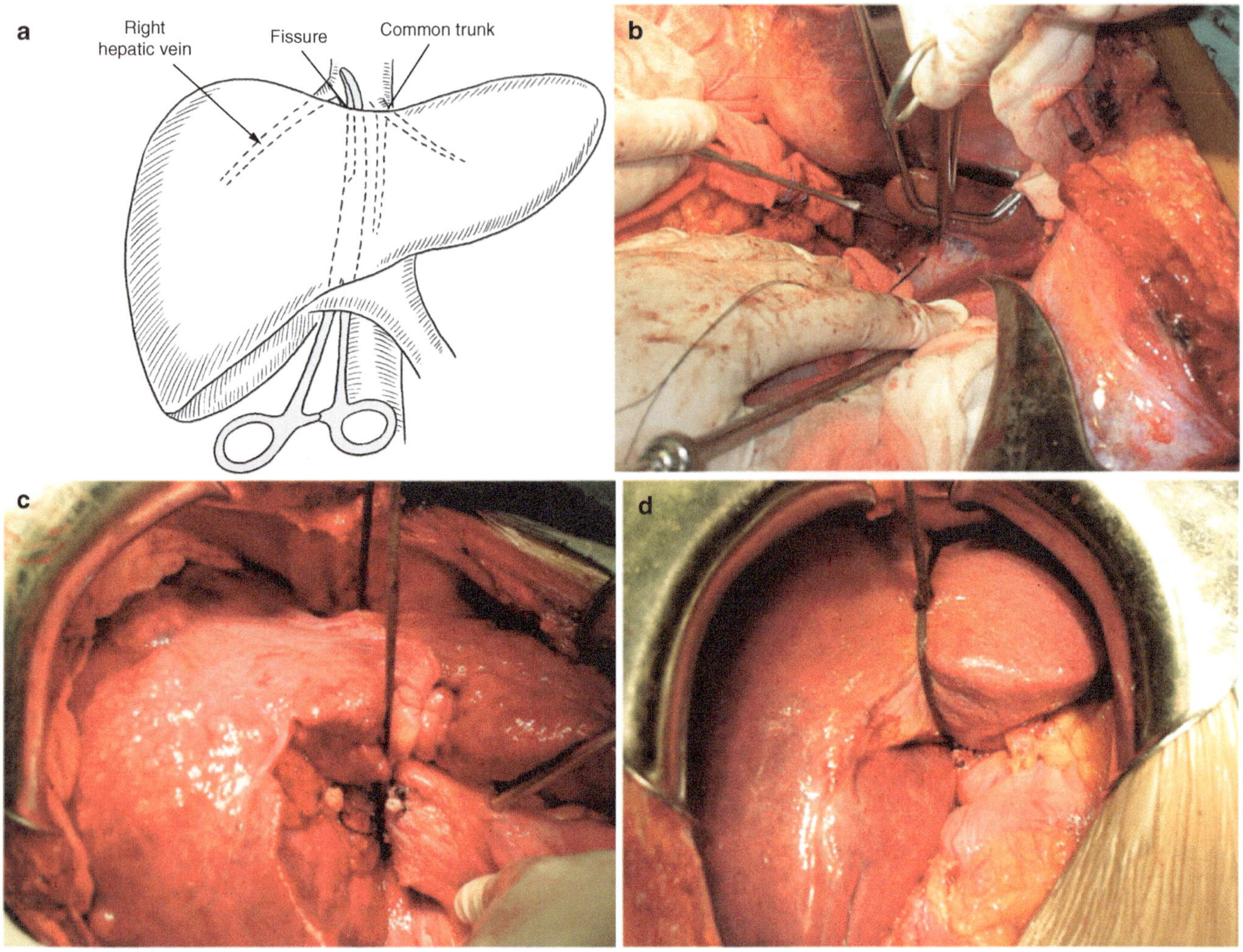

Fig. 8.7 (**a**) A dissecting forcep to establish a tunnel. (**b**) Establishing the tunnel. (**c**) Sling. (**d**) Using Belghiti's hanging technique in the liver section

8.3 (II) Hepato-Caval Ligament

The detailed anatomy is described in Sect. 3.5.

The Japanese surgeons call this ligament the Makuuchi ligament. Although Professor Makuuchi is not the first one who described this ligament, he was the first one to point to the important application of this ligament in the identification and isolation of the right hepatic vein.

8.3.1 Clinical Importance of Hepato-caval Ligament

For extrahepatic dissection, isolation and control of the right hepatic vein, the safest method is to first dissect the hepatic intervenous fossa and fissure so as to dissect the anterior and the medial sides of the right hepatic vein (see Sect. 8.2). This is followed by dissection and division of the hepato-caval ligament on the lateral and inferior sides of the right hepatic vein. The final step is to dissect between the anterior wall of the inferior vena cava and the right hepatic vein to put a sling around the right hepatic vein. Division of the small hepatic veins along the right border of the inferior vena cava facilitates this final step of isolation (Fig. 8.8).

8.4 (III) Ligamentum Venosum

The detailed Anatomy is described in Sect. 3.1, and Sects. 6.6 and 6.10.

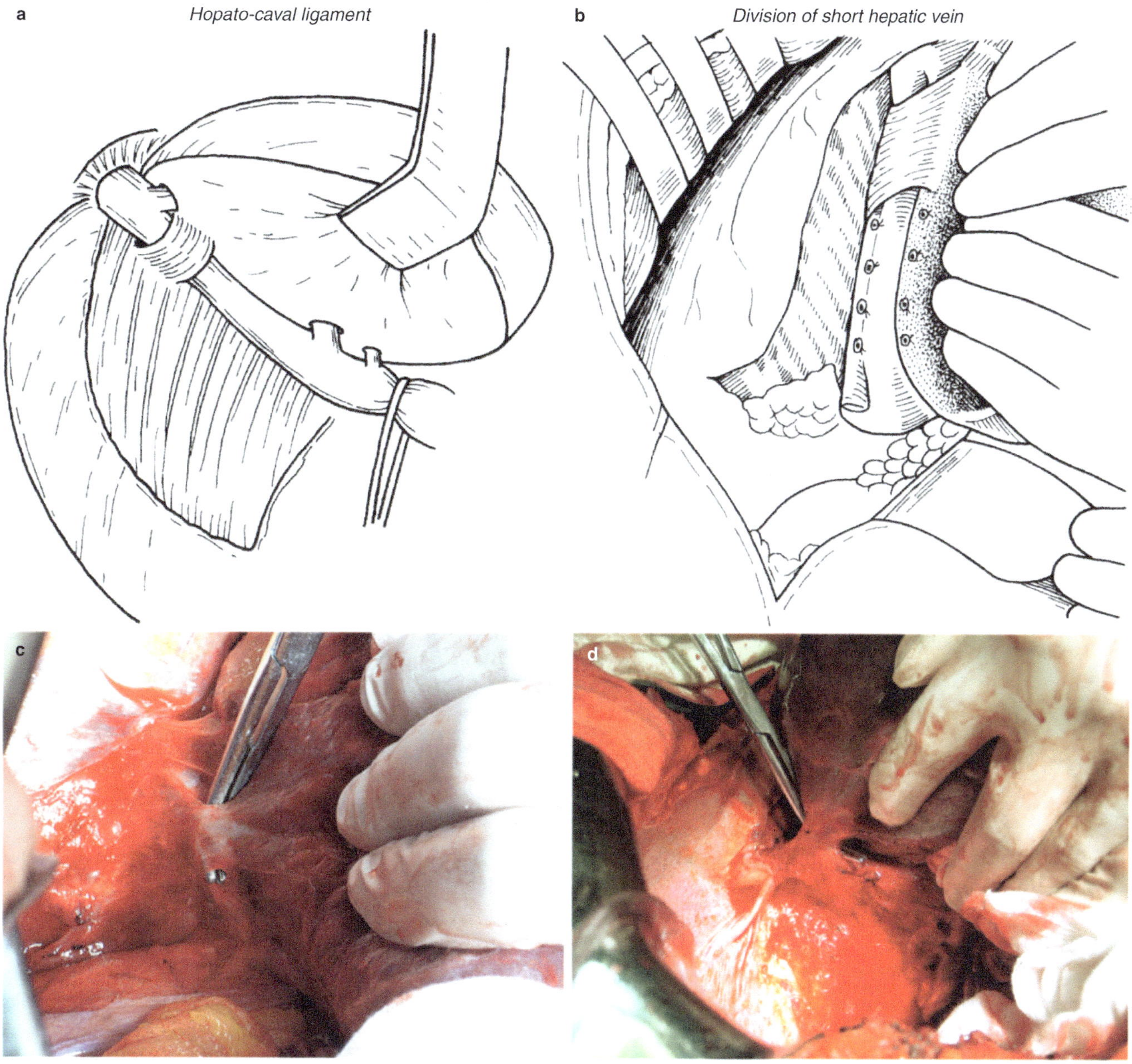

Fig. 8.8 (**a**) Hepato-caval ligament. (**b**) Division of short hepatic vein. (**c**) Dissection of hepato-caval ligament. (**d**) Isolation of right hepatic vein

The ligamentum venosum is also called the Arantius ligament as it was first described by the Italian anatomist, Giulio C. Arantius (1530–1589). It is the remains of the ductus venosus of the foetus (Fig. 8.9). After birth, it is represented by a thin fibrous cord lying in the fissure of the ligamentum venosum. Its position is therefore constant and in every individual, it connects the left portal vein with the posterior of the common trunk of the middle/left hepatic veins (Fig. 8.10).

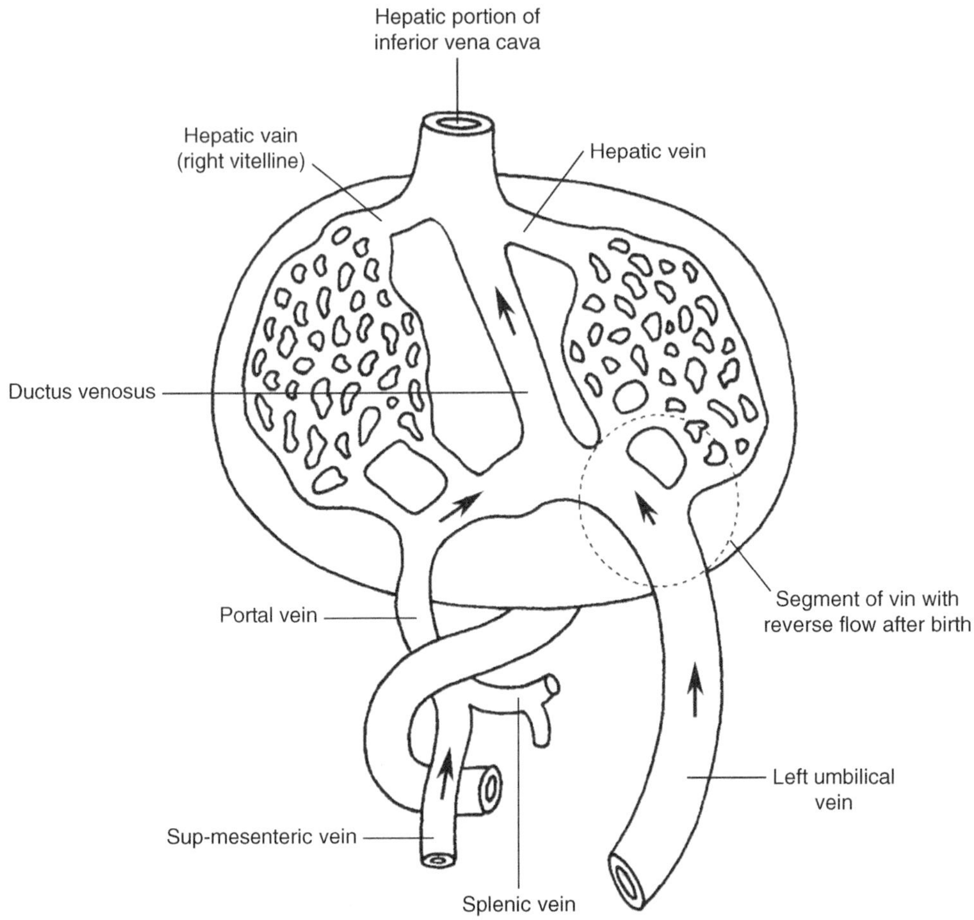

Fig. 8.9 Ductus venosus in the foetus

8.4.1 Clinical Importance of Ligamentum Venosum

For extrahepatic dissection, isolation and control of the common trunk of the middle/left hepatic veins, the safest method is to first dissect the hepatic intervenous fossa and fissure so as to dissect the anterior and the medial sides of the common trunk (see Sect. 8.2). The left triangular ligament is dissected to free the lateral side of the left hepatic vein. This is followed by dissection of the ligamentum venosum from the liver and the lesser omentum. The ligamentum verosum is divided, the upper end is lifted upwards and dissected towards the head of the patient while the lower end is left behind. The dissection of the ligmentum venosum superiorly leads to dissection of the back of the common trunk of the middle/left hepatic veins. After this dissection, it becomes easy to put a clamp around the common trunk of the middle/left hepatic veins or to put a sling around the common trunk (Fig. 8.11).

An easier way to control the extrahepatic right hepatic vein or the common trunk of the middle/left hepatic veins is to free the coronary and left triangular ligaments to the edges

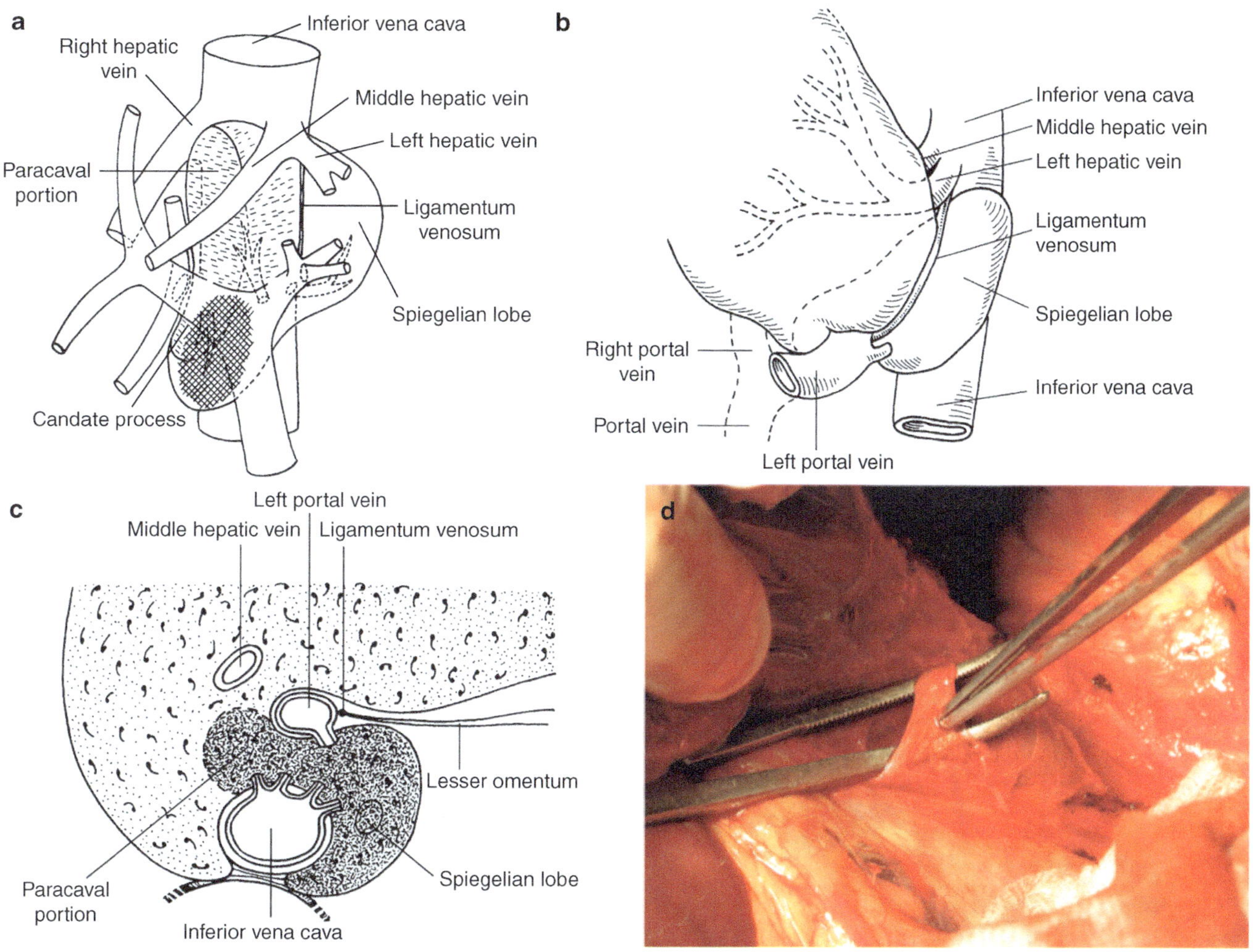

Fig. 8.10 (**a**, **b**) Ligamentum venosum. (**c**) Relationship of ligamentum venosum and lesser omentum. (**d**) Operative photograph of ligamentum venosum

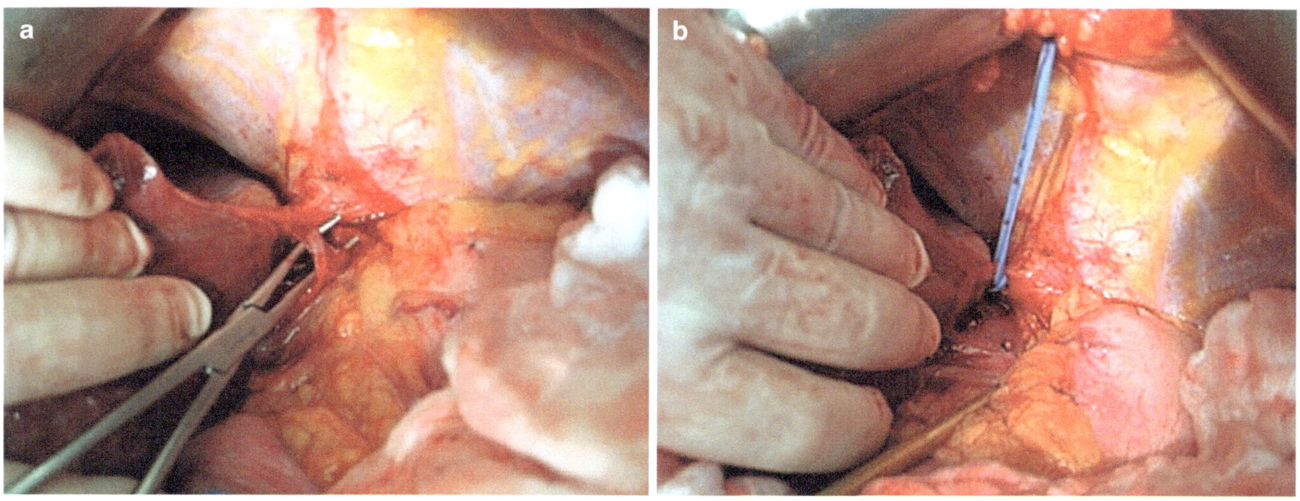

Fig. 8.11 Dissection of the common trunk of middle/left hepatic vein by tracing the ligamentum venosum

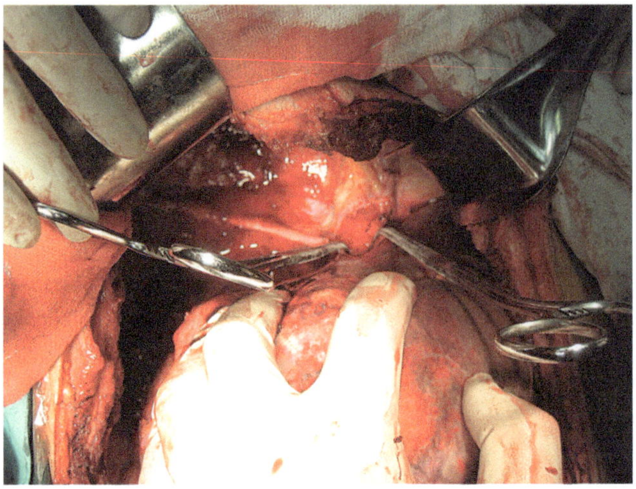

Fig. 8.12 A simpler way to extrahepatic control the right hepatic vein and the common trunk of the middle/left hepatic vein using vascular clamps

of the right hepatic vein and the left hepatic vein, dissect the hepatic intervenous fossa and fissure and to use two vascular clamps to separately control the right hepatic vein and the common trunk of the middle/left hepatic veins (Fig. 8.12).

8.5 (IV) Hepatic Hilar Plate

The detailed anatomy is described in Chap. 4.

The hepatic hilar plate consists of four parts: the hilar plate, the cystic plate, the umbilical plate and the plate of Arantius (Fig. 8.13).

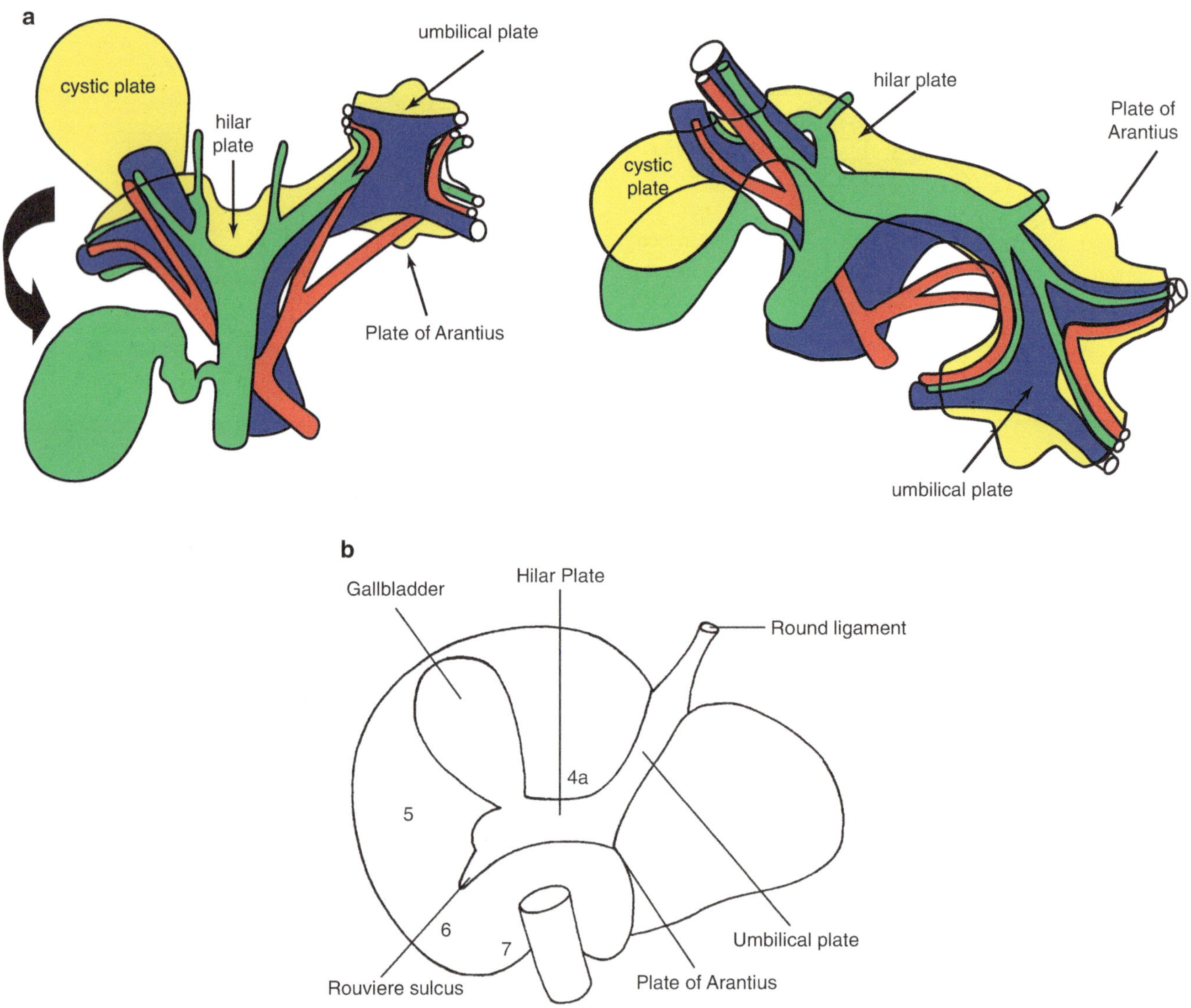

Fig. 8.13 (a) Hepatic Hilar plate. (b) The four parts of the hepatic Hilar plate

8.5.1 Clinical Application of the Hepatic Hilar Plate

The hilar plate can be lowered (Fig. 8.14):

1. To extrahepatically isolate the left duct for bypass in patients with biliary stricture or carcinoma affecting the confluence of the hepatic ducts
2. To either isolate the right, the left, or both the right and the left hepatic pedicles after lowering of the hilar plate (see Sect. 4.3). This facilitates right or left hemihepatectomy or the use of hemihepatic vascular inflow control (see Chap. 13) (Fig. 8.15).

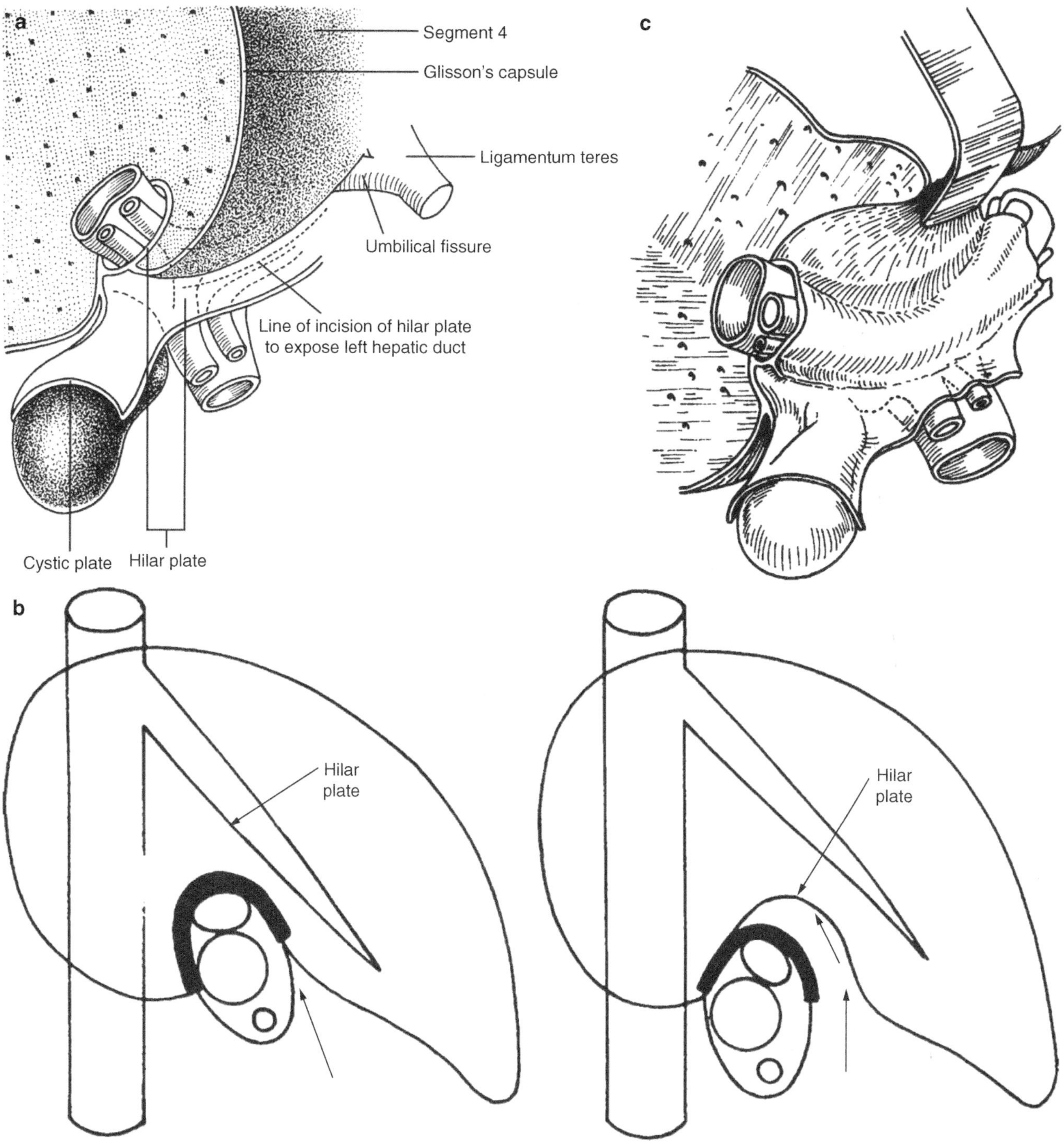

Fig. 8.14 Lowering of hilar plate

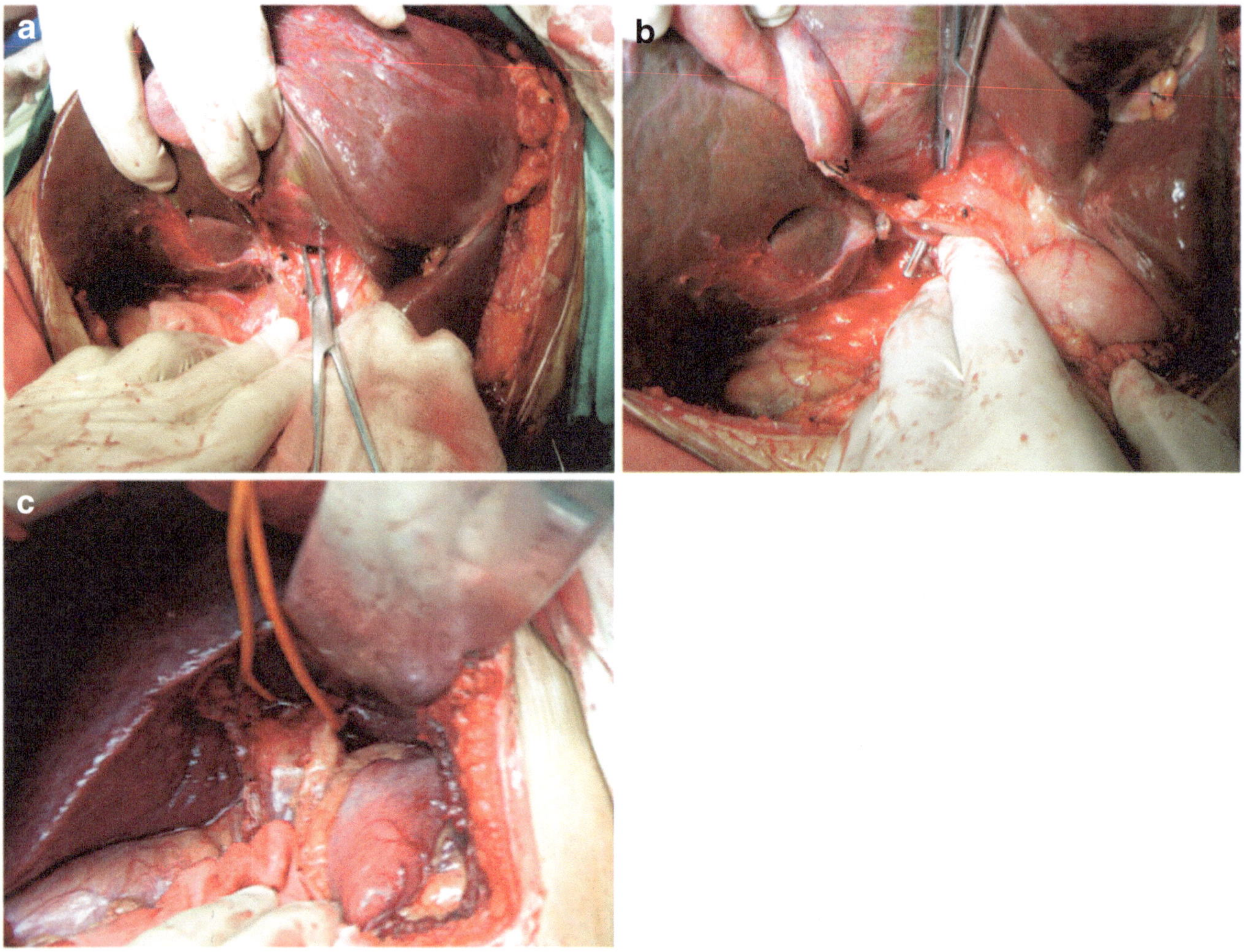

Fig. 8.15 (**a**) Lower of Hilar plate. (**b**) Isolation of right hepatic pedicle. (**c**) Sling around left hepatic pedicle

8.6 (V) Short Hepatic Veins

The short hepatic veins drain directly from the back of the liver into the inferior vena cava. The detailed anatomy is described in Sect. 3.3.

8.6.1 Clinical Application of the Anatomy of the Short Hepatic Veins

1. To extrahepatically dissect, isolate and control the right hepatic vein, we have described that it is easier to first dissect the hepatic intervenous fossa and fissure (Sect. 8.2) and to divide the hepato-caval ligament (Sect. 8.3). Dissection and division of the small hepatic veins along the right border of the inferior vena cava facilitates isolation of the hepato-caval ligament.
2. The short hepatic veins mainly drain from the back of the caudate lobe of the liver into the inferior vena cava along its right and left borders. This explains why the Belghiti liver hanging technique can go through a tunnel that is avascular (Fig. 8.16).
3. In liver resection, if one of the short hepatic veins is torn because of too much anterior traction onto the liver in an attempt to better expose the liver, torrential bleeding can result.

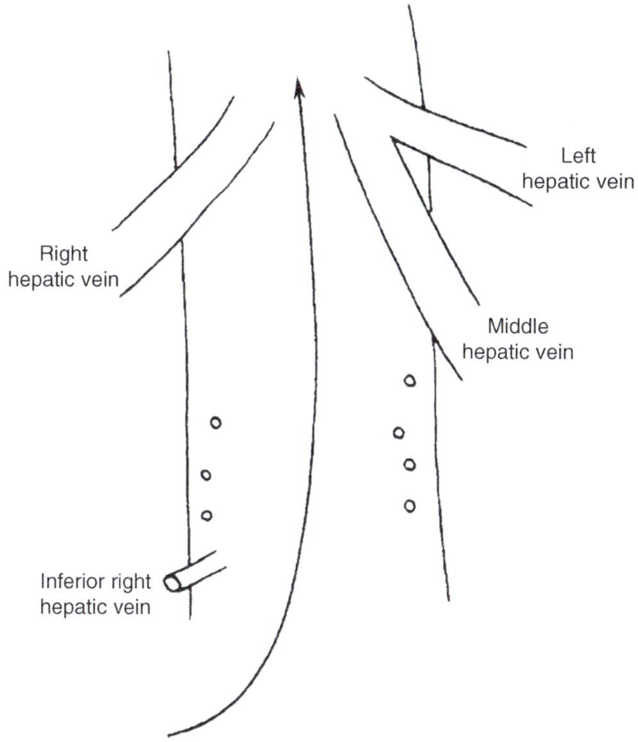

Fig. 8.16 Belghiti's avascular tunnel

4. In about 10% of patients, there is a big right inferior hepatic drain (Fig. 8.17). If this vein is divided, part of the right liver (mainly segment 6) can become congested because of compromised outflow drainage. On the other hand, in the case of right side liver resection, even when the right hepatic vein has to be resected, segment 6 can still be preserved by preserving the inflow blood supply from the right hepatic pedicle and outflow blood supply by this vein.

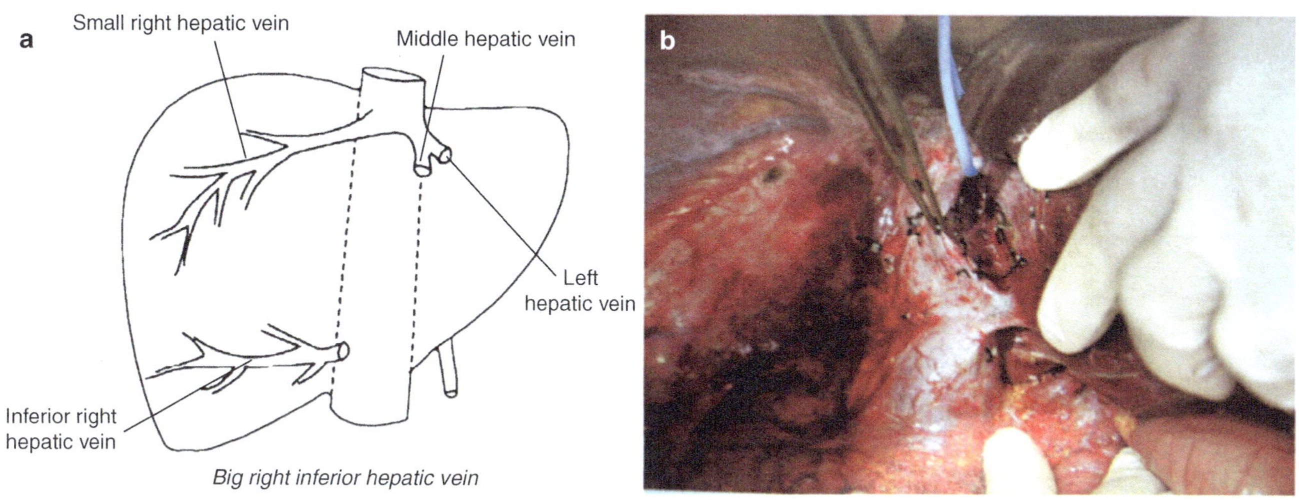

Fig. 8.17 (a) Big right inferior hepatic vein. (b) Big right inferior hepatic vein

8.7 Conclusion

In conclusion, an adequate understanding of the five important anatomical structures helps surgeons to carry out safe and better liver operations.

Further Reading

Blumgart LH, editor. Liver, biliary & pancreatic surgery. Video atlas. Philadelphia, USA: Elsevier Saunders; 2011.

Jamieson G, Launois B. Chapter 2: Living liver resections and liver transplantation: the anatomy of the liver and associated structures. In: Jamieson GG, editor. The anatomy of general surgical operation. 2nd ed. Edinburgh: Elsevier Churchill Livingstone; 2006. p. 8–23.

9 Anatomy Relevant to Liver Transplantation

9.1 Types of Liver Transplantation

There are many classifications of liver transplantation. A simple classification of the types of liver transplantation can be seen in Table 9.1.

9.2 Reconstruction of Anatomical Structures

In liver transplantation, there are four important anatomical structures that need to be reconstructed: the hepatic artery, the portal vein, the common hepatic duct and the hepatic veins (or the inferior vena cava). Adequate knowledge of the extrahepatic and intrahepatic prevailing patterns (the most commonly seen patterns) and their anomalies of these four important structures are important in liver transplantation operations as anomalies can make the operation even more complicated.

Table 9.1 Types of liver transplantation

Classification of Liver Transplantation
(The donated liver can be the whole liver or part of the liver)
Source of Donated Liver:
Cadaveric (brain death, cardiac death)
– Whole liver
– Reduced size
– Split liver
Living (Living-related, living non-related):
– Extended right hemiliver
– Right hemiliver
– Left hemiliver
– Left lateral section
Recipient:
Orthotopic
– Conventional technique
– Piggyback technique
Auxillary
– Orthotopic
– Temporary
– Heterotopic
Domino liver transplantation
Dual graft liver transplantation

9.3 Whole Liver Transplantation (in Cadaveric Liver Donation Only)

As the whole liver is used in donation, intrahepatic anomalies of any of the four important anatomical structures have no impact on the transplant operation. However, extrahepatic anomalies will affect the operation to a varying degree, depending on the type and the location of the anomaly.

9.3.1 Operative Steps of Whole Liver Transplantation (See Chap. 19)

The operative steps in whole liver transplantation are shown in Fig. 9.1. The steps involve reconstruction of the hepatic

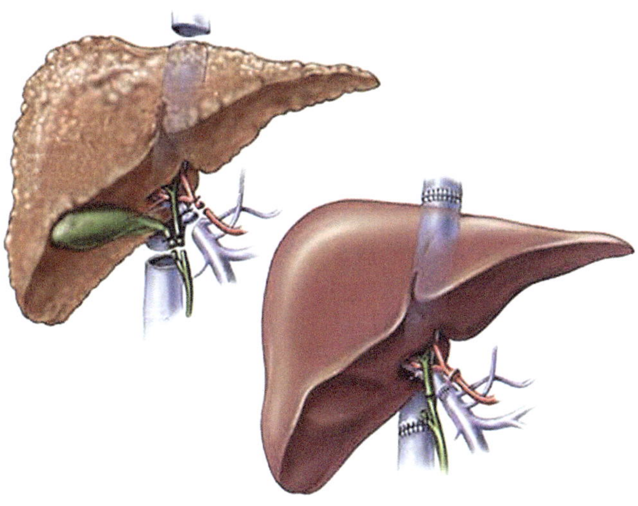

Fig. 9.1 Diagrammatic representation of the steps in whole liver transplantation

W. Y. Lau, *Applied Anatomy in Liver Resection and Liver Transplantation*, https://doi.org/10.1007/978-981-16-0800-1_9

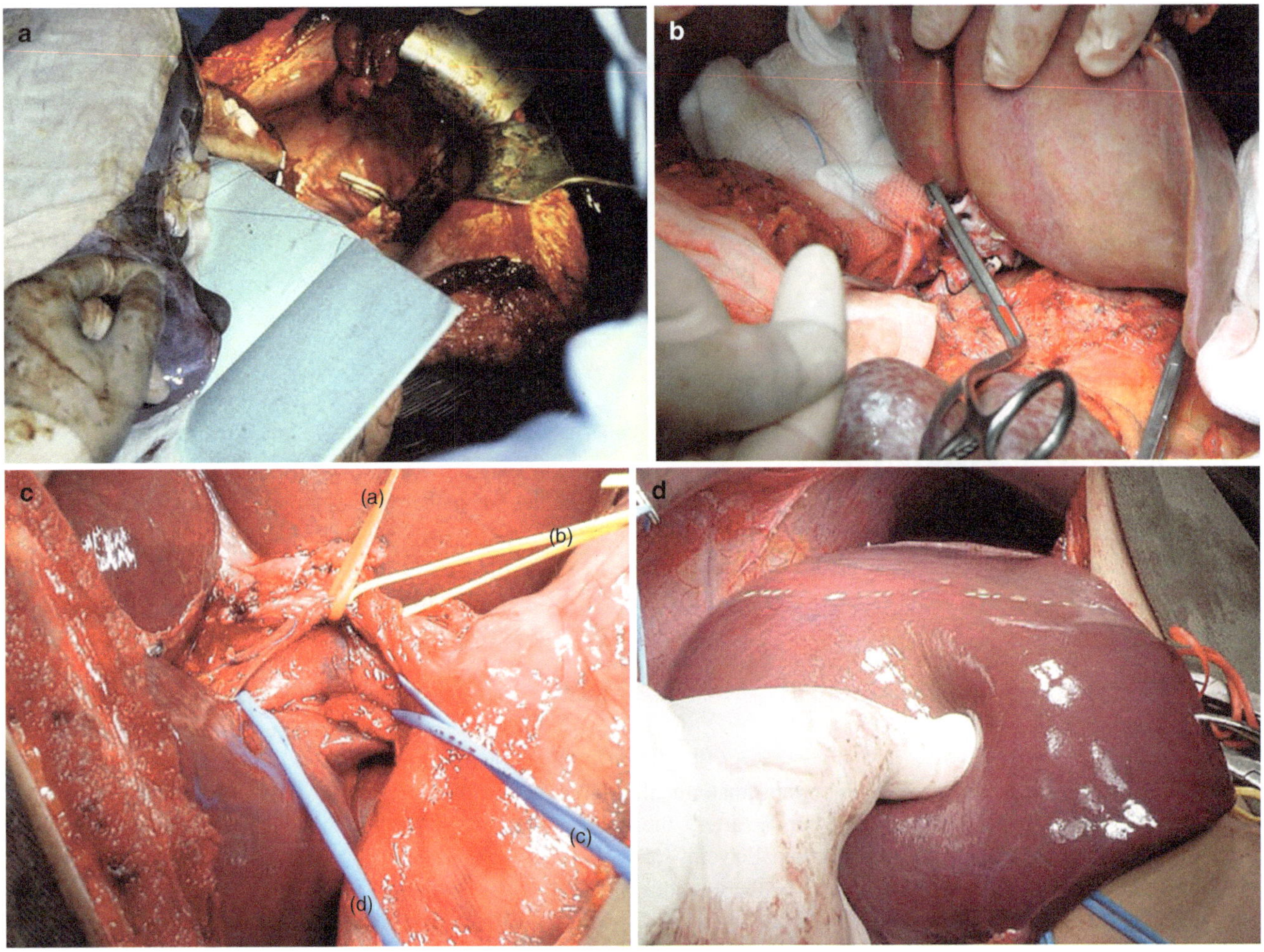

Fig. 9.2 Whole liver transplantation. (**a**) Anastomosis of supra- and infra-hepatic inferior vena cava. (**b**) Anastomosis of portal vein (slings **c**, **d**), hepatic artery (sling **a**) and bile duct (sling **b**). (**d**) Reperfusion of liver

veins (or inferior vena cava), the portal vein, the common hepatic artery (usually under a microscope) and finally the common hepatic duct.

The actual operation is shown in Fig. 9.2.

9.4 The Prevailing Anatomical Structures Are the Structures Commonly Seen in an Individual. The Less Commonly Seen Patterns Are Anomalies

9.4.1 Hepatic Artery

The prevailing pattern of the common hepatic artery is shown in Fig. 9.3.

However, extrahepatic anomalies of the common hepatic artery are common, and many anomalies have been described (Fig. 9.4).

It is very difficult to remember all these anomalies. For practical reasons, these anomalies can be simplified into the following two types:

(a) Anomalies arising from a single common hepatic artery in origin (Fig. 9.5)
(b) The right and the left hepatic arteries arising from two separate arteries from two different origins (Fig. 9.6)

Real examples of such anomalies are shown in Fig. 9.7.

The management of these arterial anomalies is more complicated. The approaches are either anastomosing two

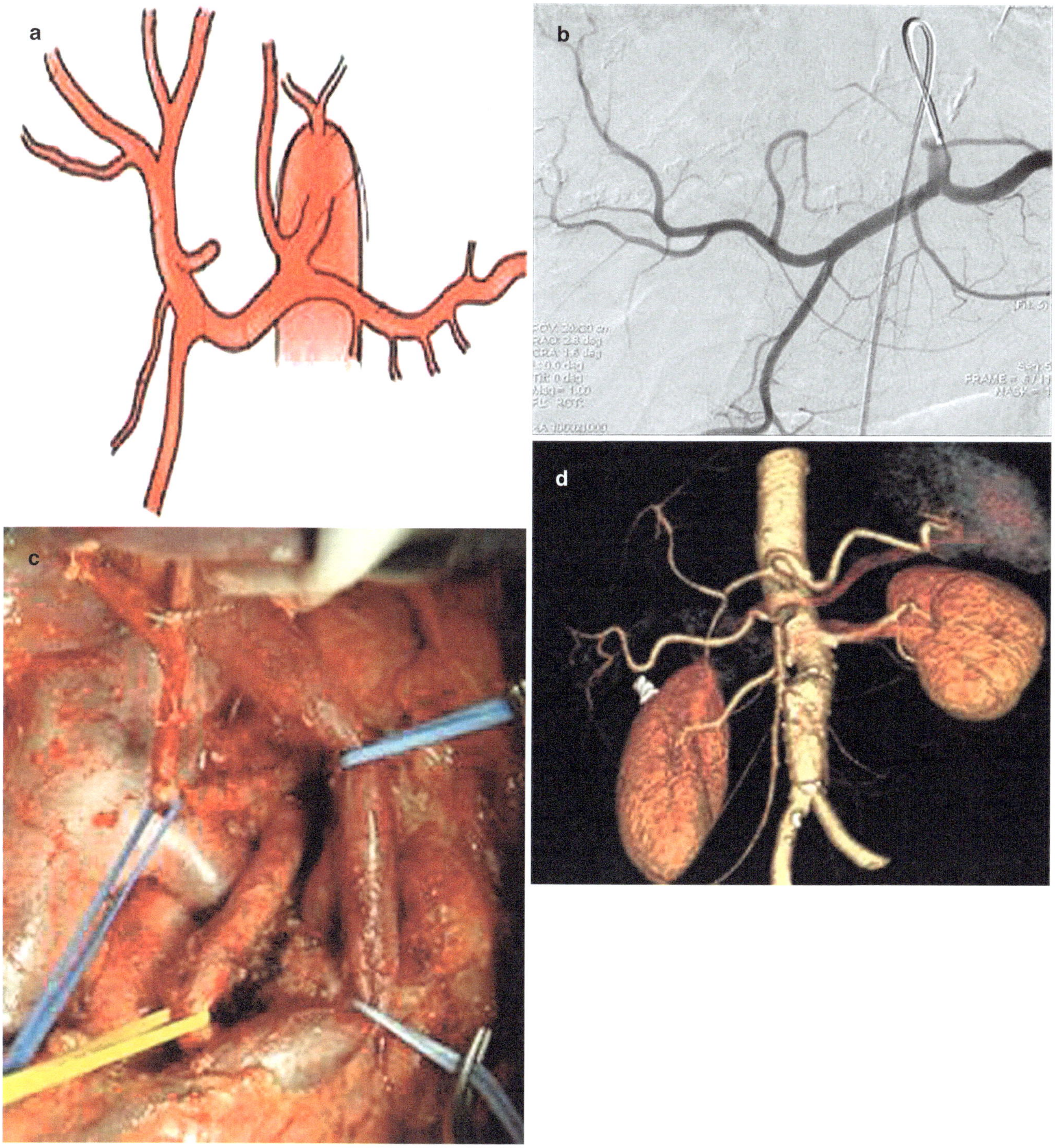

Fig. 9.3 Extrahepatic hepatic artery. (**a**) Diagram. (**b**) Angiogram. (**c**) Operation. (**d**) 3D reconstruction

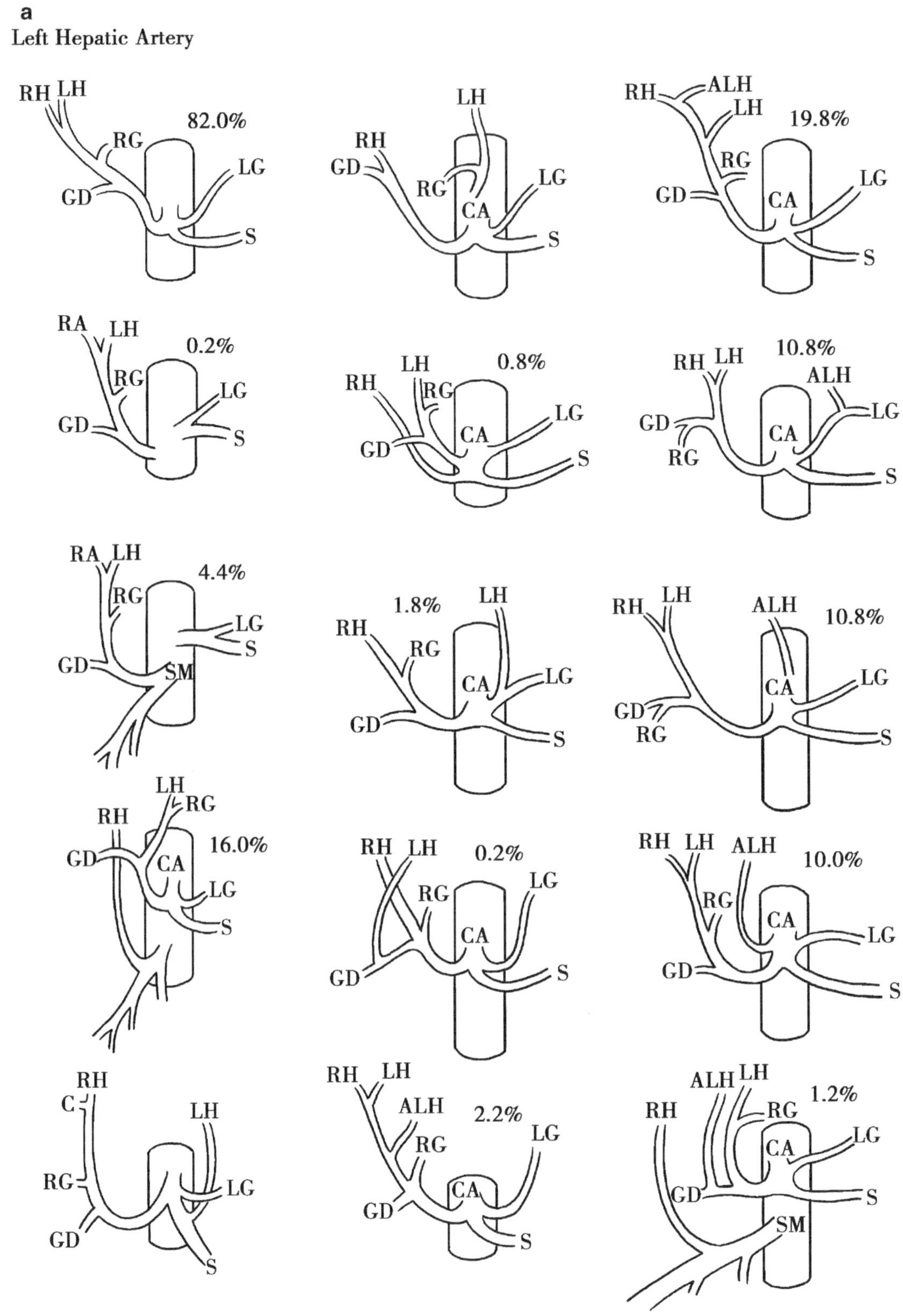

Fig. 9.4 Anomalies of extrahepatic common hepatic artery. *RH* right hepatic artery, *LH* left hepatic artery, *C* cystic artery, *GD* gastroduodenal artery, *RG* right gastric artery, *CA* coeliac artery, *LG* left gastric artery, *S* splenic artery

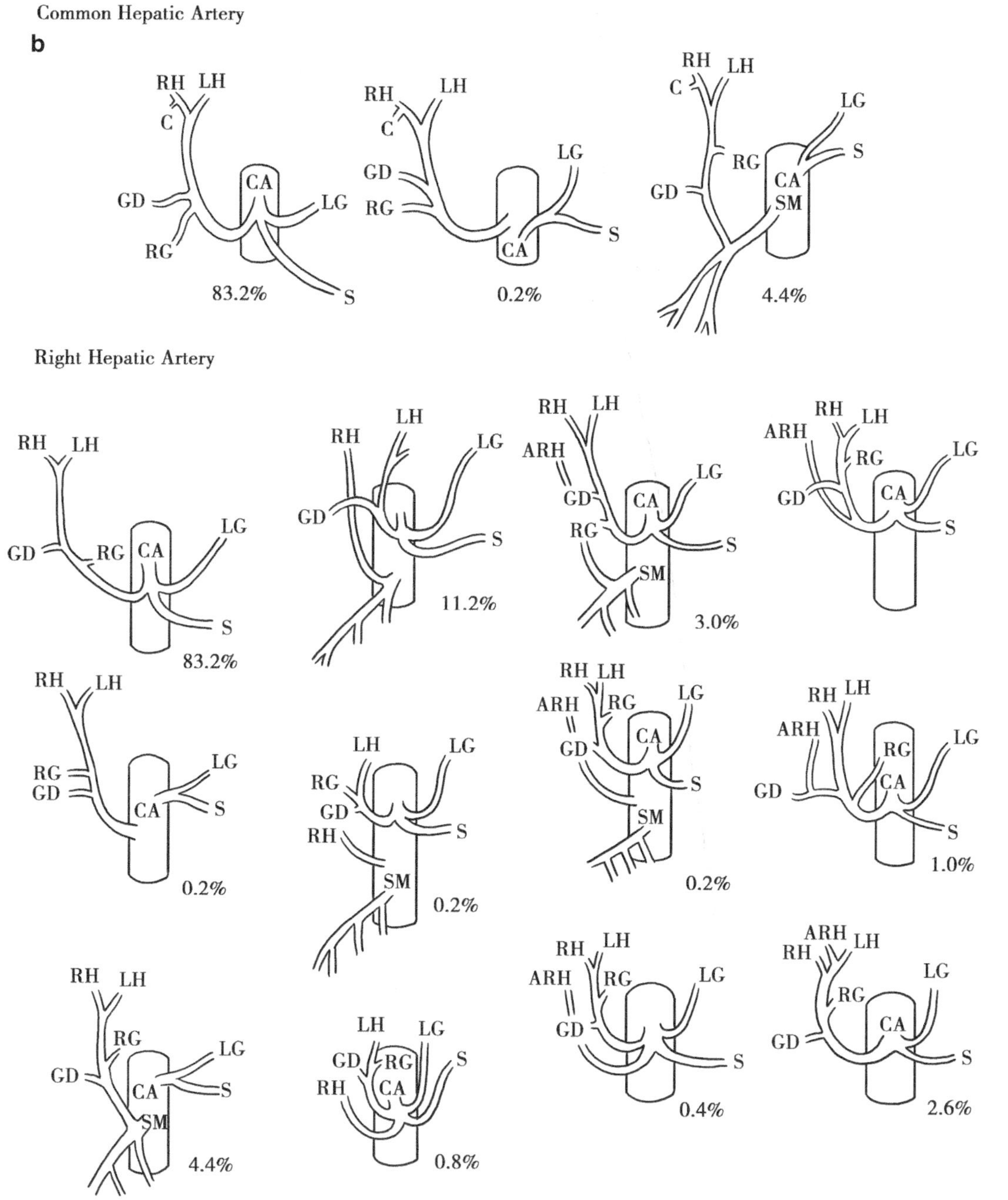

Fig. 9.4 (continued)

vessels into a single vessel, or using a carrel patch to do the reconstruction (Fig. 9.8).

9.4.2 Common Hepatic Duct

Anomalies affecting the extrahepatic biliary system are common (Fig. 9.9).

However, these anomalies are not important provided that enough care is taken to recognise these anomalies and to divide the biliary system distal enough so that a single bile duct can be used in biliary reconstruction, similar to the prevailing pattern with a single common hepatic duct.

9.4.3 Portal Vein

Anomalies of the extrahepatic portal system are rare. However, in cirrhotic patients, portal vein thrombosis can make liver transplantation technically more difficult. The

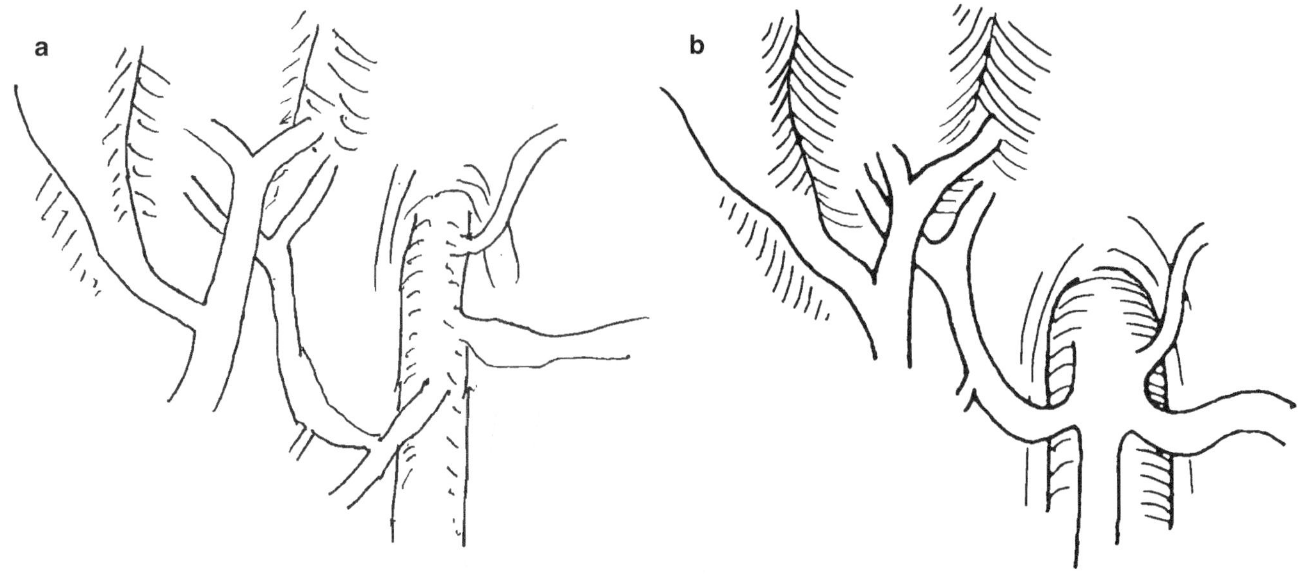

Common hepatic artery arises from superior mesenteric artery

Common hepatic artery arises from the common trunk together with the coeliac artery and the superior mesenteric artery

Fig. 9.5 Anomalies arising from a single hepatic artery. (**a**) Common hepatic artery arises from superior mesenteric artery. (**b**) Common hepatic artery arises from the common Trunk together with the coeliac artery and the superior mesenteric artery

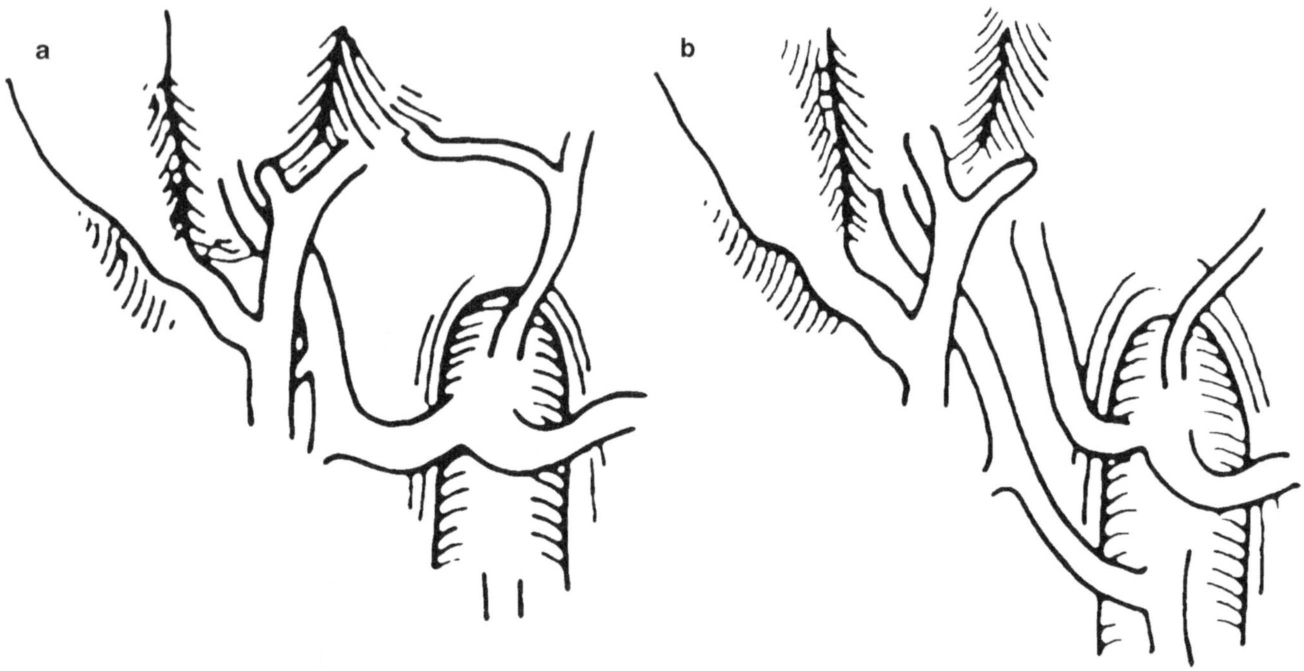

Fig. 9.6 The right and the left arteries arise from two separate arteries from two different origins. (**a**) Left hepatic artery from left gastric artery. (**b**) Right hepatic artery from superior mesenteric artery. (**c**) Left hepatic artery from left gastric artery, right hepatic artery from superior mesenteric artery. (**d**) Right and left hepatic arteries from different origins of aorta. (**e**) Left hepatic artery from the aorta, right hepatic artery from coeliac artery

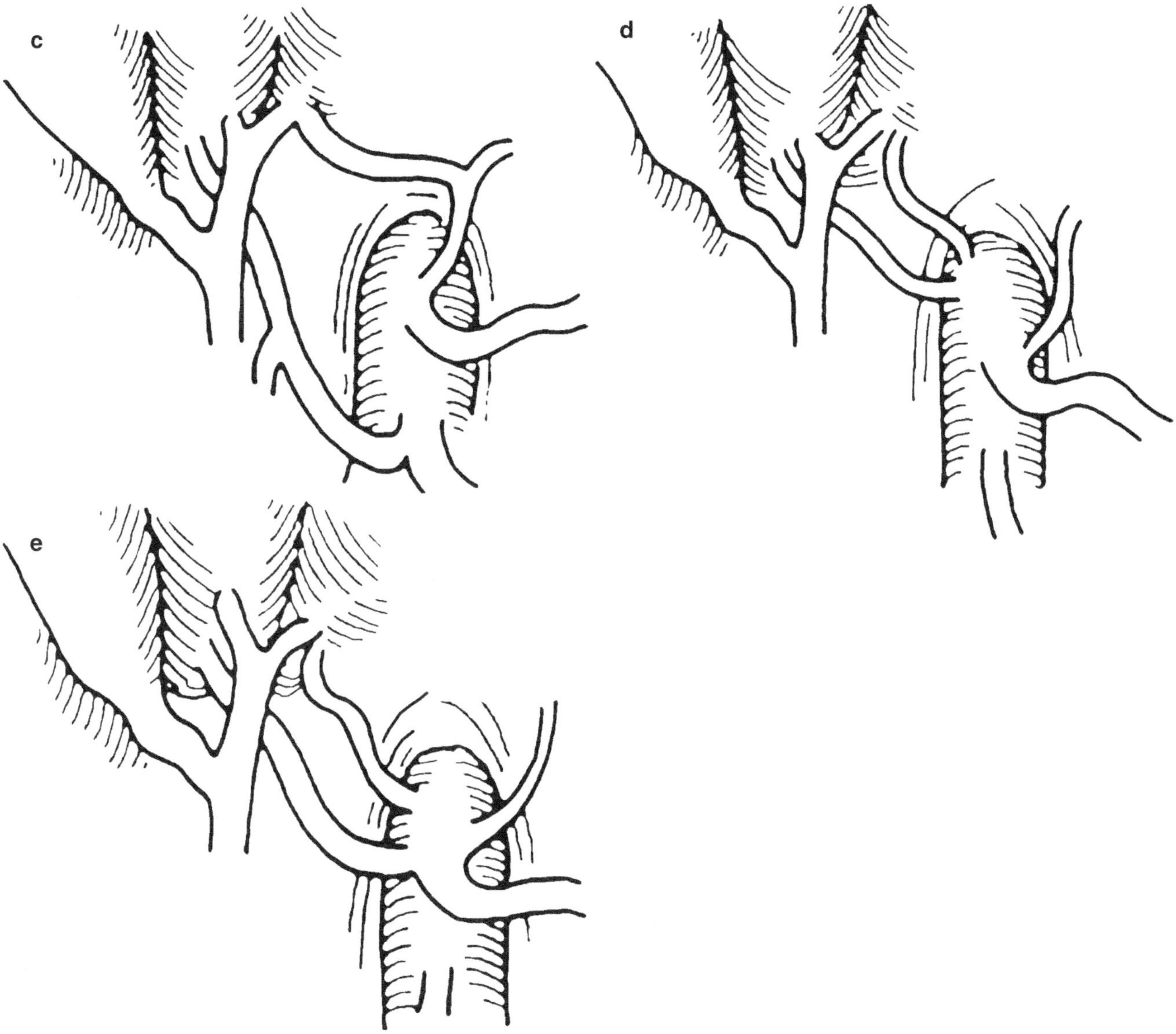

Fig. 9.6 (continued)

donor portal vein may have to be anastomosed to the superior mesenteric vein of the recipient, sometimes a graft is required to bridge the gap.

9.4.4 Hepatic Veins

Extrahepatic anomalies of the hepatic vein are rare. In such rare cases, a piggyback liver transplant can solve the technical problems (Fig. 9.10).

9.4.5 Partial Liver Transplant

With the exception of reduced-size liver transplantation which is seldom carried out nowadays, it is important to have adequate liver volumes and function on the two sides of the liver after the liver has been transected into two parts. In split liver transplantation, the right and the left sides of the liver will be used on two liver recipients. In living donor liver transplantation, a part of the liver will be transplanted onto the recipient and the remaining part of the liver remains in

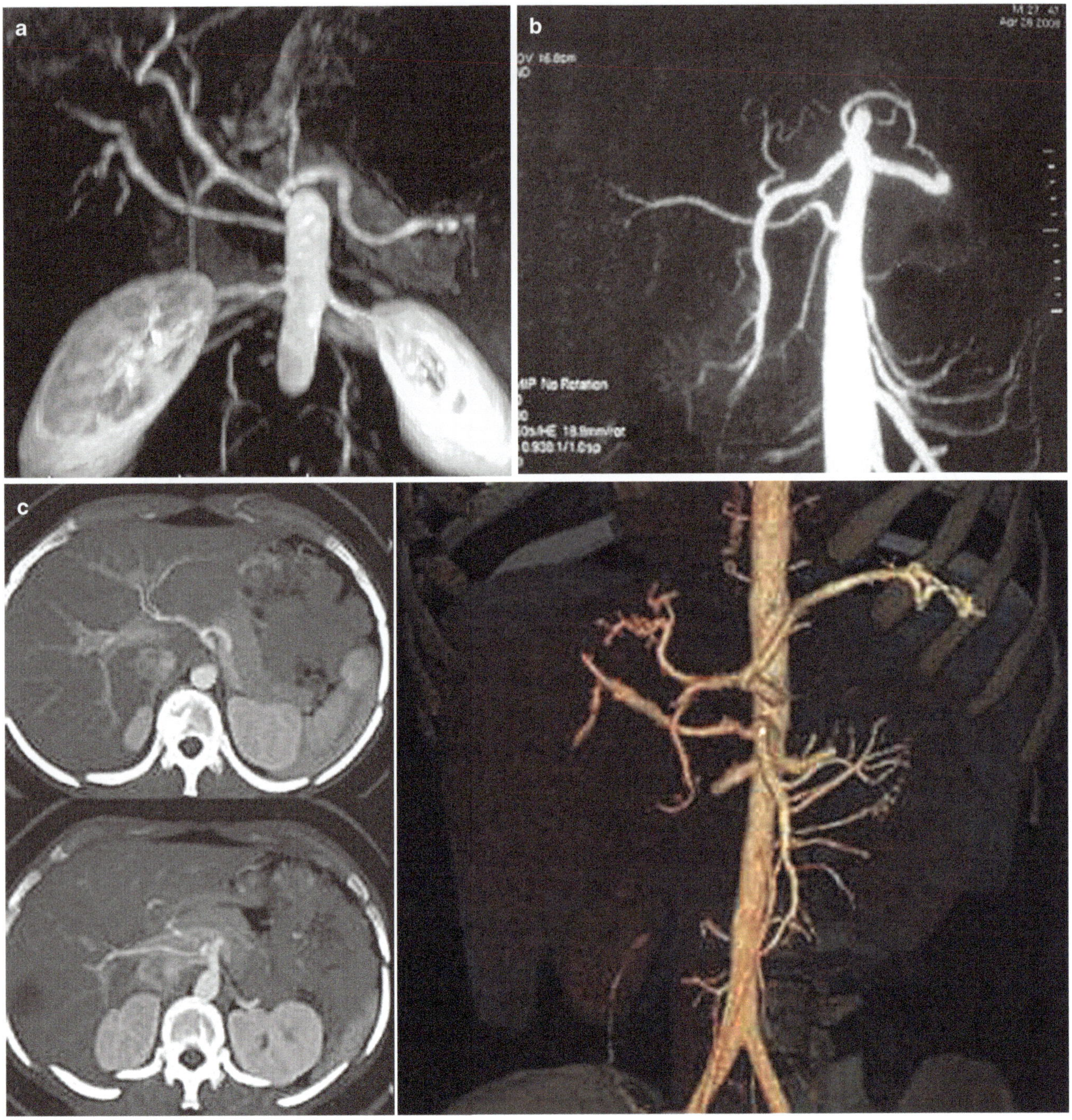

Fig. 9.7 Examples of anomalies

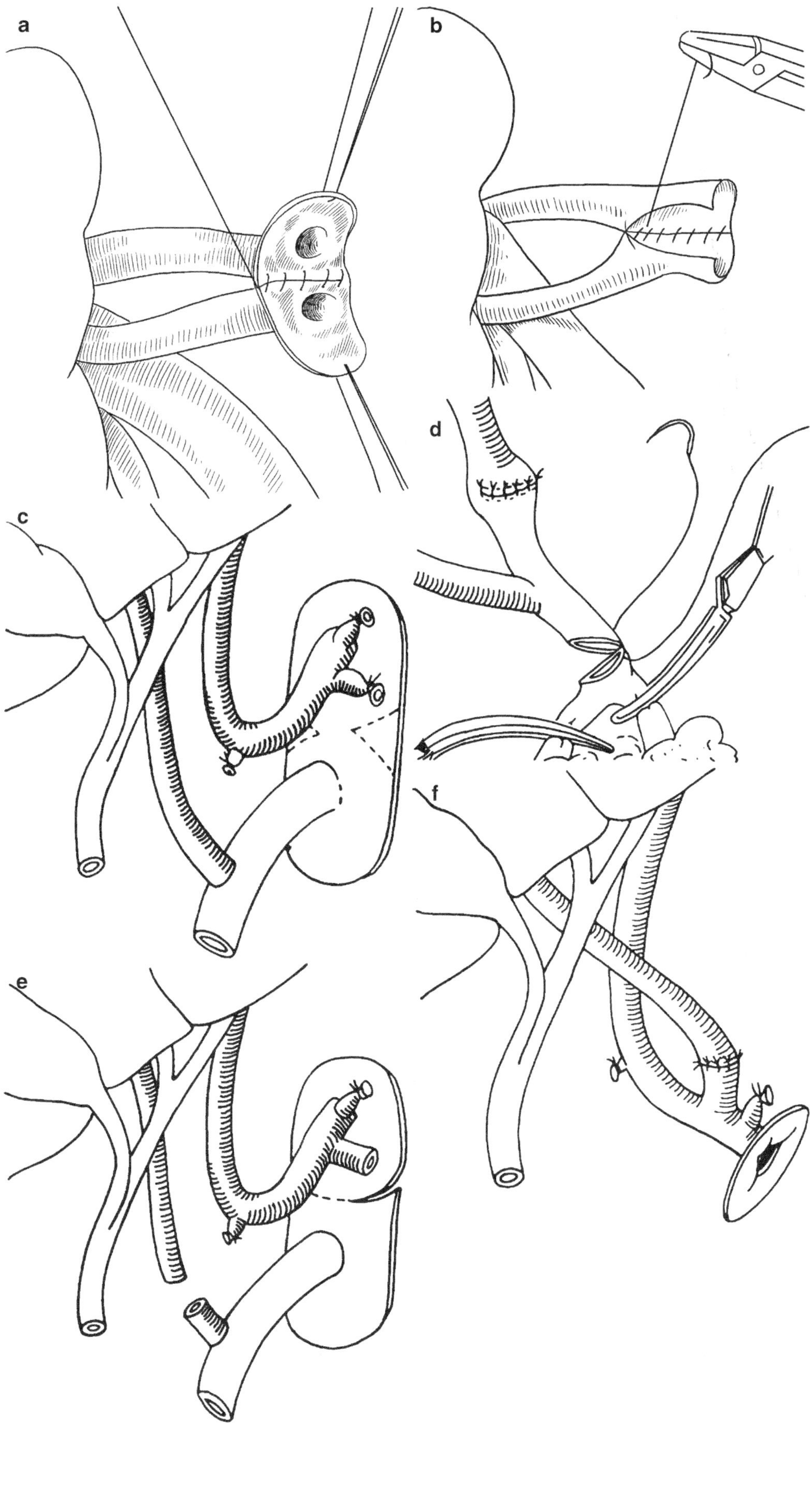

Fig. 9.8 Using arterioarteries anastomosis or a carrel patch before reconstruction. (**a**, **b**) The two carrel patches are sutured together to form one opening to anastomose with the recipient superior mesenteric artery or sutured as in (**c**, **d**). The alternative method is shown in (**e**, **f**) the divided hepatic artery is anastomosed to the splenic artery is anastomosed to the stump of the spleen artery, or to the gastroduodenal artery

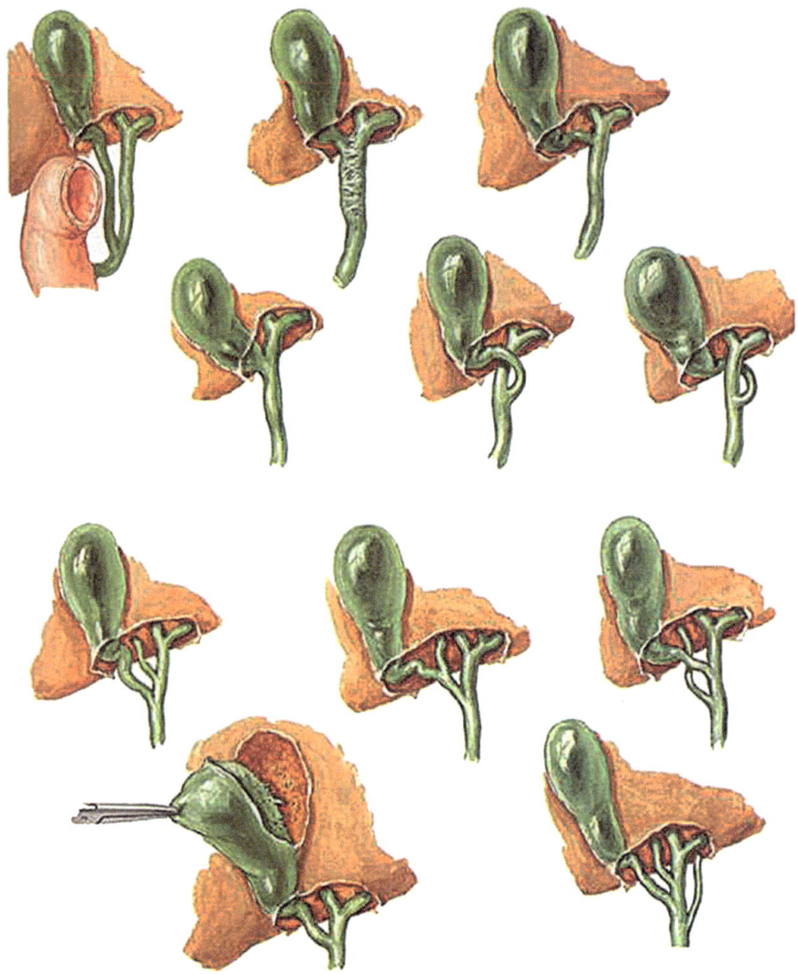

Fig. 9.9 Extrahepatic anomalies of the biliary system

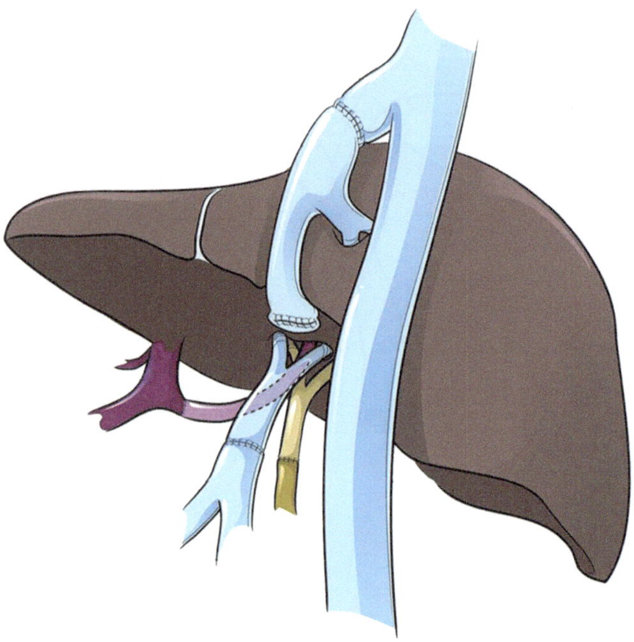

Fig. 9.10 Piggyback liver transplantation

the donor to support his health and his life. Thus, liver transection needs to be carried out not only precisely along the pre-planned plane so as to avoid leaving behind necrotic liver tissues, but also to leave behind enough lengths of the hepatic arterial, portal venous, hepatic venous and biliary structures for reconstruction (Fig. 9.11).

9.5 Intrahepatic Structures: Prevailing Pattern and Anomalies

It is also important to know the important intrahepatic vascular and biliary structures in the prevailing pattern and in the common anomalies to avoid getting into problems after liver transection.

In the use of a right hemiliver as a donor organ for living donor liver transplantation, it is still controversial as to whether the whole, part or no part of the middle hepatic vein should be kept in the donor's or the recipient's part of the liver. Those who advocate keeping the middle hepatic vein for the donor argue that segment 4 of the remnant liver of the donor will not

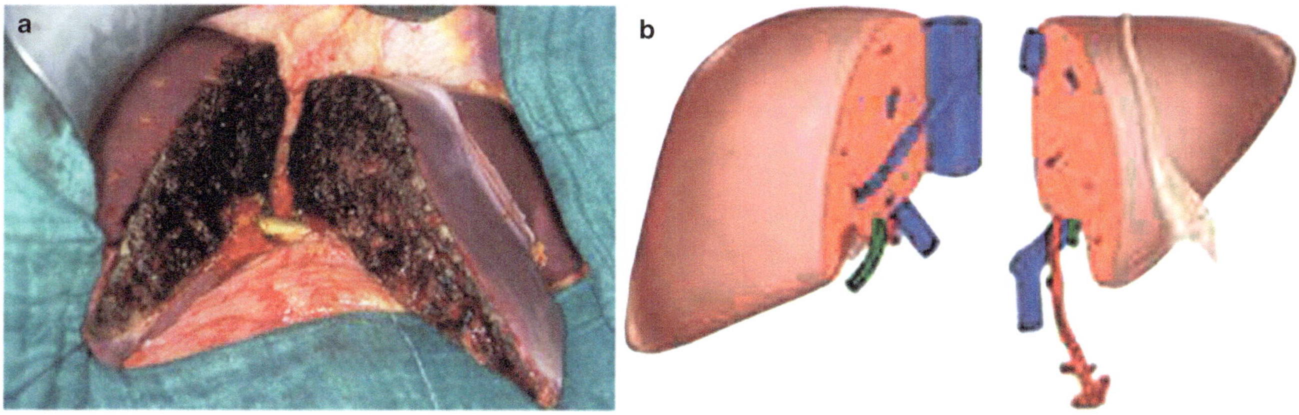

Fig. 9.11 Liver transection in liver transplantation. (**a**) Operative photograph; (**b**) Diagrammatic representation

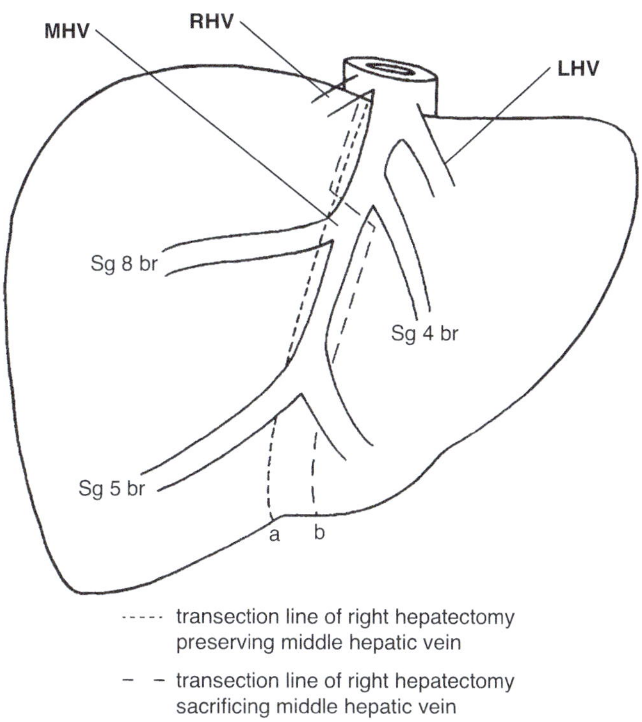

Fig. 9.12 Preserving part of the middle hepatic vein to preserve drainage of segment 4 through the segment 4 branch

become congested. However, the recipient may receive a liver with congested segments 5 and 8. The solution is to connect a vein graft to drain into the inferior vena cava to relieve the congestion. A compromise is to transect the middle hepatic vein as shown in Fig. 9.12 to preserve the segment 4 branch of the middle hepatic vein which drains the liver segment 4.

9.5.1 Intrahepatic Anomalies of the Hepatic Venous System

In some patients with anomalies of the intrahepatic venous branches (57.4% of individuals have segment 4 vein anomaly, and 60% of individuals have umbilical vein anomaly), the middle hepatic vein can be kept in the donated part of the right liver without causing congestion of segment 4 in the donor because these anomalous veins drain the liver segment 4 (Fig. 9.13).

Another important anomaly is the right inferior hepatic vein. This vein can be seen on CT scans in 10% of individuals because it is large. Transection of a large right inferior hepatic vein can lead to congestion of segment 6 in the donor liver. The solution is to do a vascular drainage into the inferior vena cava using a vascular graft if the congested area in segment 6 is large (Fig. 9.14).

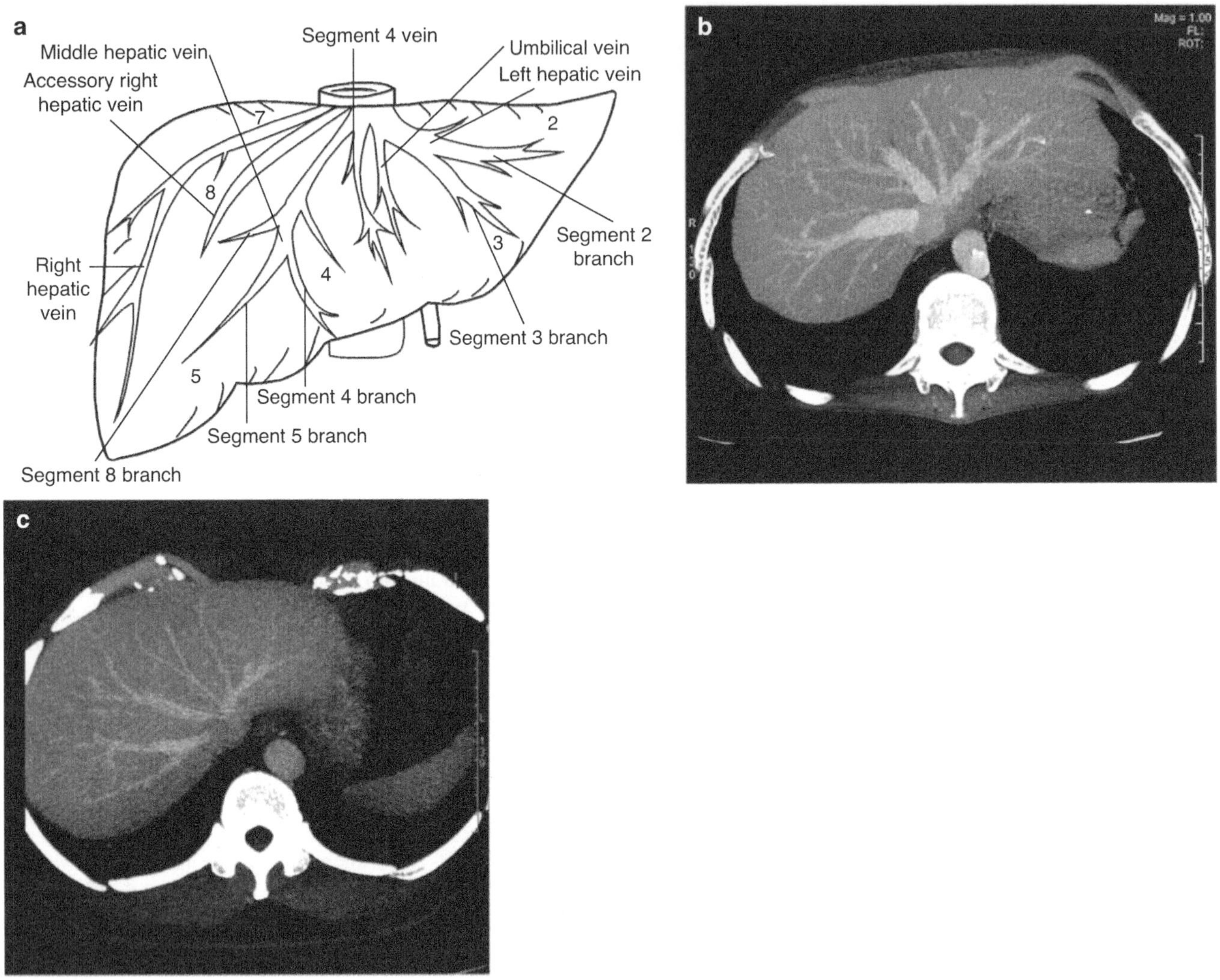

Fig. 9.13 (**a**) Diagrammatic representation; (**b**) Anomalous segment 4 vein; (**c**) Anomalous segment 4 vein and umbilical vein

9.5.2 Intrahepatic Anomalies of the Portal Venous System

The prevailing and the common anomalies of the intrahepatic portal venous system are shown in Fig. 9.15.

For trifurcation of the portal venous system, it is important not to transect the portal vein too near to the trifurcation, or a stricture will develop after operation (Fig. 9.16).

For the anomaly with the anterior sectoral portal venous branch draining into the left portal vein (Fig. 9.17) or the other less commonly seen anomalies (Fig. 9.18), they are not suitable for living donor liver transplant.

9.5.3 The Prevailing and Common Anomalies of the Intrahepatic Biliary System (Fig. 9.19)

For the trifurcation type, transection of the intrahepatic duct should not be carried out too close to the trifurcation, or a stricture will develop after operation (Fig. 9.20). The anomaly with the right posterior sectoral duct draining into the left hepatic duct is not suitable for living donor liver transplantation (Fig. 9.21).

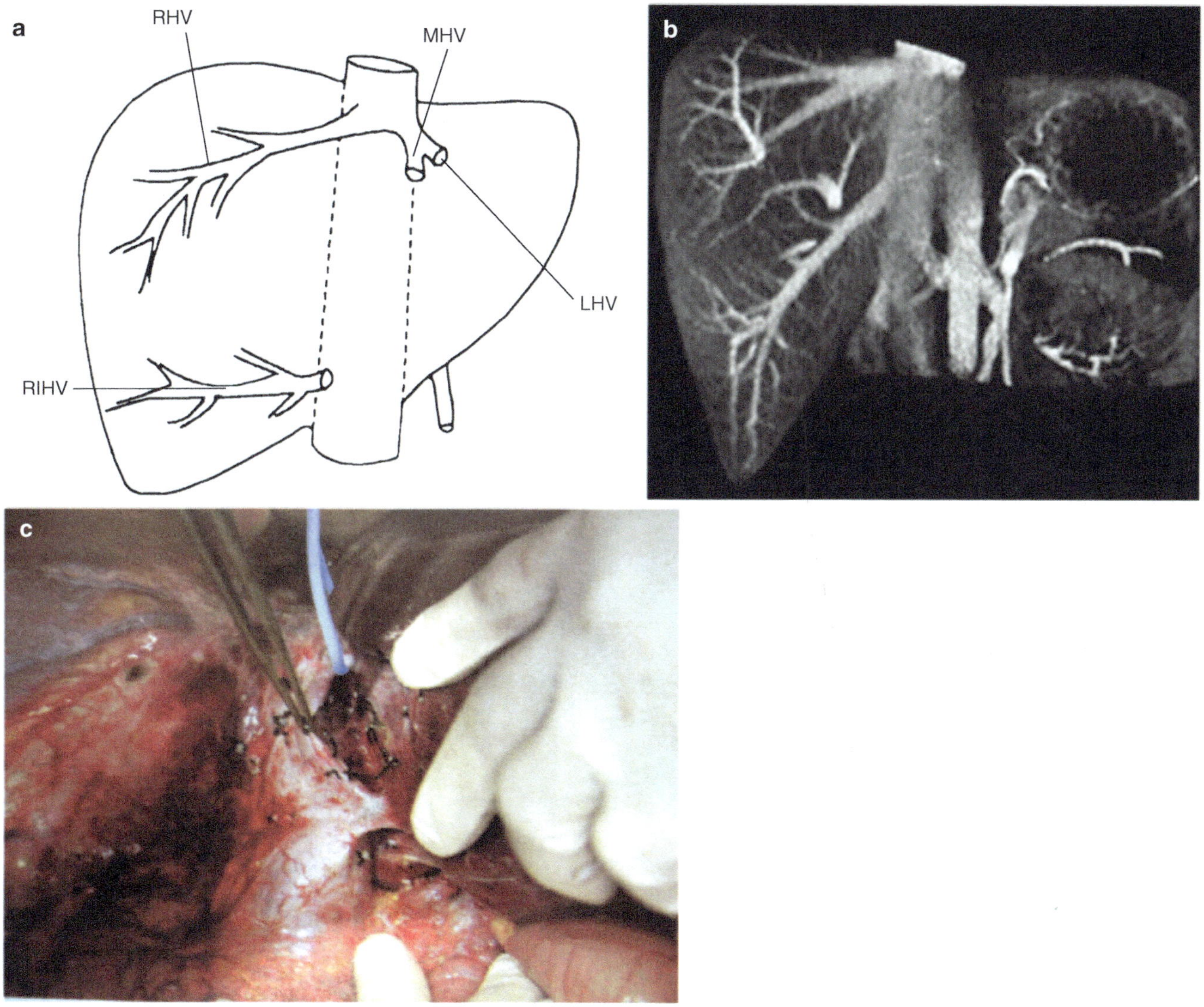

Fig. 9.14 Large right inferior hepatic vein (RIHV). (**a**) Diagrammatic representation; (**b**) Radiological feature; (**c**) Operation findings. *RIHV* right inferior hepatic vein. *RHV* right hepatic vein. *MHV* middle hepatic vein, *LHV* left hepatic vein

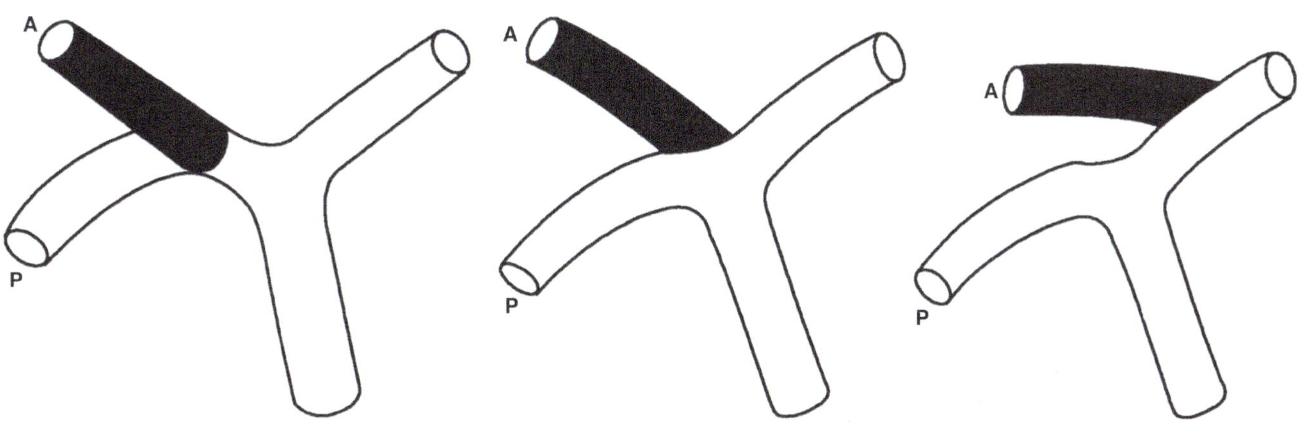

Fig. 9.15 Prevailing and common anomalies of the intrahepatic portal vein

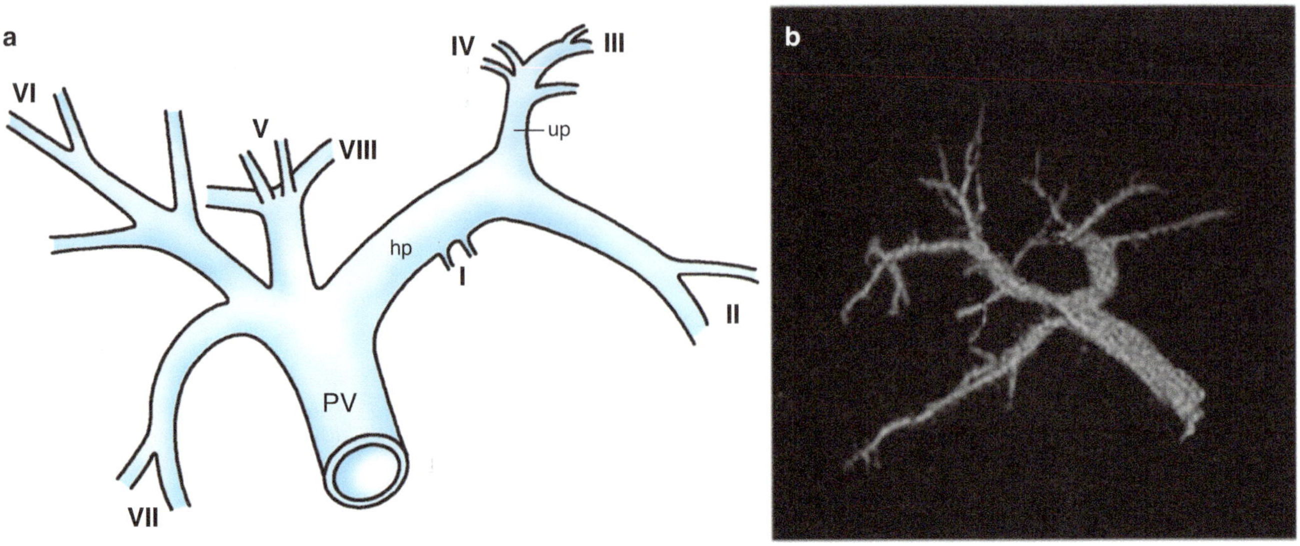

Fig. 9.16 Trifurcation of portal vein. *PV* portal vein, *hp* transverse portion of portal vein, *up* umbilical portion of portal vein. *I, II, III, IV, V, VI, VII, VIII* portal vein branches to live segments 1, 2, 3, 4, 5, 6, 7, 8. (**a**) Diagrammatic representation; (**b**) Radiological feature

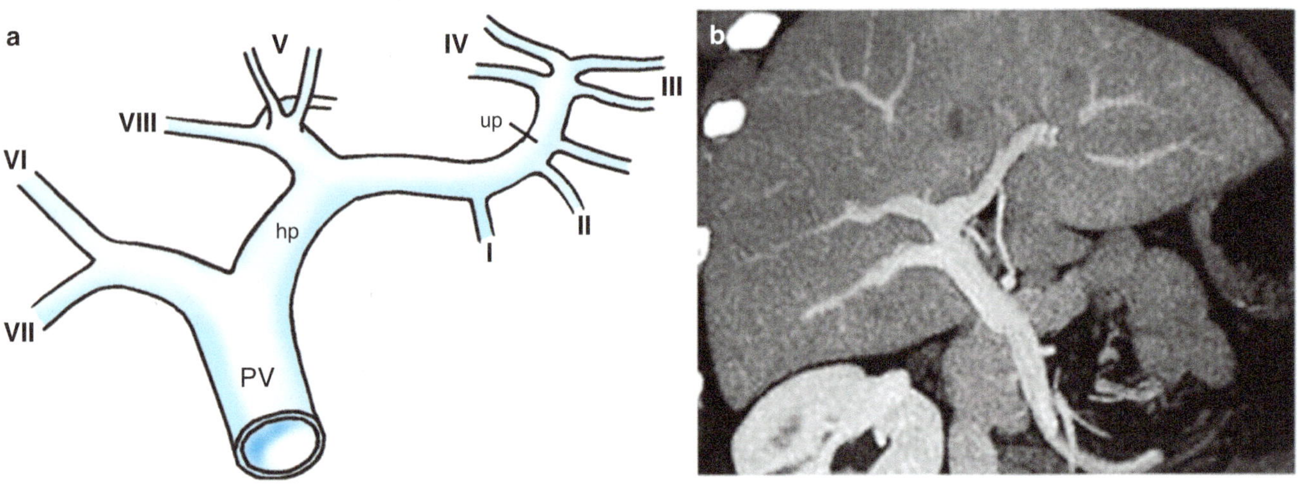

Fig. 9.17 Anterior sectoral portal vein branch into left portal vein. (**a**) Diagrammatic representation; (**b**) Radiological feature

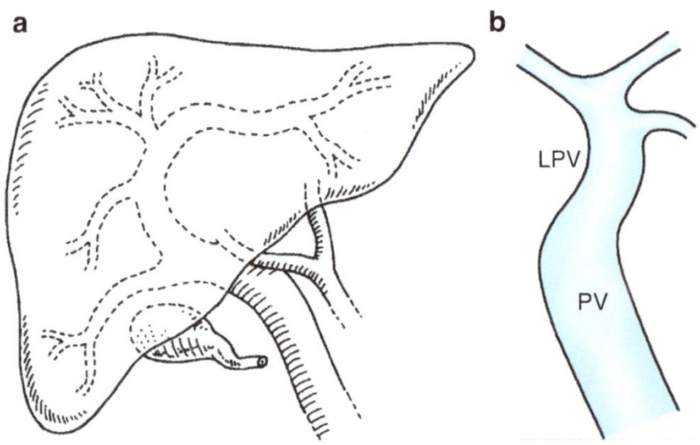

Fig. 9.18 Other less commonly seen anomalies. (**a**) Single right portal vein supplying whole liver. (**b**) Single left portal vein supplying whole liver

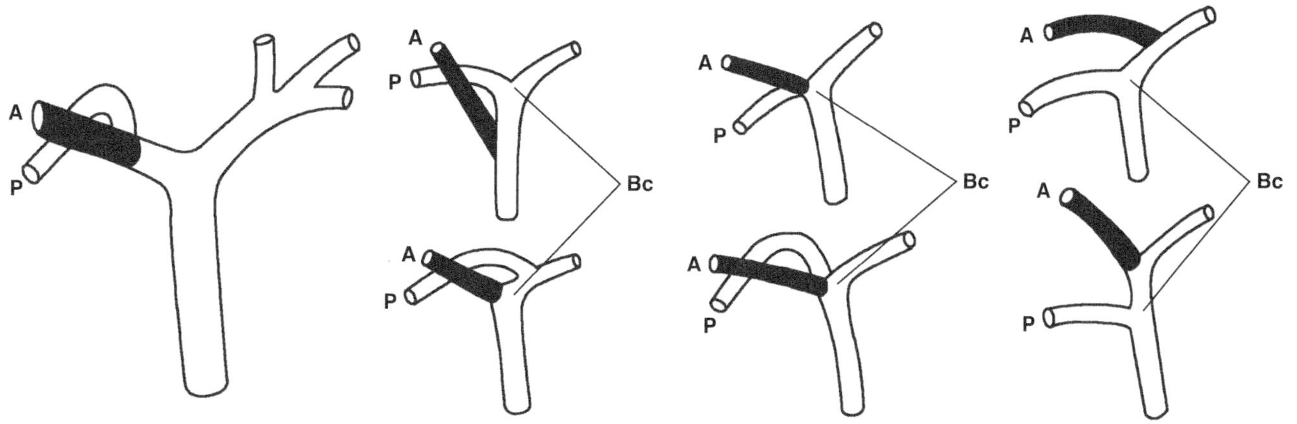

Fig. 9.19 Prevailing and common anomalies of the intrahepatic biliary system

Fig. 9.20 Trifurcation of intrahepatic ducts

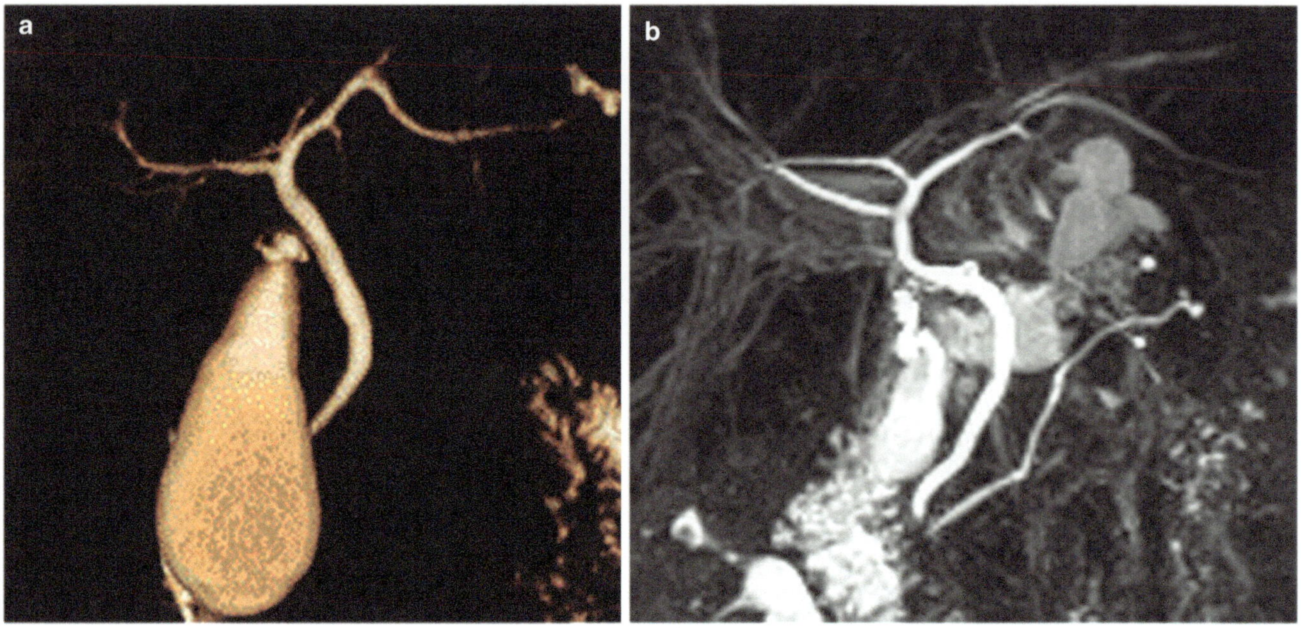

Fig. 9.21 The anomaly with the right posterior sectoral duct draining into the left duct and is not suitable for living donor liver transplantation. (**a**) A small right posterior sectoral duct; (**b**) Anomalous right posterior sectoral duct

9.6 Conclusion

Extra- and intrahepatic anomalies can affect the results of liver transplantation. It is important to identify these anomalies before and during liver transplantation. Three-dimensional reconstruction of vessels and bile ducts helps to understand the important extra- and intrahepatic anatomical structure better.

Further Reading

Blumgart LH, Hann LE. Chapter 1: Surgical and radiologic anatomy of the liver and biliary tract. In: Blumgart LH, Fong Y, editors. Surgery of the liver and biliary tract, vol. 1. 3rd ed. London: W.B. Saunders; 2000. p. 3–33.

Fan ST. Living donor living transplantation. Hong Kong, China: Takungpao Publishing Co., Ltd.; 2007.

Jamieson G, Launois B. Chapter 2: Liver resections and liver transplantation: the anatomy of the liver and associated structures. In: Jamieson GG, editor. The anatomy of general surgical operation. 2nd ed. Edinburgh: Elsevier Churchill Livingstone; 2006. p. 8–23.

10 Defining the Couinaud's Liver Segments Clinically

Couinaud divided the liver into two hemilivers, four sectors and eight segments (or nine segments). Each segment is an individual unit with its own arterial and portal venous supply, and biliary and hepatic venous drainage. Each segment can be resected individually or in combination with adjacent segment(s) (see Chaps. 2 and 3).

10.1 Surgical Exposure of the Liver

Adequate exposure of the liver during surgery requires the use of a good incision. Commonly used incisions are shown in Fig. 10.1. For patients who are thin, a right subcostal incision with midline extension is adequate. A right thoracoabdominal incision is seldom required, even with a big tumour with diaphragmatic involvement. The bilateral subcostal incision with midline extension can have the midline incision extended to split the sternum in operations for patients with hepatocellular carcinoma with tumour thrombus extended up the inferior vena cava into the right atrium.

The modern retractor retracts the rib cages in an upward, outward and forward direction, thus fully exposing the diaphragmatic and superior surfaces of the liver. Access to the suprahepatic inferior vena cava is not a problem with the use of modern incisions and retractors (Fig. 10.2).

10.2 Preoperative Imaging to Define the Couinaud's Segments

Preoperative ultrasound defines the inferior vena cava and the right, middle and left hepatic veins with ease. The middle hepatic vein (MHV) divides the liver into the right and left hemilivers. On the right side, the right hepatic vein (RHV) divides the right hemiliver into the right anterior sector (segments 5, 8) and the right posterior sector (segments 6, 7) (Fig. 10.3).

The main portal vein (MPV) can be seen easily. By tracing the right portal vein branches, the right anterior portal vein (RAPV) can be traced which divides into the segment 5 and 8 branches with segment 8 being superior, and segment 5 inferior. Similarly, the right posterior portal vein (RPPV) divides into the branch to segment 7 superiorly and the branch to segment 6 inferiorly (Fig. 10.4).

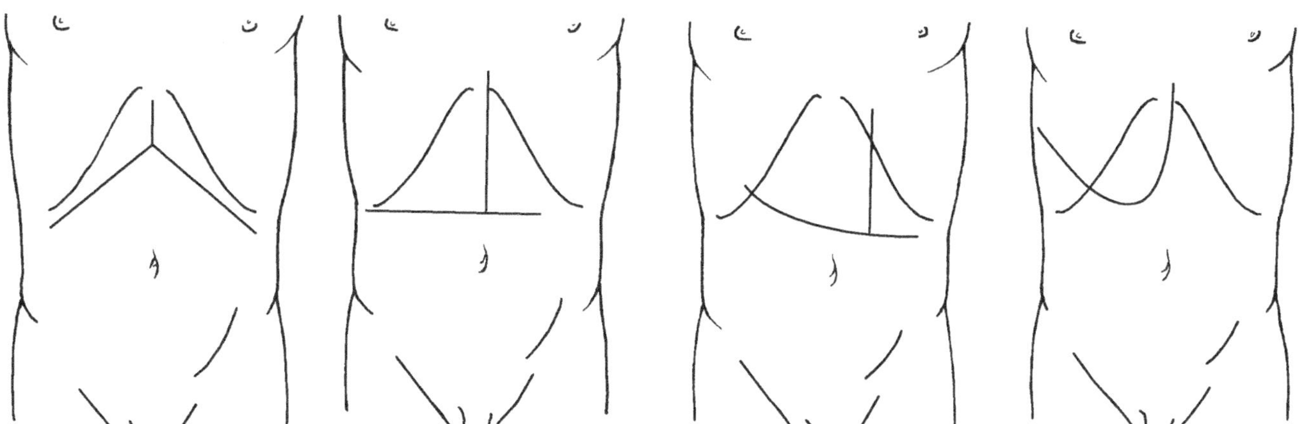

Fig. 10.1 Common incisions for liver surgery

W. Y. Lau, *Applied Anatomy in Liver Resection and Liver Transplantation*, https://doi.org/10.1007/978-981-16-0800-1_10

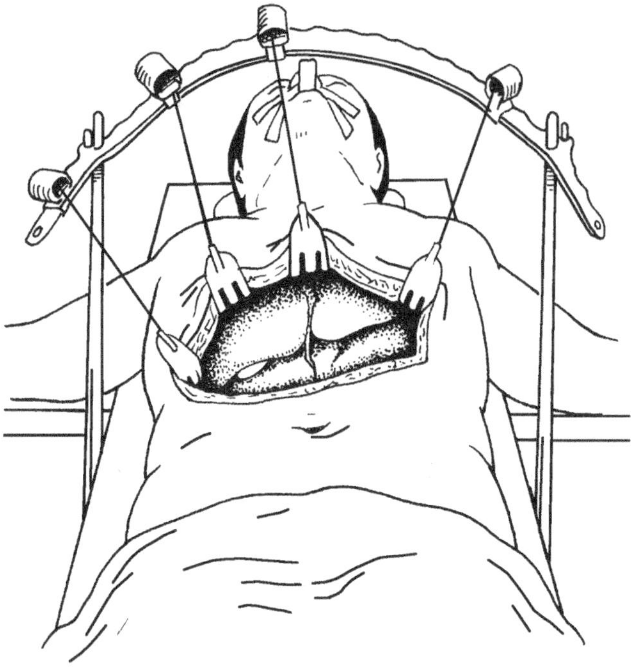

Fig. 10.2 Adequate exposure of liver with a retractor

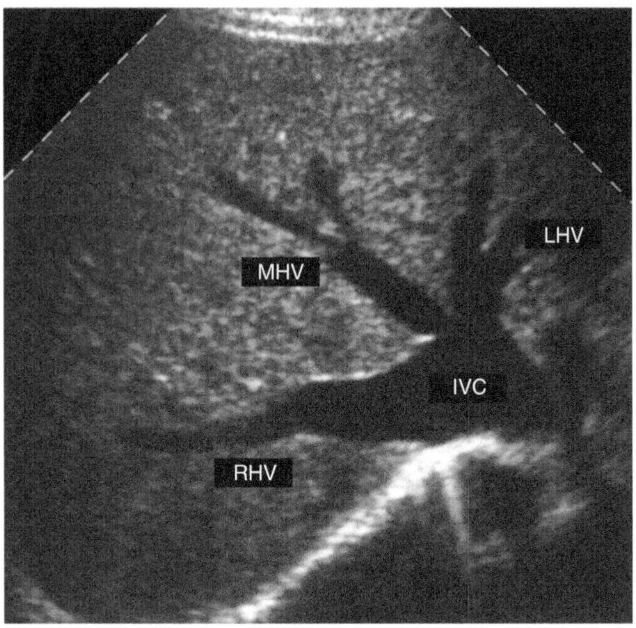

Fig. 10.3 Ultrasound showing inferior vena cava and the hepatic veins. *IVC* inferior vena cava, *LHV* left hepatic vein, *MHV* middle hepatic vein, *RHV* right hepatic vein

In the left hemiliver, the left hepatic vein (LHV) divides the left hemiliver into the left lateral sector (segment 2) and left medial sector (segments 3 and 4) (Fig. 10.3).

The left portal vein (LPV) can be traced from its transverse portion (T) to the umbilical portion (U). Branches then come off from the umbilical portion to supply segment 4 on the right side and segments 3 and 2 on the left side (Fig. 10.5).

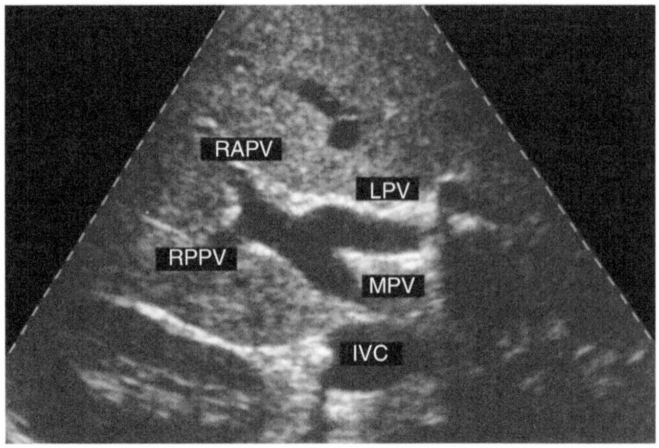

Fig. 10.4 Ultrasound showing the Main Portal Vein and its Right Sectoral Branches. *IVC* inferior vena cava, *MPV* main portal vein, *LPV* left portal vein, *RAPV* right anterior portal vein, *RPPV* right posterior portal vein

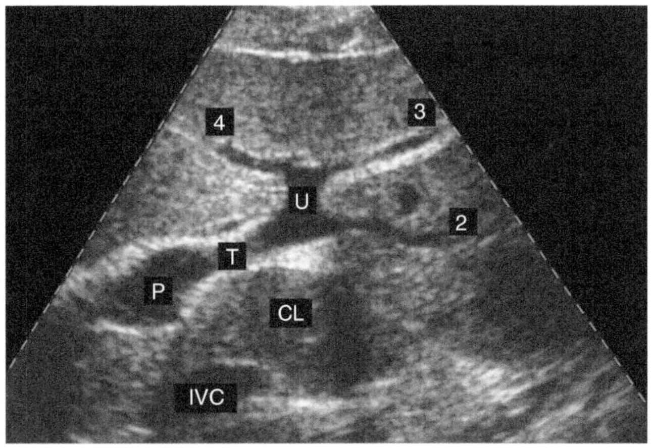

Fig. 10.5 Ultrasound showing left portal vein and its branches. *CL* Caudate Lobe, *IVC* inferior vena cava, *P* portal vein, *T* transverse portion of portal vein, *U* umbilical portion of portal vein, 2,3,4 portal vein branches to live segments 2, 3, 4

The caudate lobe (CL) can be seen on the transverse section to be surrounding the inferior vena cava (IVC) (Fig. 10.6). The lesser omentum (LV) can be seen as a white hyperechoic line in front of the caudate lobe (CL). A vein draining the caudate lobe (white arrows) directly into the inferior vena cava can be seen in Fig. 10.7.

Thus, the three hepatic veins divide the liver into four sectors and the portal vein branches define the seven segments (2–8). The caudate lobe can be defined easily on ultrasound to be around the inferior vena cava.

Similarly, on preoperative computed tomography or magnetic resonance imaging, the middle hepatic vein divides the liver into the right and left hemilivers, and the right and the left hepatic veins divide the liver into four sectors. The portal vein defines segments 2–8, and the caudate can be identified because of its size and shape (Figs. 10.8, 10.9 and 10.10).

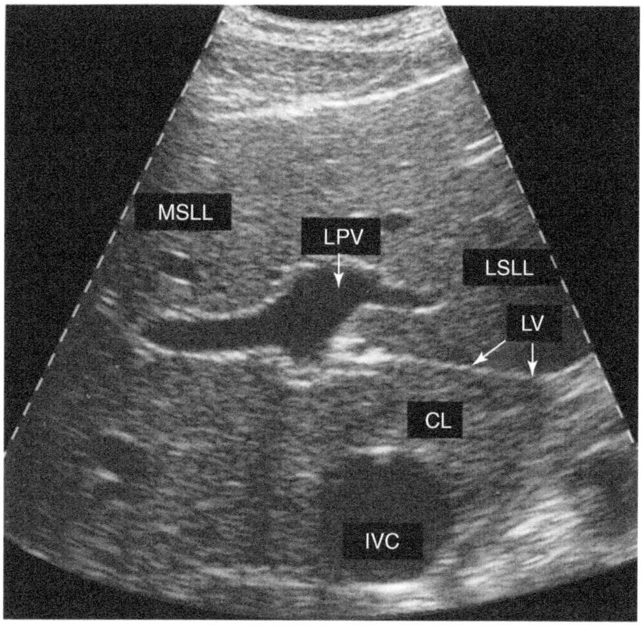

Fig. 10.6 Transverse section of ultrasound showing caudate lobe. *CL* Caudate Lobe, *IVC* inferior vena cava, *LPV* left portal vein, *LSLL* lateral section of left liver, *LV* lesser omentum, *MSLL* medial section of left liver

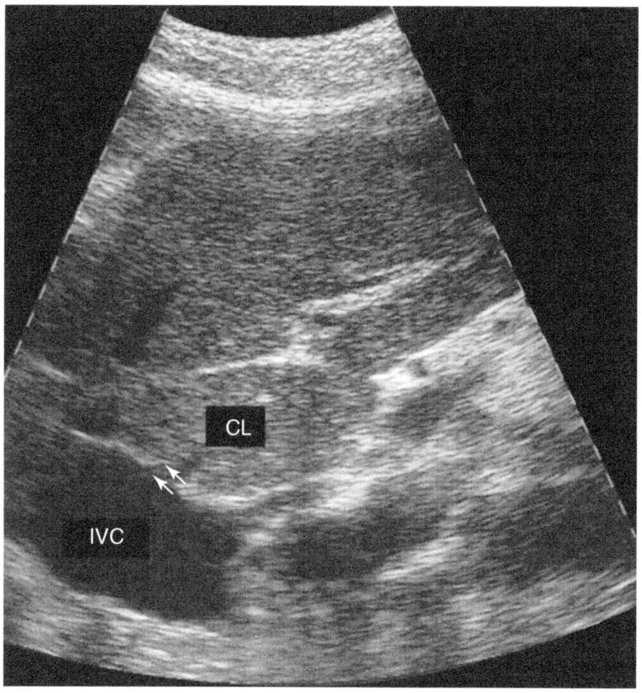

Fig. 10.7 Sagittal section of ultrasound showing caudate lobe (CL). *CL* Caudate Lobe, *IVC* inferior vena cava. Arrow points to a vein draining into IVC

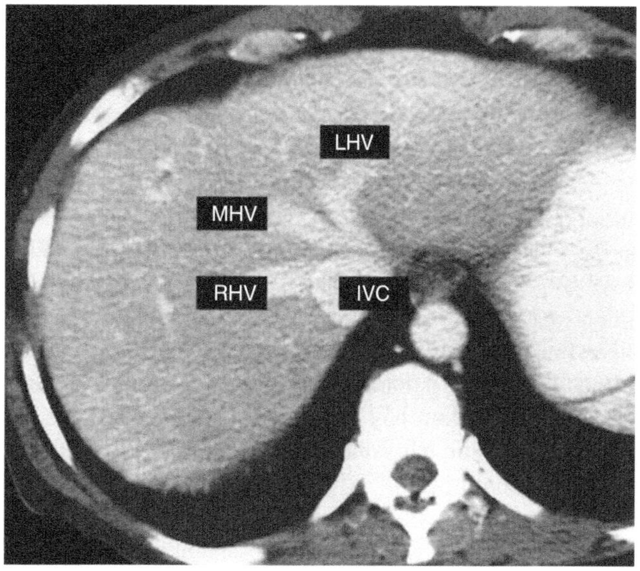

Fig. 10.8 CT scan showing inferior vena cava (IVC) and the right hepatic vein (RHV); middle hepatic vein (MHV); and the left hepatic vein (LHV)

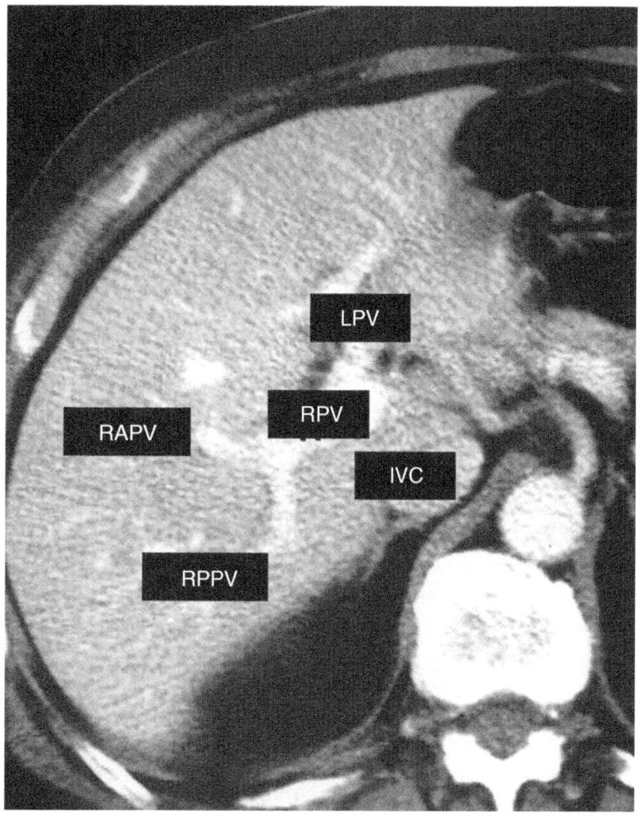

Fig. 10.9 CT scan showing right portal vein (RPV) and its right anterior portal vein (RAPV) and posterior portal vein (RPPV) branches. *LPV* left portal vein. *IVC* inferior vena cava

On preoperative imaging, we should have a good idea of the number of tumours, their relationship to the major hepatic veins and portal venous branches, and the liver segments in which these tumours are located. Thus, a preoperative decision can be made to allow the best surgical procedure to be carried out to resect the tumour(s) and to determine the plane of liver transection so as to get an adequate resection margin without sacrificing too much non-tumourous liver tissues.

10.3 Intraoperative Ultrasound

Intraoperative ultrasound is indispensable in modern liver surgery (Fig. 10.11). It is especially important in cirrhotic liver surgery because a soft hepatocellular carcinoma is difficult to see and to feel within a hard and nodular cirrhotic liver. In metastatic carcinoma to the liver, the situation is better as the tumour is hard and the liver is soft and smooth. Furthermore, the atrophy–hypertrophy complex present in a cirrhotic liver can make the gross anatomy of the liver distorted and confusing.

Intraoperative ultrasound allows the detection of small tumourous lesions which are otherwise not detectable on preoperative imaging. In the case shown in Fig. 10.12a, a 2.5-cm hepatocellular carcinoma was present in the right liver. Intraoperative ultrasound showed a 1 cm hepatocellular carcinoma in the left liver which was not seen on preoperative computed tomography (Fig. 10.12b).

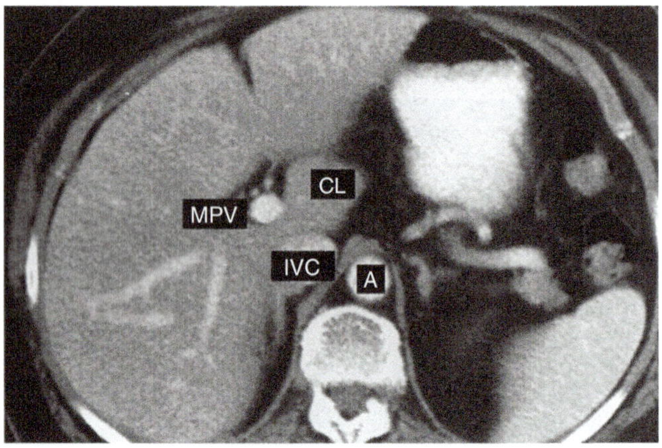

Fig. 10.10 CT showing caudate lobe. *MPV* main portal vein, *IVC* inferior vena cava, *A* aorta, *CL* caudate lobe

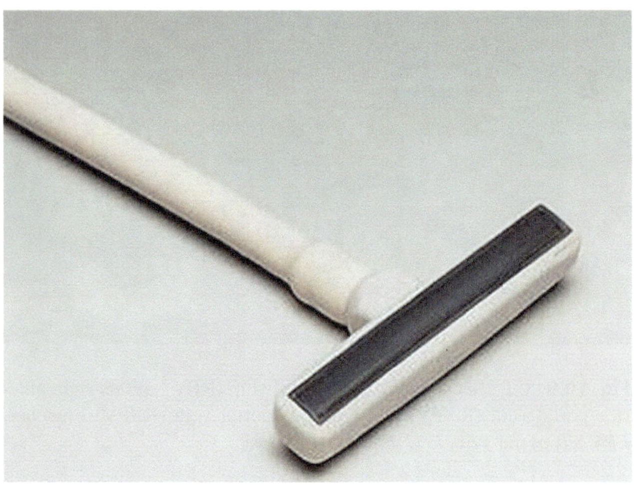

Fig. 10.11 Linear probe used in Intraoperative Ultrasound

Furthermore, intraoperative ultrasound in other patients showed compression of the right portal vein (Fig. 10.13), and tumour thrombus in the right hepatic vein and inferior vena cava (Fig. 10.14). We reported that intraoperative ultrasound decisively altered the preoperatively planned surgical treatment strategy in 32% of patients who underwent laparotomy for hepatocellular carcinoma, and significantly decreased the rate of positive tumour margin involvement after liver resection when compared with patients who received no intraoperative ultrasonography (0% vs. 16%).

Intraoperative ultrasound can help to determine the transection plane (TP) in relation to the tumour (HCC) and to a large vessel (V) (Fig. 10.15). With fingers (F) at the back of the liver, the relation of the transection plane (TP) to the edge of the tumour (HCC) can be shown even better (Fig. 10.16).

10.4 Operative Steps in Intraoperative Ultrasound

There are six operative steps in intraoperative ultrasound: (1) general inspection of the whole liver to detect unexpected lesions not detected preoperatively; (2) a systematic anatomical study to trace the three hepatic veins, the main portal vein and its branches so that individual Couinaud's liver segments can be determined; (3) location of the tumour in the liver segment(s); (4) determination of the liver segment(s) to be resected, this depends on the segments in which the tumour is located, the vasculobiliary sheaths and the hepatic venous drainage that need to be sacrificed to get an adequate resection margin; (5) marking of the liver parenchymal transection plane on the surface of the liver; and (6) redetermination of the distance of the transection plane to the edge of the tumour (the planned resection margin).

10.5 Surface Anatomy of the Liver (See Chap. 1)

The surface anatomy can be used to identify some anatomical parts of the liver (Figs. 10.17 and 10.18). However, not all liver segments and sectors have surface landmarks. Hepatic atrophy–hypertrophy complex can make surface markings to define sectors and segments difficult and inaccurate.

The midplane between the right and left livers lies in a plane that crosses the gallbladder bed anteriorly to the inferior vena cava posteriorly. The falciform ligament divides the

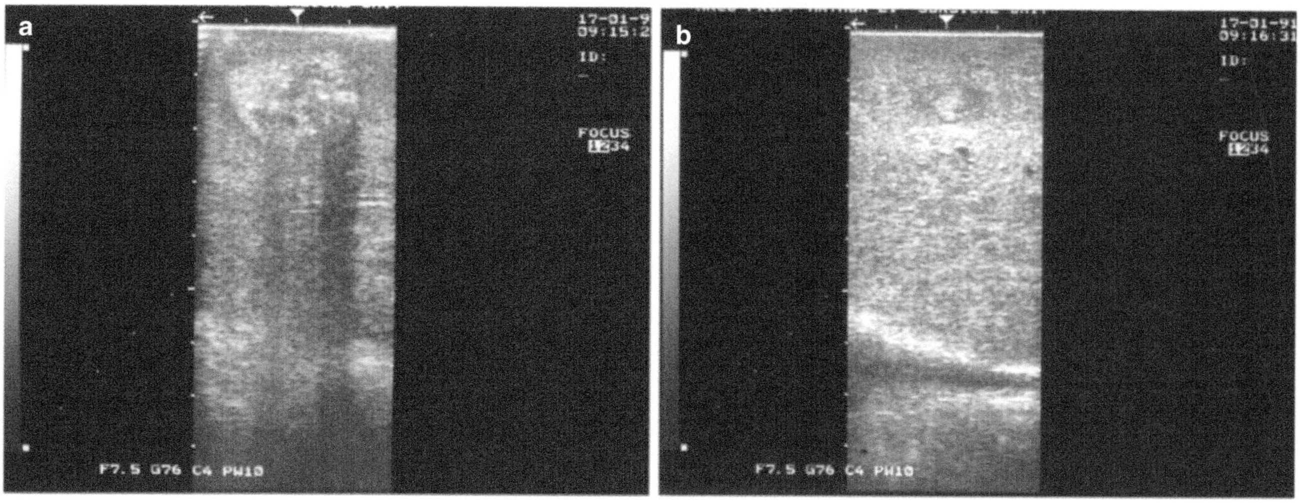

Fig. 10.12 Intraoperative ultrasound. (**a**) Showing a 2.5-cm hepatocellular carcinoma. (**b**) A 1 cm hepatocellular carcinoma not shown on preoperative CT

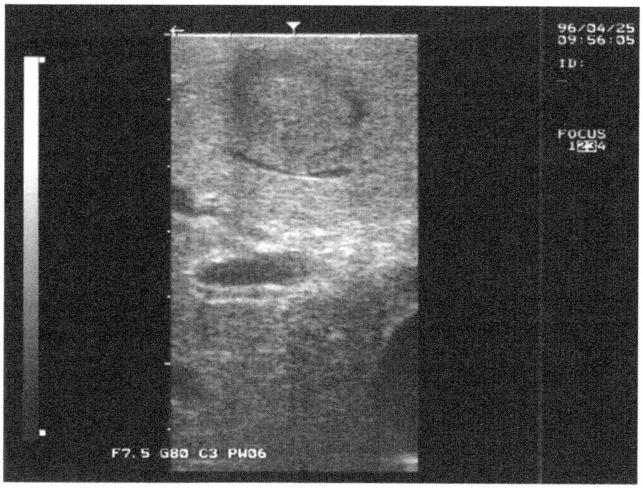

Fig. 10.13 Intraoperative ultrasound showing compression of right portal vein by tumour

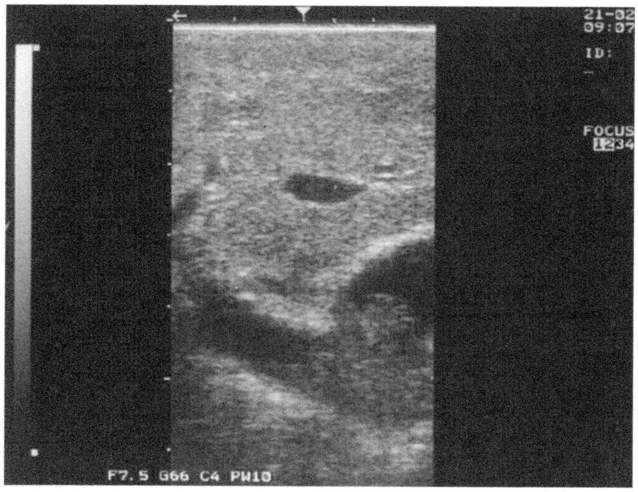

Fig. 10.14 Intraoperative ultrasound showing tumour thrombus in IVC and right hepatic vein

left liver into the medial section (segment 4) and the lateral section (segments 2 and 3). The quadrate lobe represents segment 4b while the rest of segment 4 is 4a.

In the right liver, there is no surface landmark except the Rouviere sulcus which marks where the right posterior sectional pedicle enters into the liver (see Sect. 4.2).

The caudate lobe and process stand out clearly on the surface anatomy.

10.6 Intraoperative Ultrasound to Define the Liver Segments

Like preoperative ultrasound, intraoperative ultrasound is used to define the individual hemilivers, sectors and segments. Intraoperative ultrasound has the advantage over preoperative ultrasound in that there is no abdominal wall or ribs to interfere with the ultrasound waves.

Again the middle hepatic vein divides the liver into the right and the left livers (Fig. 10.19).

In the right liver, the right hepatic vein divides it into the right anterior and posterior sectors. Tracing the right portal vein helps to identify the right anterior sectoral portal vein which divides into the segment 5 branch inferiorly and the segment 8 branch superiorly.

In the left liver, the left hepatic vein divides it into left lateral sector (segment 2) and left medial sector (sectors 3, 4) (Fig. 10.19). Tracing the left portal vein helps to identify segments 2, 3, and 4 (Figs. 10.20 and 10.21).

The caudate lobe stands out as it surrounds the inferior vena cava and the bright echogenic lesser omentum lies in front of it (Fig. 10.22).

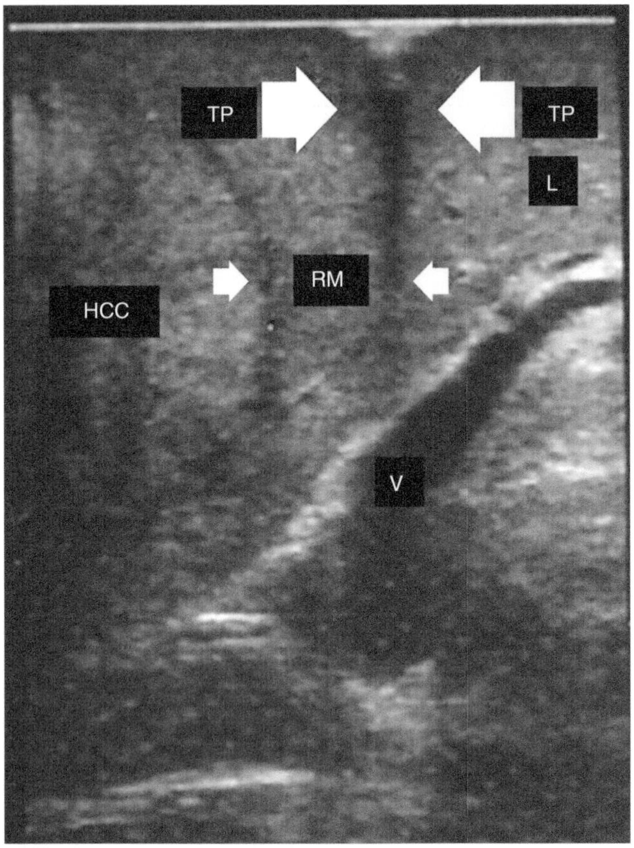

Fig. 10.15 Intraoperative ultrasound showing relationship of transection plane (TP and arrow) to tumour (HCC) and a big vessel (V). *RM* resection margin

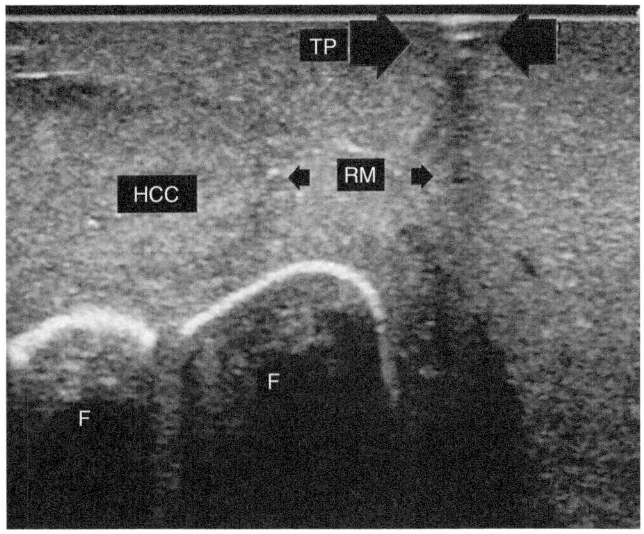

Fig. 10.16 Intraoperative ultrasound showing liver transection plane (TP and arrow) with tumour (HCC). *F* finger, *RM* resection margin. The fingers were used to guide the plane of hepatic transection

Fig. 10.17 Surface landmarks of the liver. Diaphragmatic surface

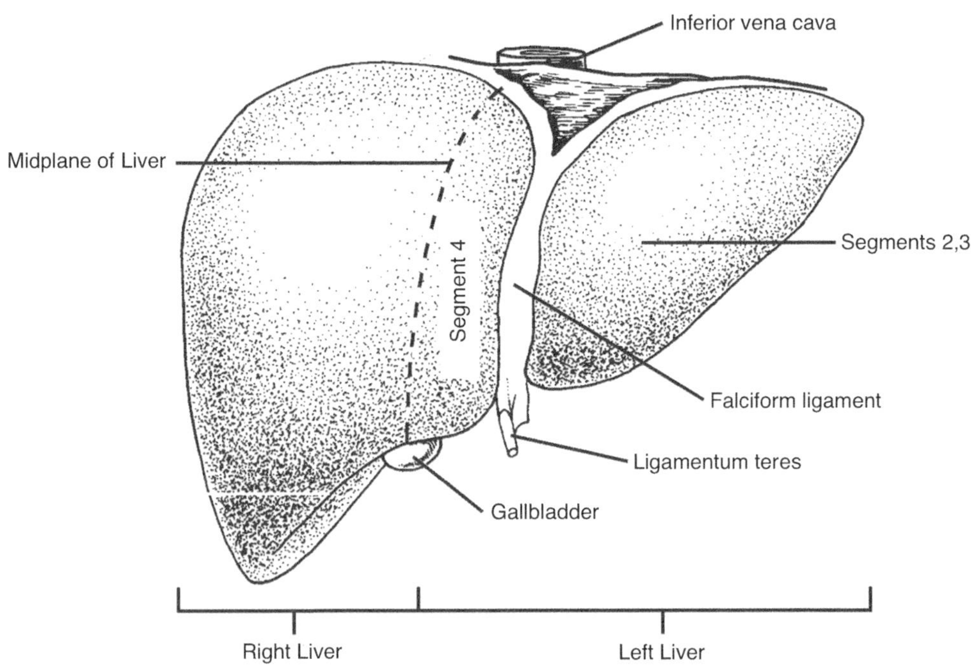

Fig. 10.18 Surface landmarks of the liver. Visceral surface

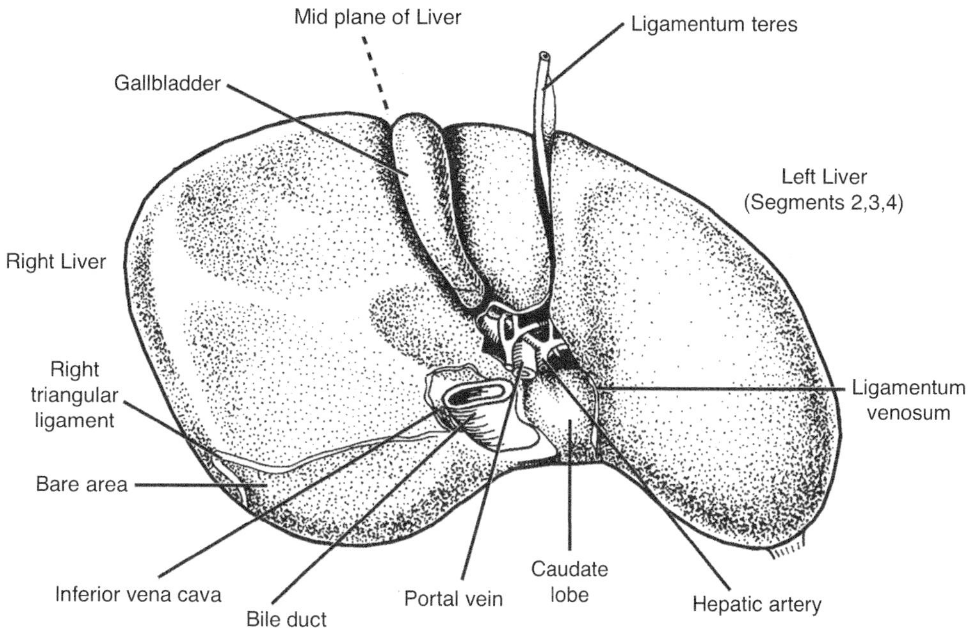

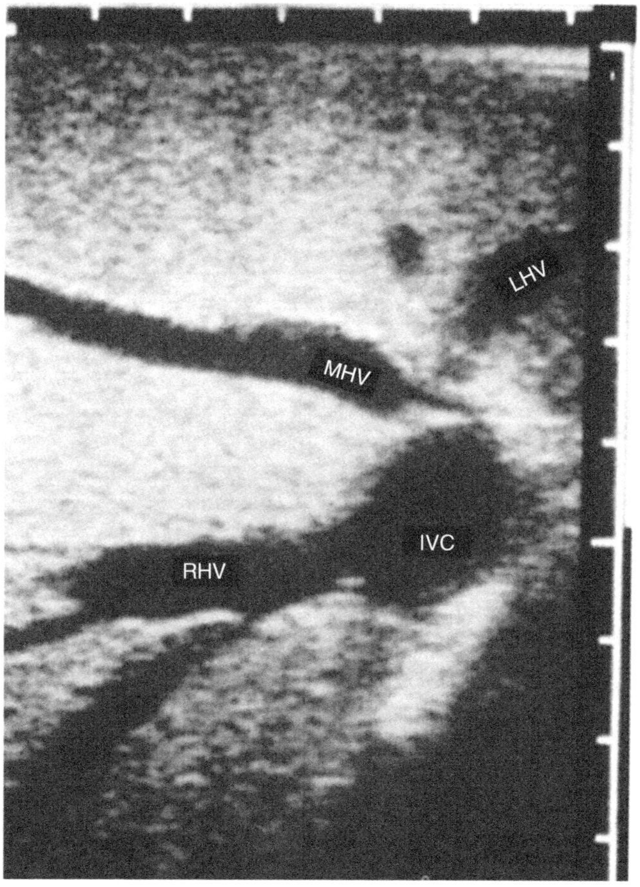

Fig. 10.19 Intraoperative ultrasound showing inferior vena cava and the three hepatic veins. *IVC* inferior vena cava, *RHV* right hepatic vein, *MHV* middle hepatic vein, *LHV* left hepatic vein

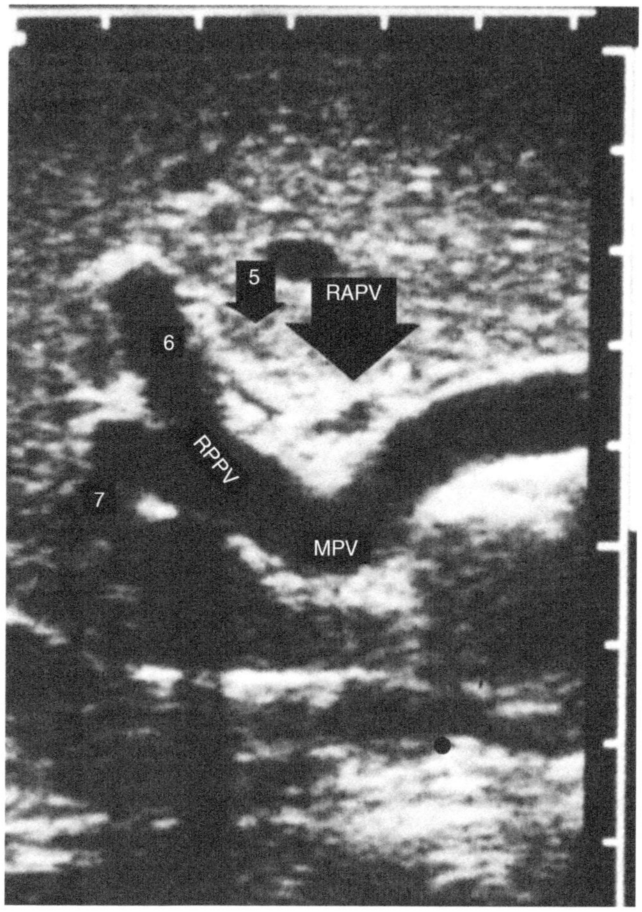

Fig. 10.20 Intraoperative Ultrasound showing MPV (main portal vein), right anterior portal vein (RAPV) and right posterior portal vein (RPPV), and their segmental branches. *5, 6, 7* portal branches to live segments 5, 6, 7

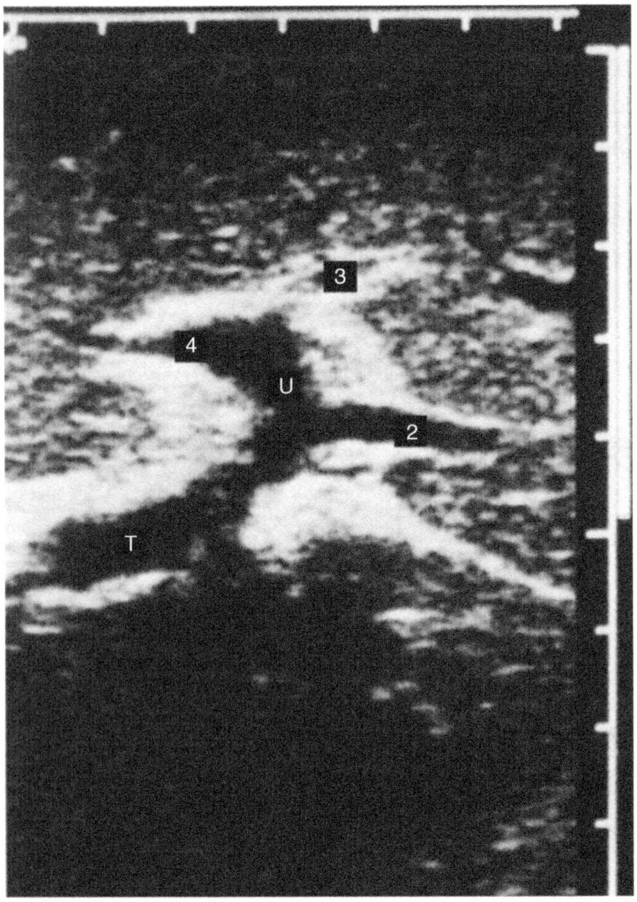

Fig. 10.21 Intraoperative Ultrasound showing the left portal vein transverse (T) and umbilical portions (U) and their branches supplying segments 2, 3, 4. *T* transverse portion of portal vein, *U* umbilical portion of portal vein, *2, 3, 4* portal vein branches to live segments 2, 3, 4

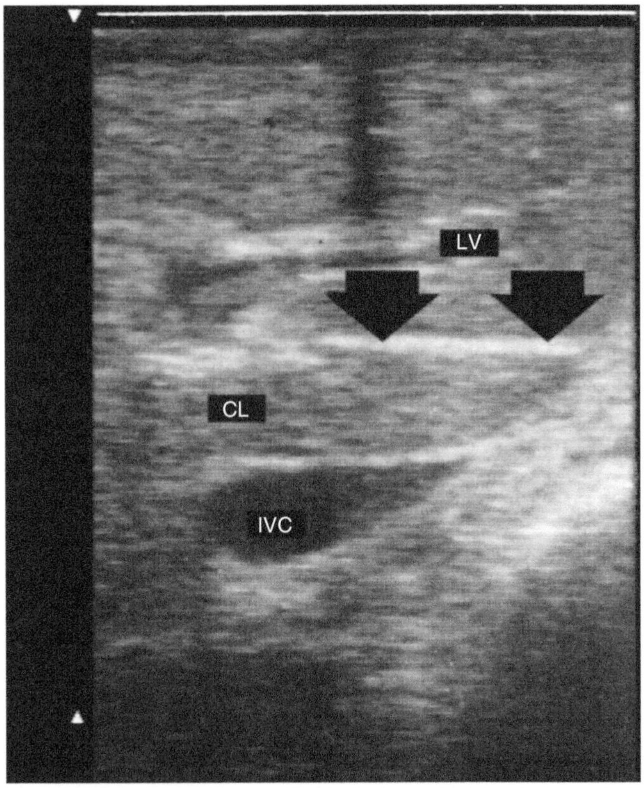

Fig. 10.22 Intraoperative Ultrasound showing the Caudate Lobe (CL) surrounding the inferior vena cava (IVC) and the lesser omentum (LV) (black arrows)

Further Reading

Lau WY, Leung KL, Lee TW, Li AKC. Ultrasonography during liver resection for hepatocellular carcinoma. Br J Surg. 1993;80:493–4.

Torzilli G, Bismuth H. Chapter 15: Intraoperative ultrasound. In: Lau WY, editor. Hepatocellular carcinoma. Singapore: World Scientific; 2008. p. 359–85.

The Three-Dimensional Body Visible System in Liver Surgery

11

11.1 Three-Dimensional Reconstruction of the Liver

Three-Dimensional-Computed Tomography (3D-CT) reconstruction has been used quite commonly in liver surgery. The 3D-reconstruction of the liver requires good quality CT (16 row, 32 row or 64 row CT), close CT cuts, intravenous injection of contrast and arterial/venous/delay phases of CT (Fig. 11.1).

CT scanners made by General Electric, Philips and Simens have 3D reconstruction.

11.2 Three-Dimensional Body Visible System of the Liver

The 3-D Body Visible System is a further development of the 3D reconstruction technique:

1. By merging the arterial/venous/delayed phases.
2. Can he made transparent so that the internal structures are visible.
3. Can have isolated reconstruction of the soft tissue systems like the arterial, portal venous, hepatic venous and biliary system, or these systems can be shown in any combinations.
4. Can spontaneously calculate the volumes of different parts of the liver.
5. Can do a simulation hepatectomy to show the large vascular/biliary structures which need to be transected.
6. Can separate out Couinaud's liver segments.
7. Can support 3D printing.

11.3 Clinical Applicability

1. Helps to understand in detail the relationship between a tumor and intrahepatic vasculo-biliary structures.
2. Helps to make the clinical decision in estimating treatment results and risks among different treatment protocols.
3. Using simulation hepatectomy, can decide on the best liver transection plane based on the vasculo-biliary structures which need to be transected.
4. Can estimate the volumes of the liver that is going to be resected and the volume of the future remnant liver.
5. Can help to identify anomalies of vasculo-biliary structures.

11.4 How Can This 3D Body Visible System of the Liver be Used?

As this System is owned by clinicians but not by the radiological department, clinicians can use this System freely for:

1. Communication between doctors with his/her patients' and relatives
2. Communication between clinicians of different specialties
3. Training of liver surgeons
4. Making video and slides to facilitate case discussion, knowledge and skills exchange, teaching, publication and research

11.5 International Development of the 3D Visible System of the Abdomen

Different such systems have been developed in various countries, e.g. EDDA in America, MeVis in Germany, Mint in Germany and Myrian in France.

All these Systems put different emphases on their own Systems, but the common features are:

1. Going after a fast automatic system.

W. Y. Lau, *Applied Anatomy in Liver Resection and Liver Transplantation*, https://doi.org/10.1007/978-981-16-0800-1_11

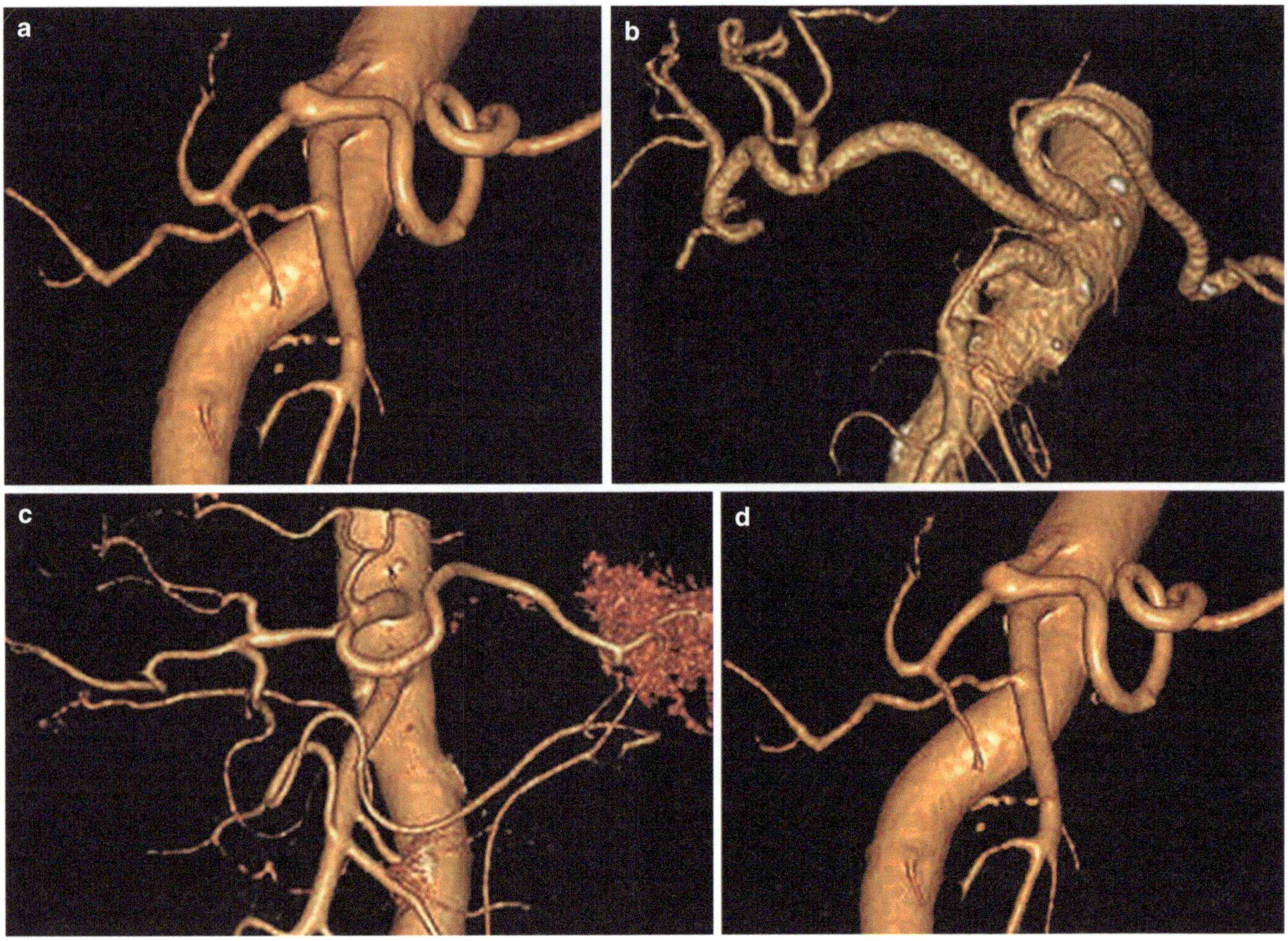

Fig. 11.1 3D-CT reconstruction of hepatic arteries

2. Requiring very high-quality CT, having difficulties in the management of more complicated cases.
3. Inadequate to support the Couinaud's liver segmentation.
4. The plane of simulation hepatectomy is flat, not 3-D.
5. Can detect down to third-degree branching of vessels only.

11.6 Development of the 3D Visible System of the Abdomen in China

The 3D Med of the Chinese Academy of Sciences is only a research project. It can only carry out basic steps in dissecting and reconstructing in the 3D visible system.

In Xiamen, a company has developed a planning system for liver resection. However, the reconstruction time is long, and the system cannot support Couinaud's liver segmentation.

The Shenzhen Yorktal Group started the 3D Visible System as an 863 research project by Professor Fang Chihua, which gradually developed into a commercialised product with the help of Professor Lau Wan Yee, Joseph.

11.7 The Yorktal 3D Body Visible System

This system has the following merits:

1. It can be used to select a particular abdominal organ to study.
2. The reconstruction is based on the original data obtained from CT scans to reconstruct different soft tissue systems like arteries, veins and bile ducts.
3. It has a fine analytical function to manage complicated cases.

4. For data obtained from not very well carried out CT scans, the system can use a vascular enhancement system to improve on the quality of the reconstruction.
5. Can reconstruct down to fourth-degree branches of vessels.
6. Can carry out Couinaud's liver segmentation.
7. Can calculate the volumes of different parts of the liver.
8. Can reconstruct dilated biliary systems, collaterals in portal hypertension, tumor thrombi in portal vein/hepatic vein.

11.8 The Basic Principles of 3D Body Visible System

Obtain basic data from CT scan
↓
Dissect 2D pictures
↓
Reconstruction of 3D pictures
↓
Clinical application: diagnosis, planning, simulation, analysis, selection

11.9 3D Body Visible System-Clinical Applications in Liver Surgery

11.9.1 Couinaud Liver Segmentation

Couinaud divided the liver into two parts: the main liver and the caudate lobe.

For the main liver, he further subdivided it into two hemilivers, four liver sectors and seven segments (Fig. 11.2).

As each liver segment has its own hepatic arterial/portal venous supply and each is drained by its own bile duct/hepatic vein branch, each liver segment is an individual unit that can be resected alone or in combination with an adjacent liver segment. Liver resection based on liver segments is called segment-based liver resection (Fig. 11.3).

To an experienced liver surgeon, after a detailed examination of 2D CT images, a 3D picture of the tumor and its relationship with the vasculo-biliary structures of the liver can be reconstructed in his mind. Such an ability takes a long time to acquire.

The main advantages of the 3D body visible system of the liver, especially to those who are less experienced are:

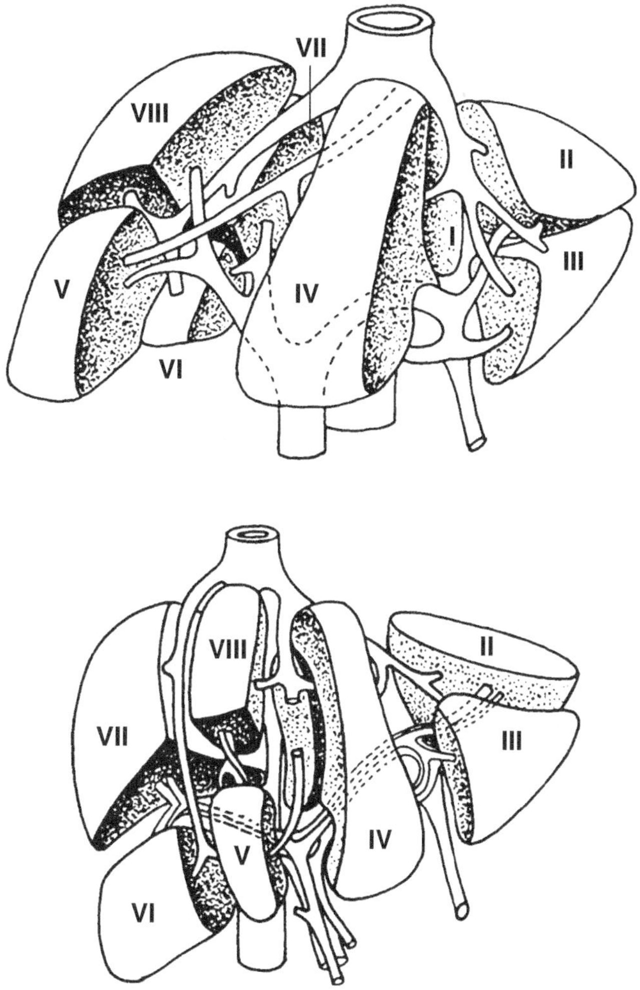

Fig. 11.2 Couinaud's liver segmentation

1. The liver can be rotated in any direction (Fig. 11.4).
2. Part or whole of the liver can be made transparent to see the inside structures.
3. Each Couinaud liver segment can be removed or reconstructed (Fig. 11.5).

11.9.2 To Determine the Position of the Tumor in the Liver

By putting the tumor back into the liver, the position of the tumor in the various segments of the liver can be determined (Fig. 11.6).

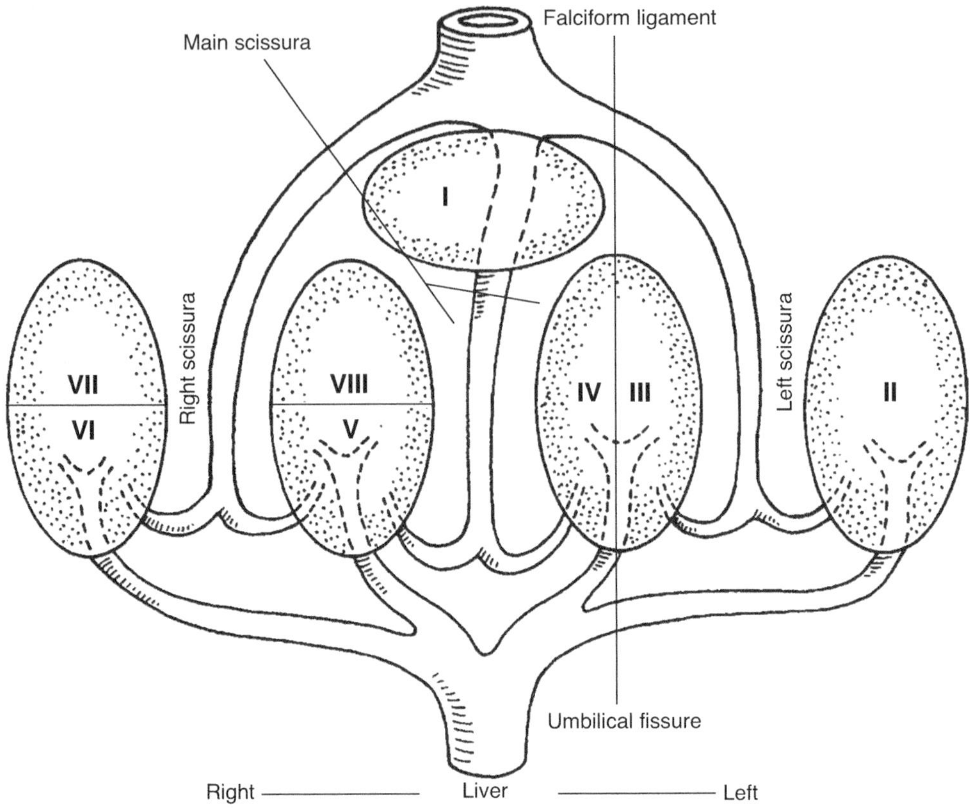

Fig. 11.3 Liver segment-based liver resection

11.9.3 To Determine the Type of Liver Resection

The type of liver resection is determined by three factors:

1. Anatomical or Non-Anatomical Liver Resection

   Figure 11.7 showed a simulation of non-anatomical liver resection. A non-anatomical liver resection, although it looks straight-forward, is actually a very complicated operation because

   (a) It is very difficult to ensure a R_0 resection margin all around the tumor, especially in the deep part of the resection.

   (b) The resection plane repeatedly goes through major vasculo-biliary structures.

   The demerits of non-anatomical liver resection can be shown easily on a simulation liver resection on a 3D body visible system of the liver (Fig. 11.8):

   (i) The liver transection plane repeatedly goes through major vasculo-biliary structures.

   (ii) The remnant liver left after liver resection contains a lot of devascularised liver tissues and liver tissues that are not drained by bile ducts. These lead to a high complication rate.

   Thus, if liver function allows an anatomical right hepatectomy is a technically easier operation than a non-anatomical resection (Fig. 11.9).

2. Is There Any Need to Sacrifice a Major Vessel?

   Figure 11.10 shows a liver cancer situated in the anterior right sector of the liver, compressing on the right hepatic vein (RHV). If liver function allows, a right hepatectomy should be carried out with a liver transection plane on the right of the middle hepatic vein.

   Figure 11.11 showed the hepatocellular carcinoma to be very near to the middle hepatic vein. If liver function allows, the transection plane should be on the left side of

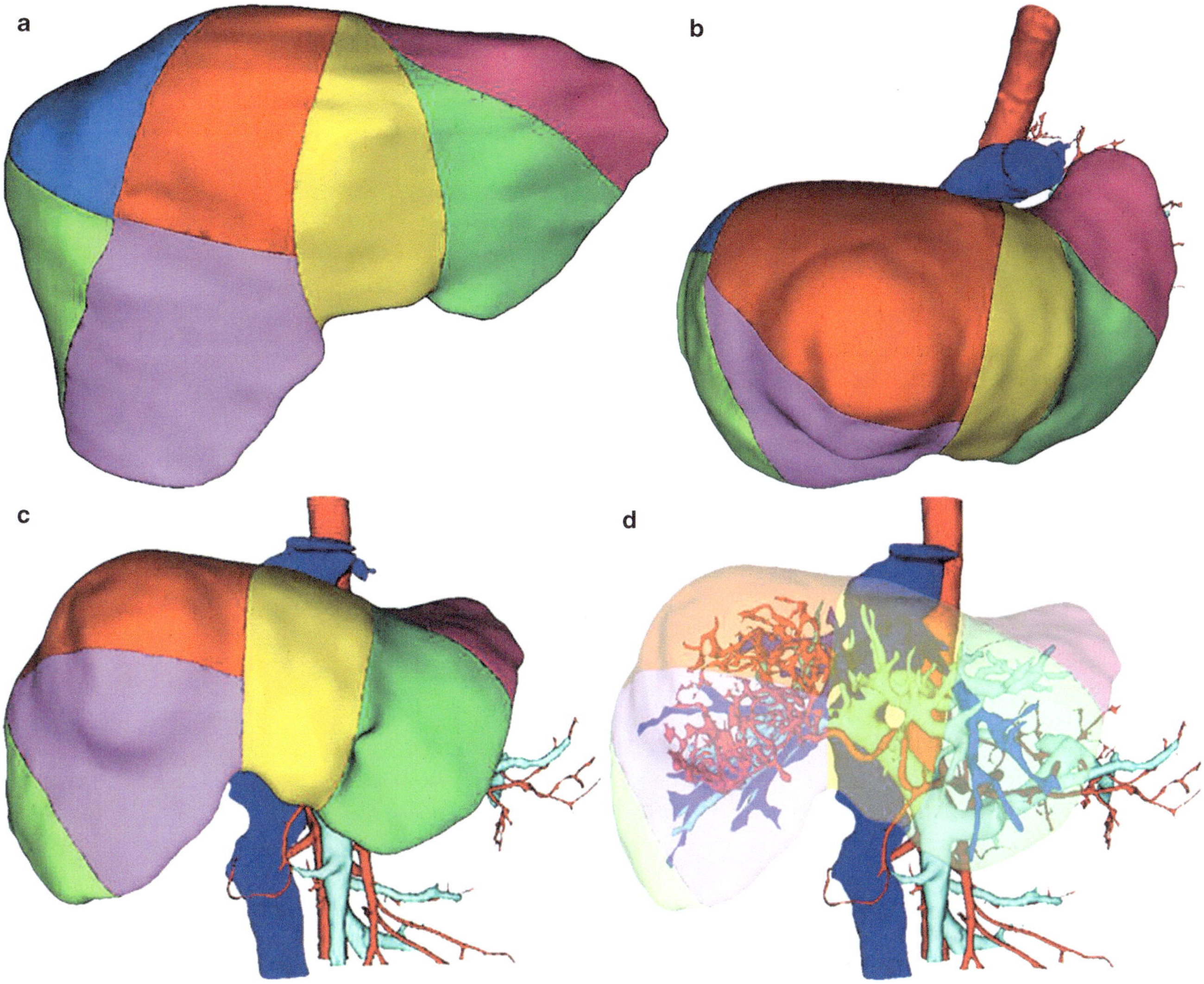

Fig. 11.4 (**a**) Each segment can be colored with a different color. (**b**) The liver can be rotated in any direction. (**c**) Vasculo-biliary structures can be put back. (**d**) The liver can be made transparent

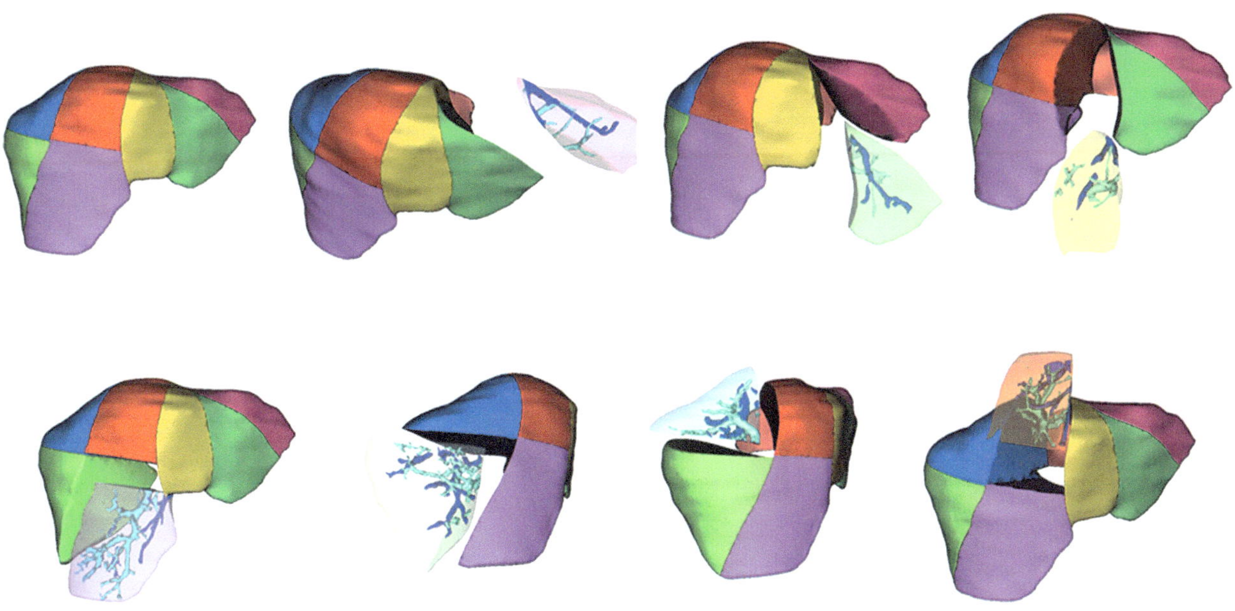

Fig. 11.5 Each liver can be made transparent and/or removed

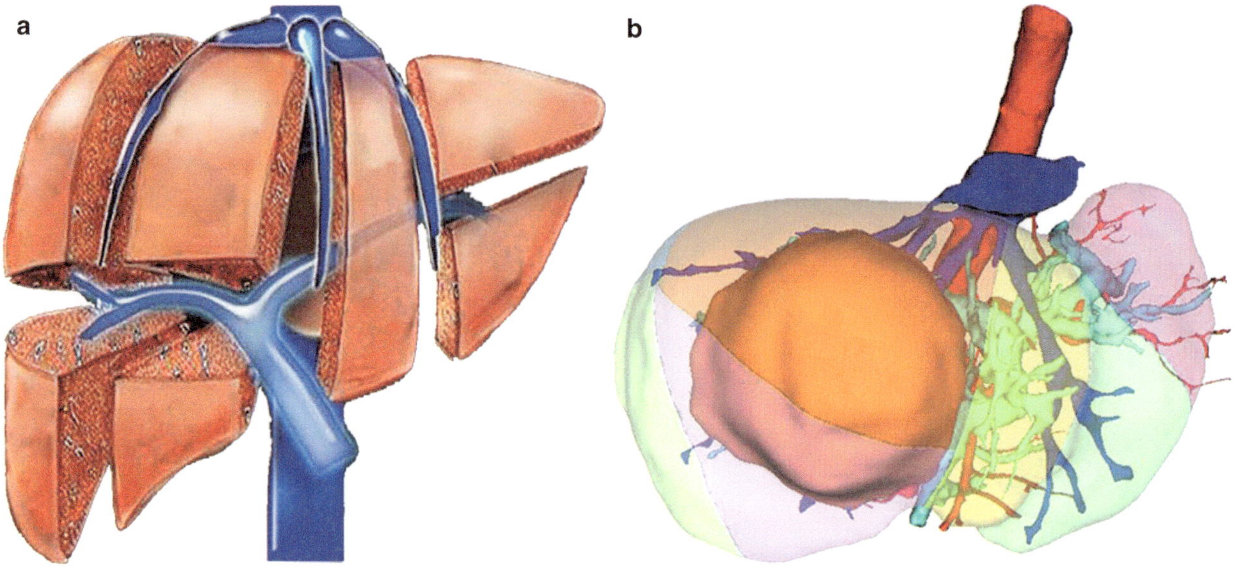

Fig. 11.6 Position of tumor in relationship to the liver segments. (**a**) Couinaud's liver segments. (**b**) Tumor in liver segments 5 and 8

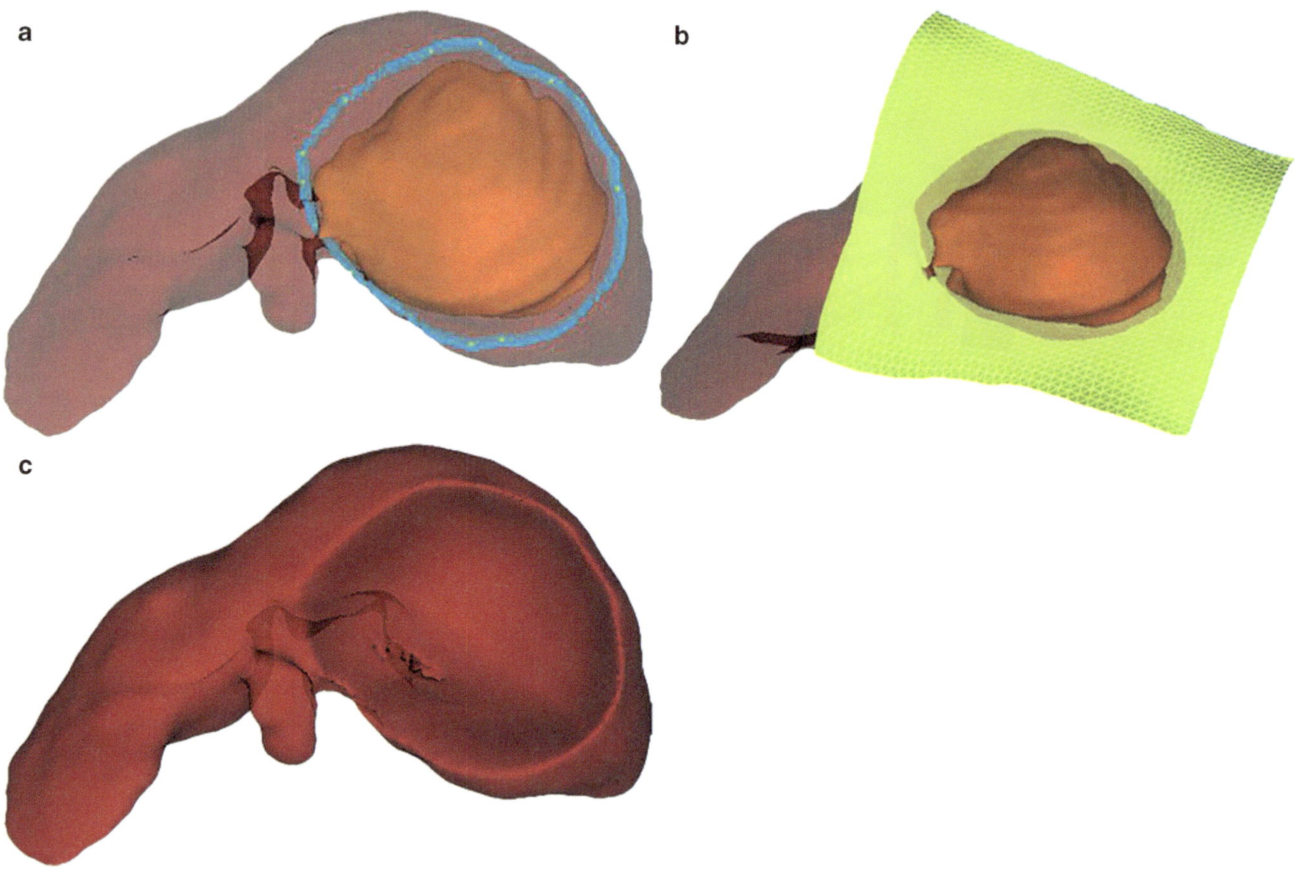

Fig. 11.7 Non-anatomical resection

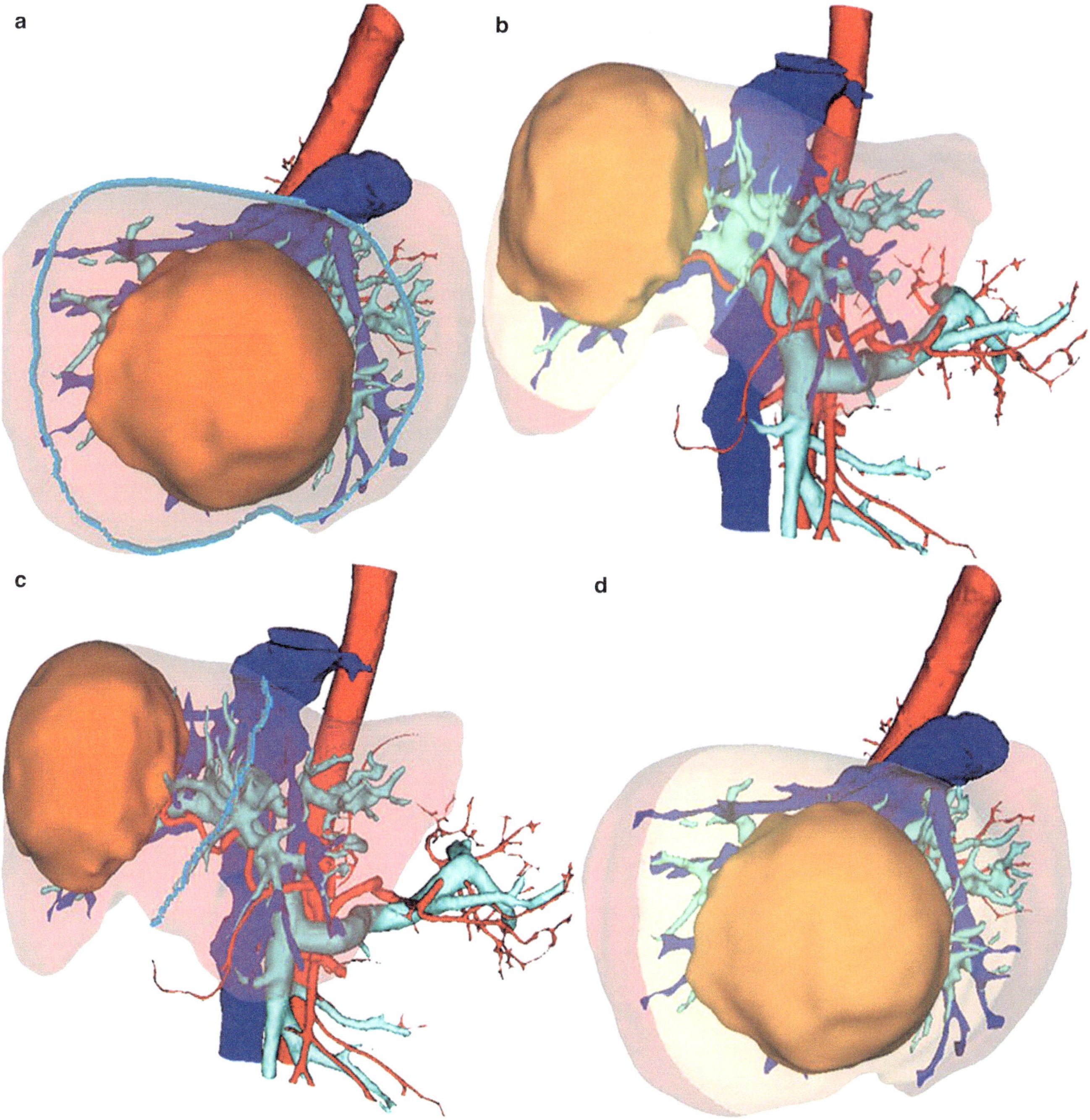

Fig. 11.8 Demerits of non-anatomical liver resection

the middle hepatic vein (MHV). Care should be taken to preserve the segment 4 vein to prevent congestion of segment 4 after liver resection.

3. Amount of Liver Resection Margin
In the study by Shi M, et al., in anatomical resection, a 2 cm resection margin resulted in less long-term recurrence than a 1 cm margin. Thus if liver function allows, a 2 cm resection margin should be aimed. Subsegmentectomy of the central liver segment (liver segments 4, 5, 8) can increase the liver resection margin.

11.9.4 Simulation Hepatectomy

Using the 3D Body Visible System of the liver, simulation hepatectomy can be carried out. Such a procedure is facilitated by:

1. The System showing the relationship of the tumor with the vasculo-biliary structures (Fig. 11.12).
2. In the transection plane of the simulated hepatectomy, all major vasculo-biliary structures that need to be transected can be shown (Fig. 11.13).
3. To estimate the volumes of the part of the liver which is going to be removed and the future remnant liver.

There are many ways to calculate:

1. Percentage
2. Absolute volume
3. Ratio of volume to Body Surface Area (BSA)
4. Ratio of volume to body weight

11.9.5 To Find Out Vasculo-Biliary Anomalies

1. Hepatic Artery

   Usually the anomalies are extrahepatic, which can make liver resectional surgery easier (Fig. 11.14).
2. Portal Vein (Figs. 11.15 and 11.16)

   Portal Vein anomalies are usually intrahepatic (Fig. 11.17).
3. Hepatic Vein Anomalies

   Segment 4 vein (Sg 4V) present in 57.4%, umbilical vein (UV) present in 60%.

   In 10% of patients, inferior right hepatic veins are so large that it can be seen on CT scan (Fig. 11.18).

   In liver transplantation, if the right liver is used as a donor graft, transection of a large right inferior hepatic vein can result in congestion of liver segment 6. The solution is to do a vascular reconstruction to improve the venous drainage of liver segment 6.
4. Bile Ducts

   Usually, bile ducts are not visible on CT scan. They can be visible when they are dilated, especially if contrast has been injected into the biliary system by a percutaneous transhepatic biliary drainage tube (Fig. 11.19).

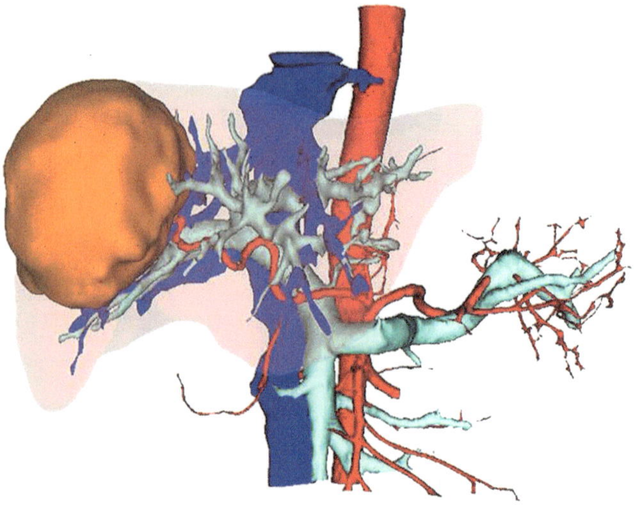

Fig. 11.9 Right hepatectomy is technically easier

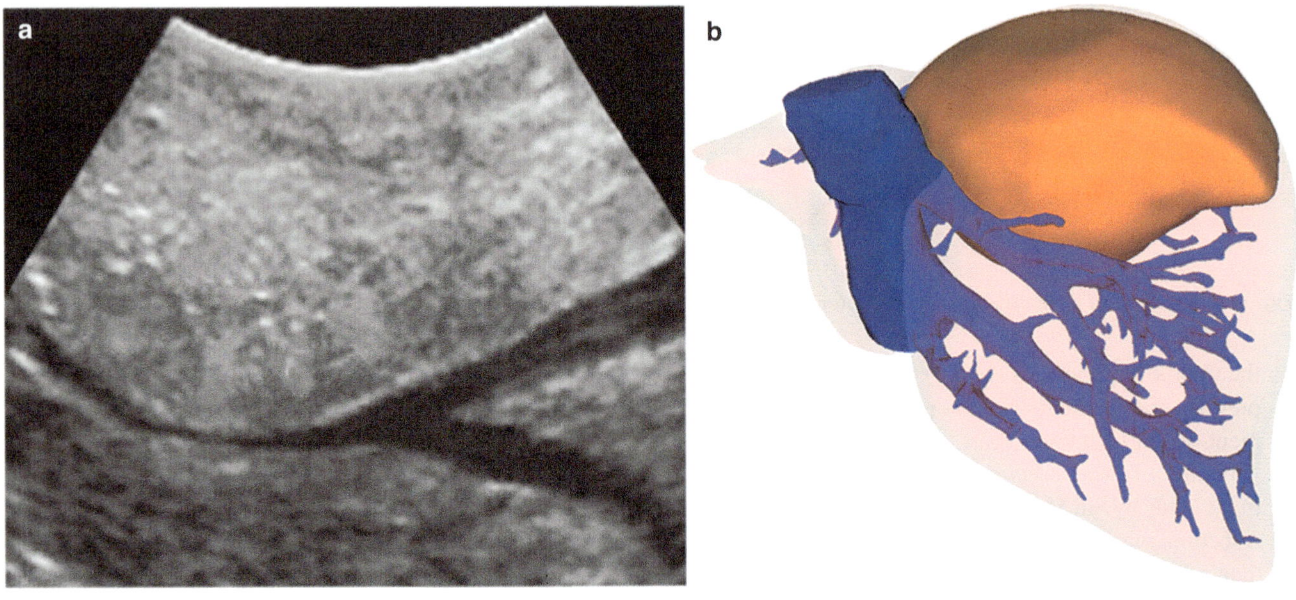

Fig. 11.10 HCC compressing on right hepatic vein (RMV)

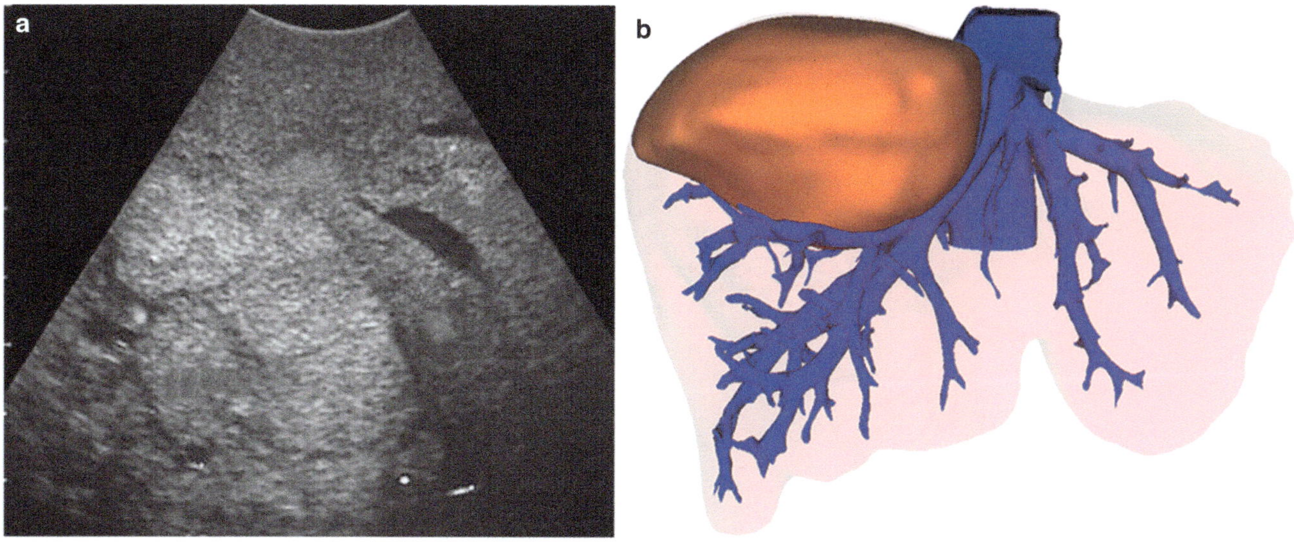

Fig. 11.11 Hepatocellular carcinoma very near to middle hepatic vein

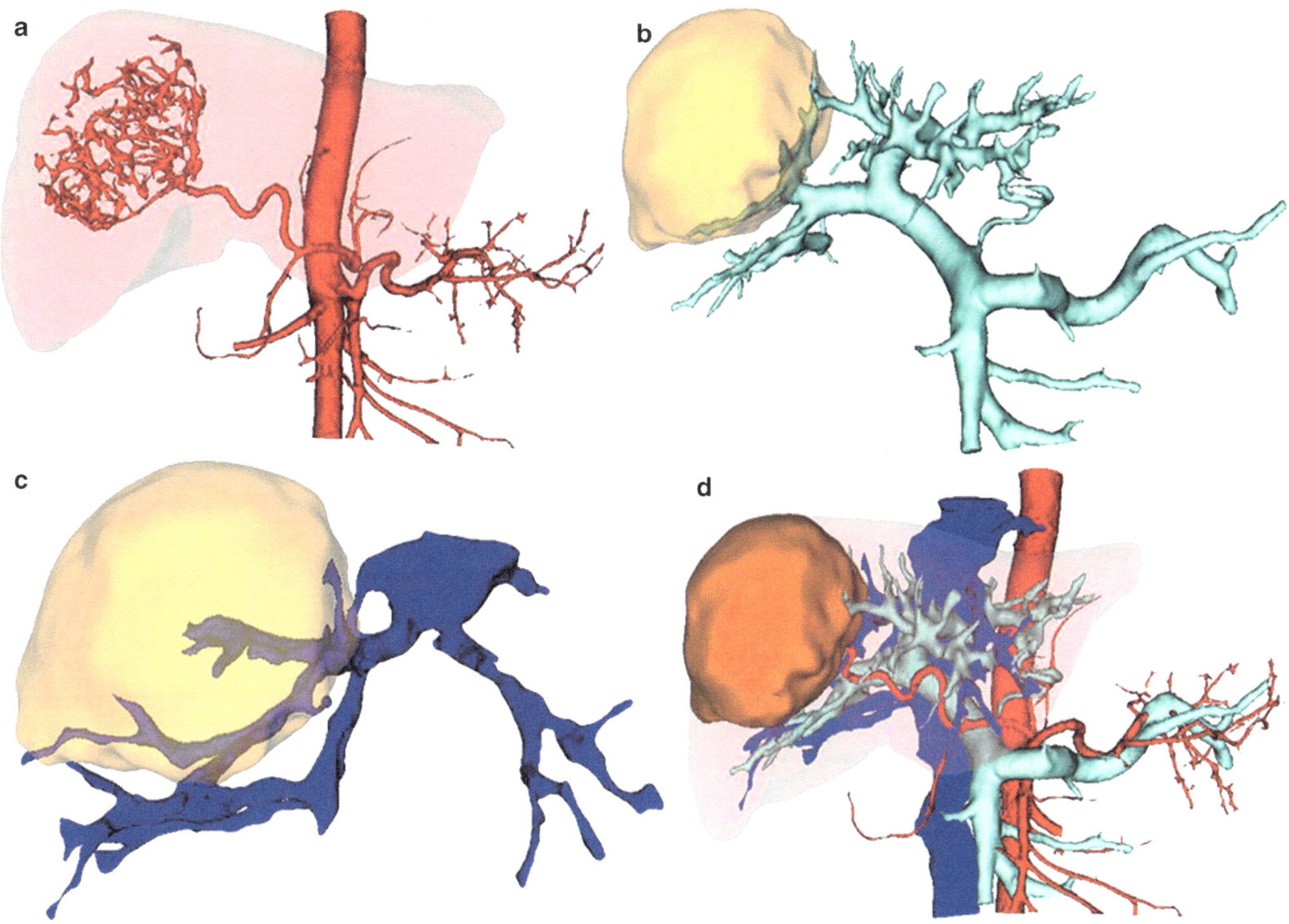

Fig. 11.12 Relationship of tumor to vasculo-biliary structures. (**a**) Hepatic artery. (**b**) Portal vein. (**c**) Hepatic vein. (**d**) Combination of vasculo-biliary structures

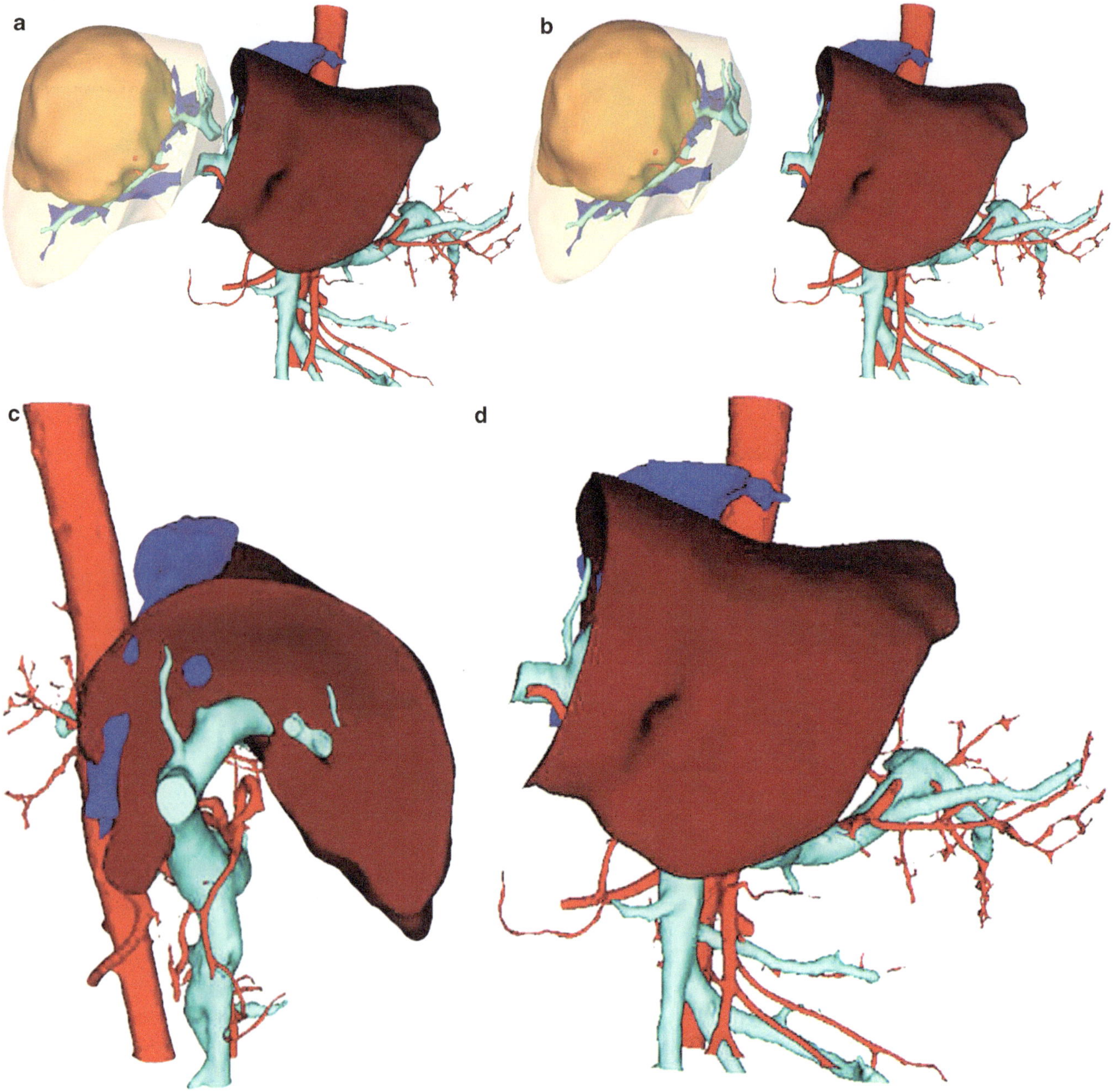

Fig. 11.13 Simulation hepatectomy. (**a**) Liver transection plane showing transected vessels and bile ducts. (**b**) Liver transection plane viewed from the front. (**c**) Liver transection plane viewed from the right. (**d**) Liver transection plane slightly oblique

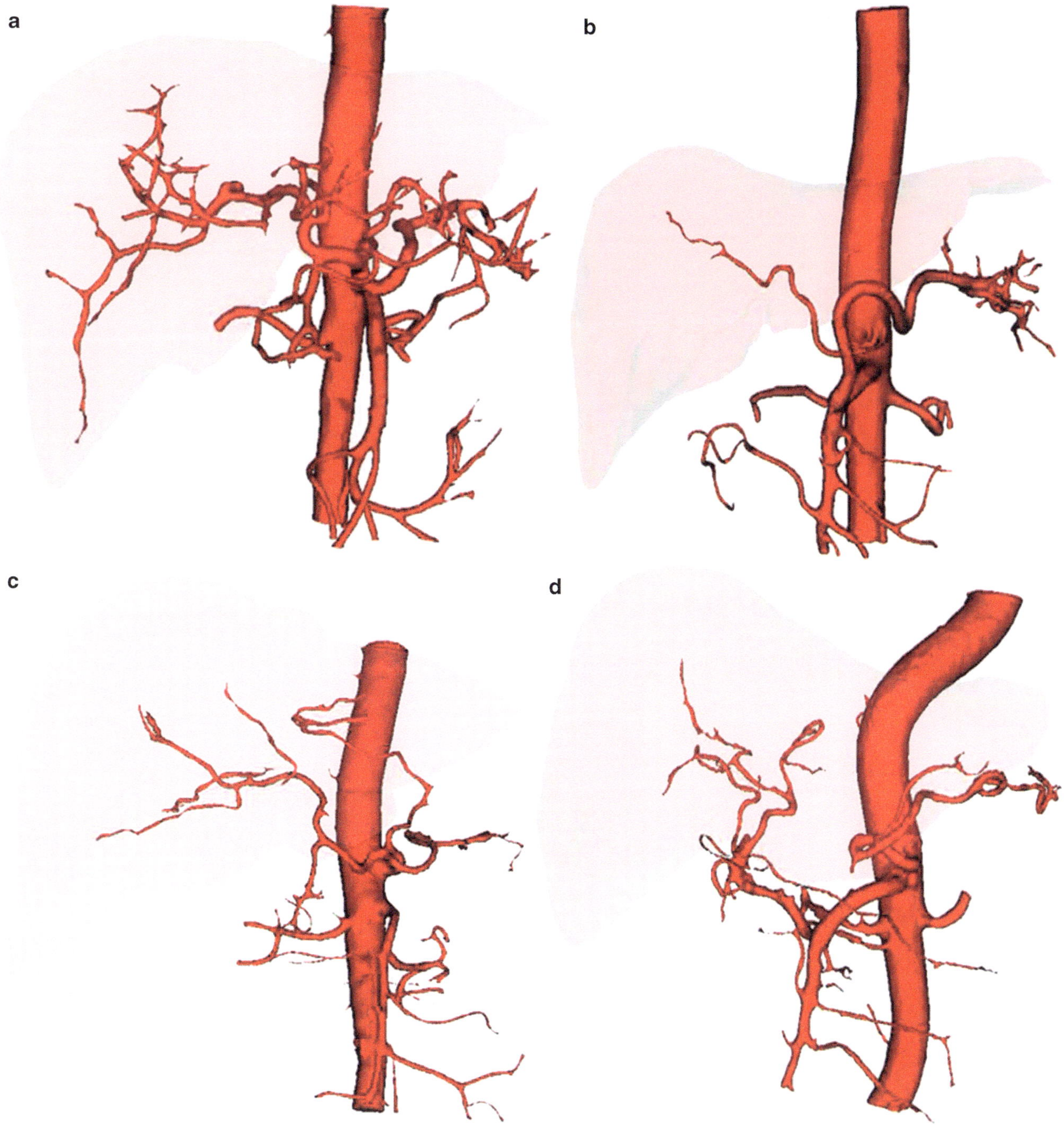

Fig. 11.14 (**a**) Normal anatomy. (**b**) Right hepatic artery from aorta, left hepatic artery from left gastric artery. (**c**) Left hepatic artery from left gastric artery. (**d**) Left hepatic artery from left gastric artery, right hepatic artery from SMA

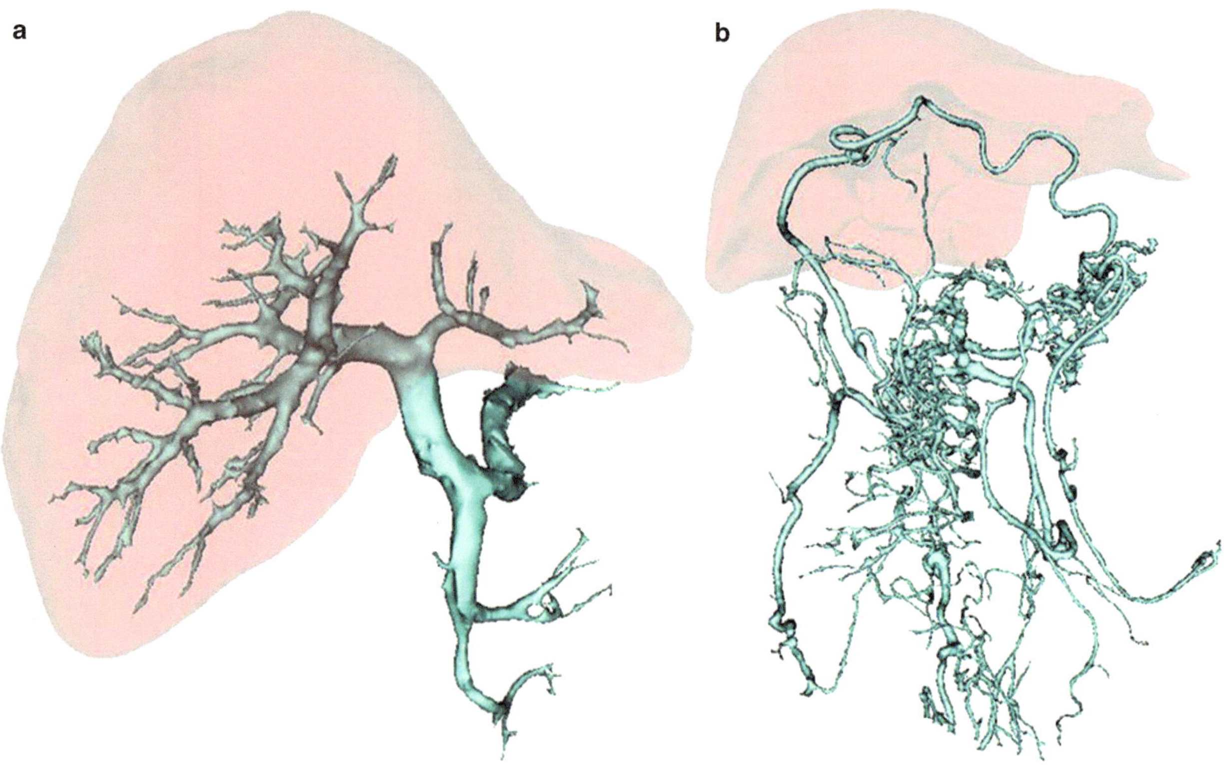

Fig. 11.15 (**a**) Normal anatomy. (**b**) Blocked main portal vein with development of collaterals

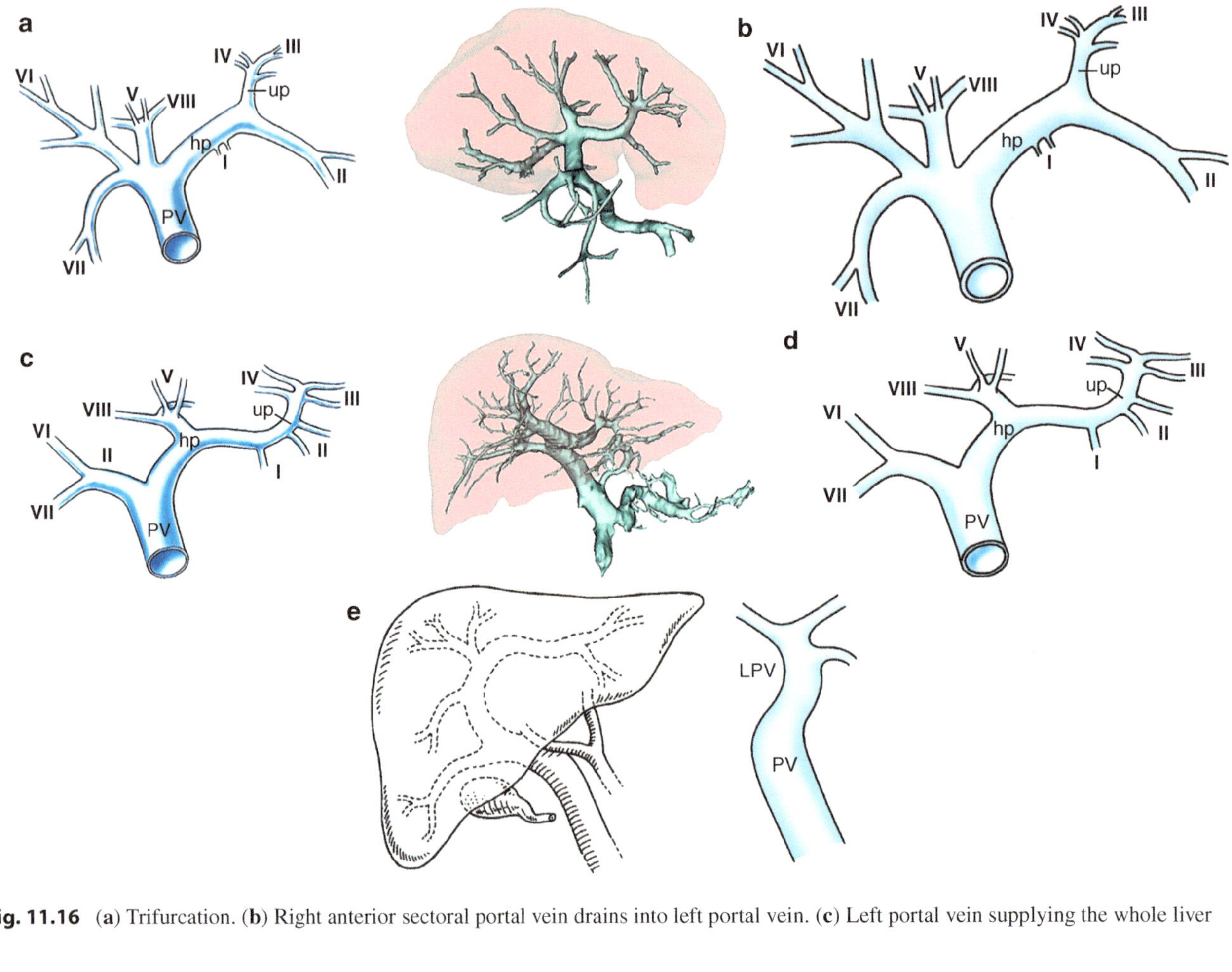

Fig. 11.16 (**a**) Trifurcation. (**b**) Right anterior sectoral portal vein drains into left portal vein. (**c**) Left portal vein supplying the whole liver

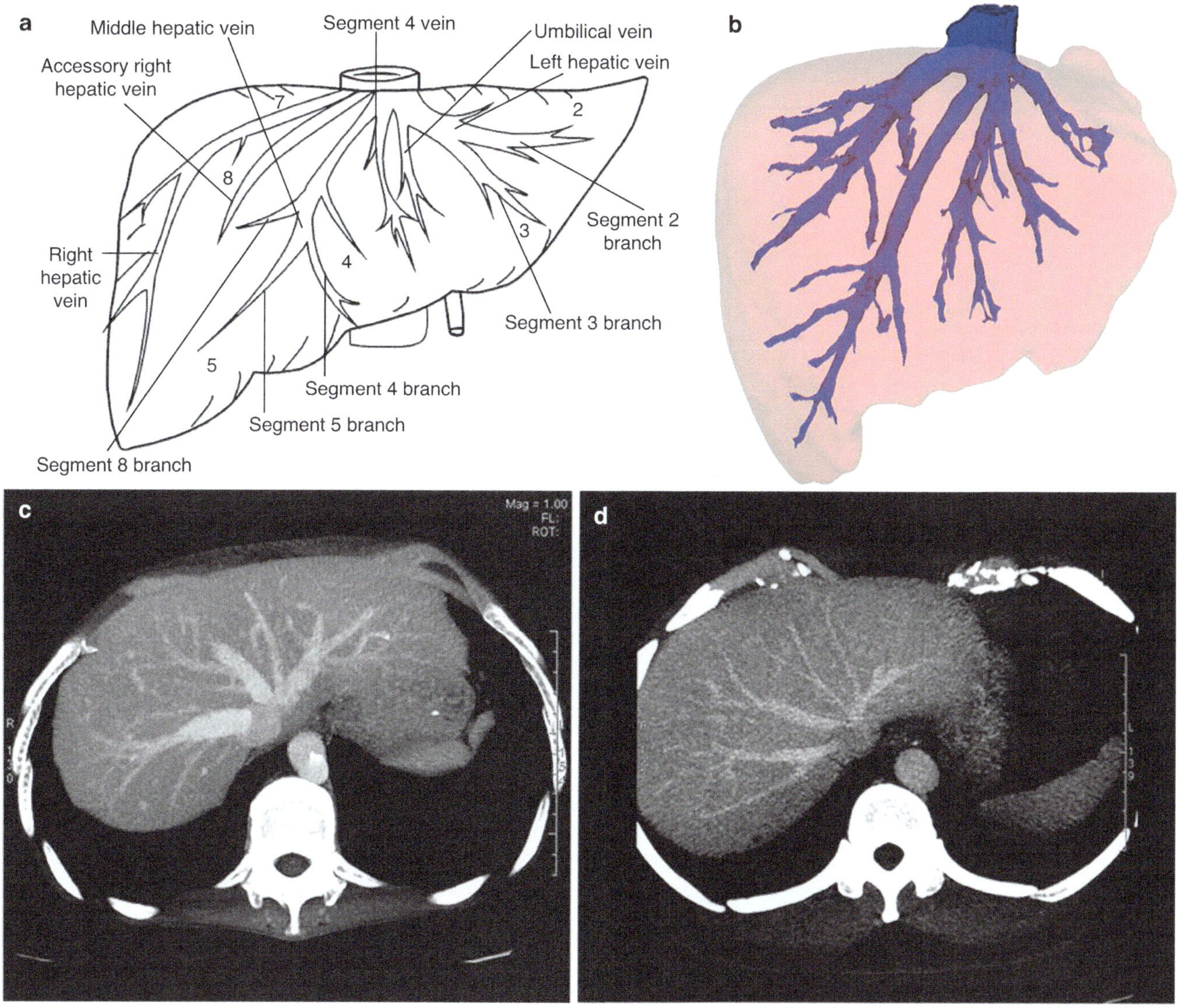

Fig. 11.17 Portal vein anomalies (Segment 4 vein and umbilical vein)

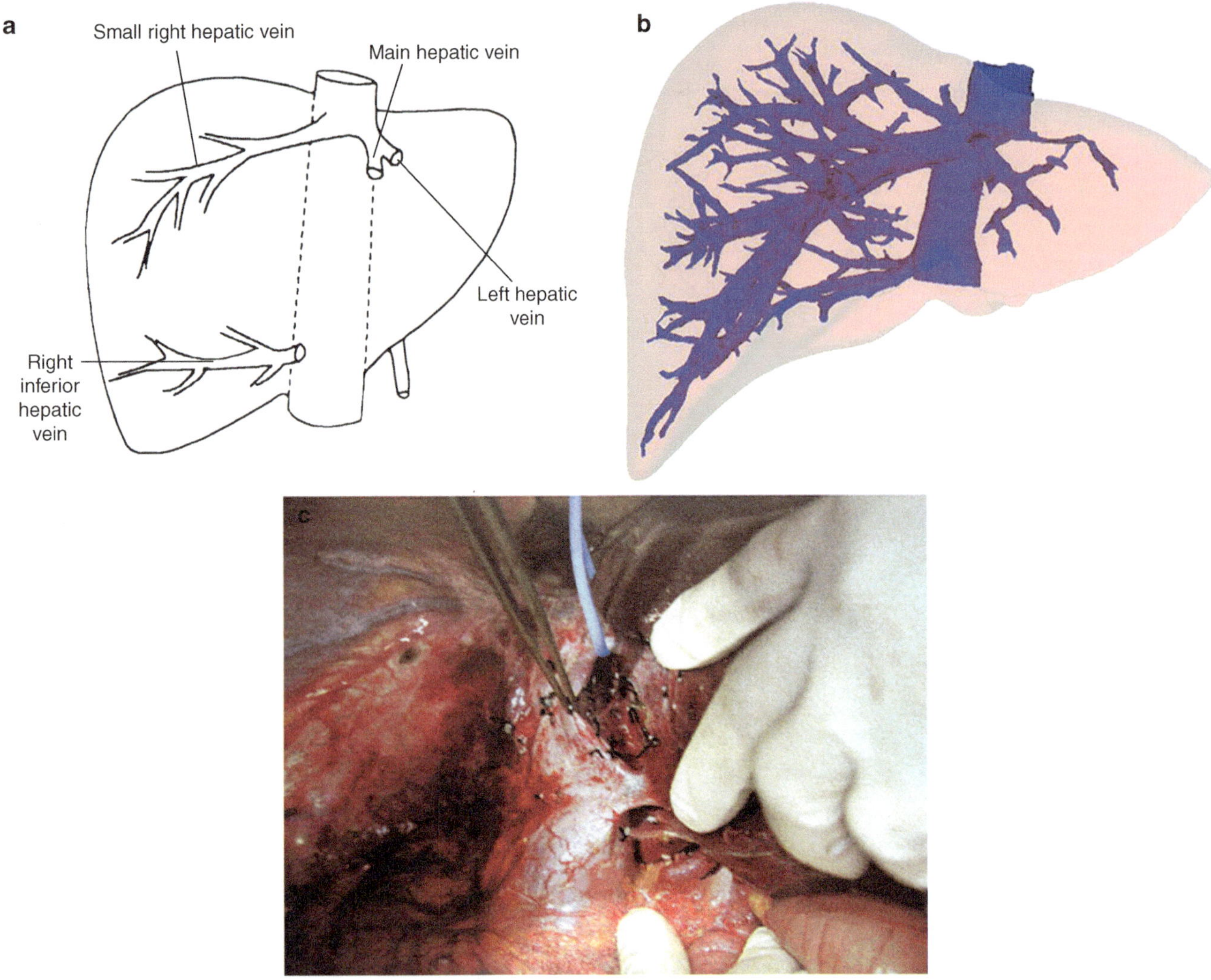

Fig. 11.18 (**a**) Large right inferior hepatic vein. (**b**) Large right inferior hepatic vein

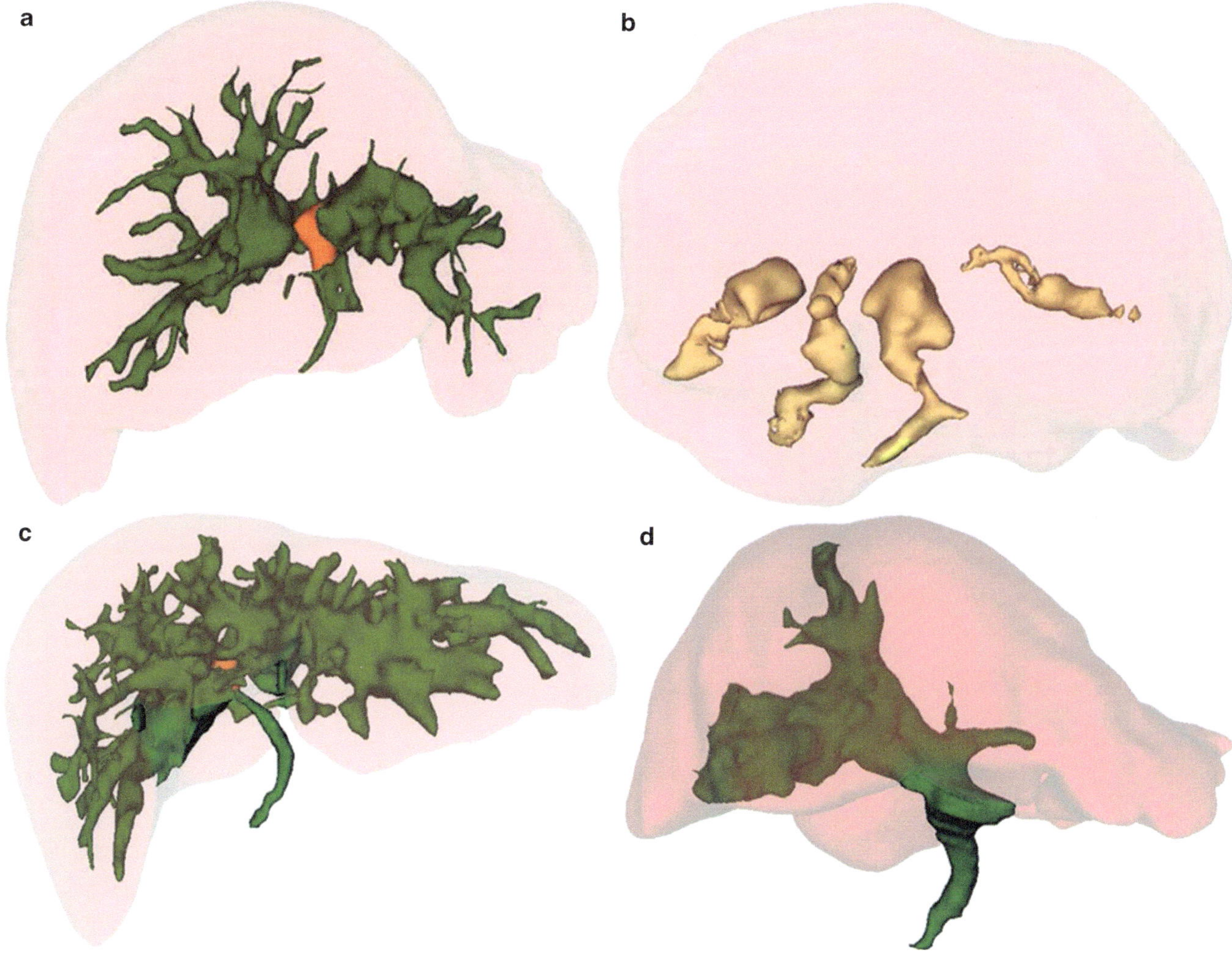

Fig. 11.19 Dilated intrahepatic ducts

11.10 Conclusion

3D Body Visible System has been used in liver resectional surgery. This System pushes the 3D reconstruction technique to a higher level of clinical application. It is especially useful for clinicians who are starting their careers as liver surgeons.

Further Reading

Fang CH, You JH, Lau WY, et al. Anatomical variations of hepatic veins: three-dimensional computed tomography scans of 200 subjects. World J Surg. 2012;36:120–4.

Shi M, Guo RP, Lin XJ, et al. Partial hepatectomy with wide versus narrow resection margin for solitary hepatocellular carcinoma: a prospective randomized trial. Ann Surg. 2007;245:36–43.

12 Low Central Venous Pressure in Liver Resectional Surgery

12.1 Blood Loss During Liver Resectional Surgery

The liver receives its abundant blood supply from both the systemic arterial and the portovenous systems. After going through the liver, the blood then drains through the hepatic venous system into the inferior vena cava and hence back to the heart. Blood loss from the raw surfaces of the liver during liver parenchymal transection is inevitable, and the amount of blood loss significantly affects post-operative morbidity, mortality and long-term survival. Minimizing intraoperative blood loss significantly improves patient outcomes.

12.2 Applied Anatomy and Physiology

Bleeding from the inflow blood coming from the hepatic artery and the portal vein can be controlled readily with porta hepatis clamping (the Pringle's manoeuvre). However, backflow bleeding still occurs from the divided branches of the outflowing major hepatic veins (the right, middle and left) and the short hepatic veins (Fig. 12.1). The rate of this bleeding is directly proportional to the presumed gradient across the divided vessels. On one side is the hepatic venous and sinusoidal pressure which is directly rated to the central venous pressure (CVP). On the other side, it is open to air which is atmospheric pressure. In other words, the rate of blood loss is proportional to the pressure gradient between the CVP and the atmospheric pressure. Another factor that determines the rate of blood loss is the diameter of the divided vessel, with a larger divided vessel losing blood at a faster rate (Fig. 12.2). Thus, it is logical to assume that haemodynamic manipulation to maintain a low CVP is one of the important techniques to decrease blood loss during liver transection. With less blood loss, there is a better view of the transected area, and larger vessels can be

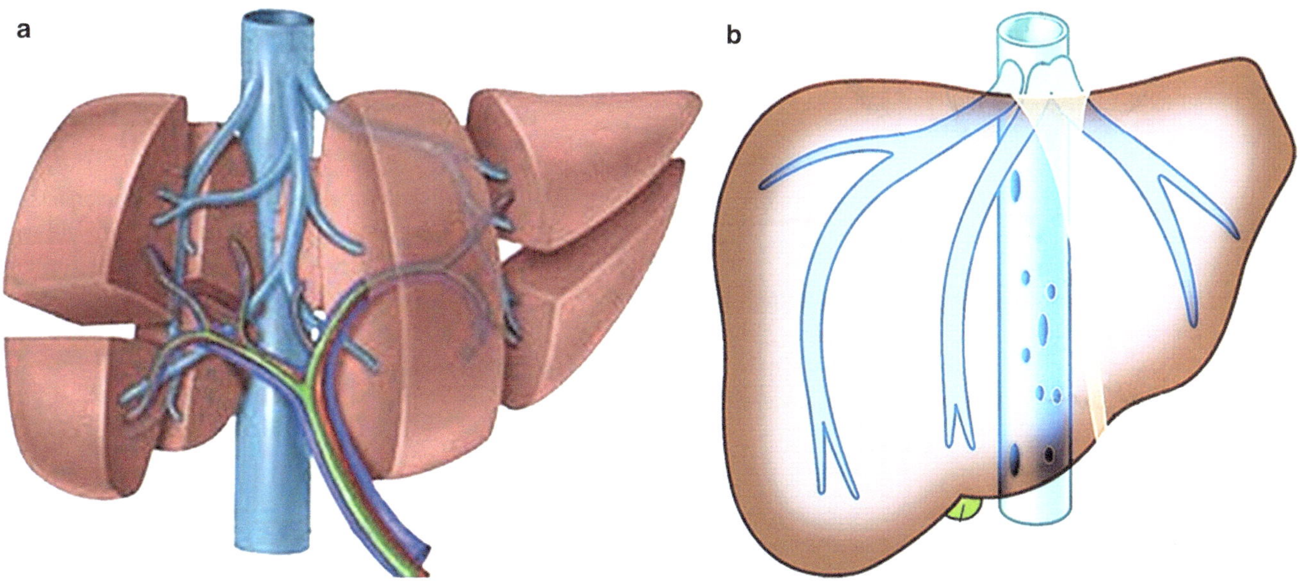

Fig. 12.1 (a) Vascular inflow and outflow of the liver. (b) After Pringle's manoeuvre, the main blood lose comes from branches of the hepatic veins

W. Y. Lau, *Applied Anatomy in Liver Resection and Liver Transplantation*, https://doi.org/10.1007/978-981-16-0800-1_12

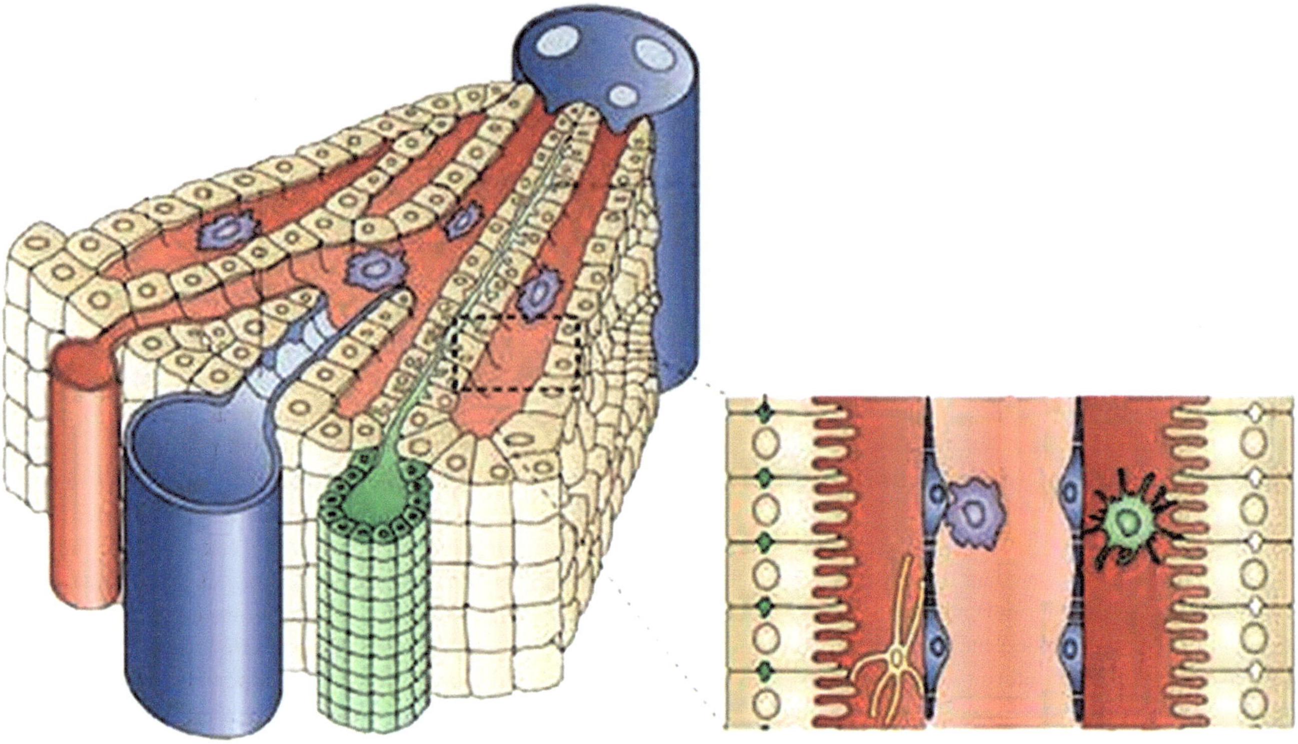

Fig. 12.2 Rate of blood loss during liver transection using Pringle's manoeuvre is related to the gradient between the venous pressure and the atmosphere and the diameters of the divided vessel

controlled and ligated before they bleed, thus entering into a virtuous cycle of less bleeding → clearer view → less bleeding → clearer view.

12.3 Advantages of a Low CVP in Liver Resectional Surgery

The advantages of a low CVP which leads to a decrease in intraoperative blood loss include:

1. Decreases the percentage of patients who requires blood transfusion
2. Shortens operative time
3. Decreases post-operative morbidities and mortalities
4. Shortens hospital stay
5. Faster recovery, especially the post-operative liver and renal functions

12.4 Why Are Some Anaesthesiologists Still Reluctant in Using Low CVP?

This mainly arises out of the lack of communication between the anaesthesiologists and the liver surgeons and in some sense, a mistrust of the anaesthesiologist on the ability of the surgeon in avoiding major bleeding during liver resectional surgery. These anaesthesiologists are worried about the narrow safety margin for the patient's cardiovascular system being able to provide adequate perfusion to the patient's major organs during an episode of major bleeding. Thus, this mutual trust has to develop gradually between the anaesthesiologist and the surgeon. Even if an anaesthesiologist declines to use low CVP, he/she should realize that a high CVP makes bleeding during liver transection worse and should avoid doing so.

The second reason is a relative lack of knowledge by the anesthesiologist in this area. Physiological studies have shown that a mean arterial blood pressure of >50 mmHg is able to maintain perfusion to major organs. Moreover, adequate oxygenation to major organs is not only dependent on the perfusion pressure in the cardiovascular system, but is also dependent on the oxygen content of the blood. This is calculated by the following equation:

$$(O_2 \text{ carried by Hb}) + (O_2 \text{ in solution}) = (1.34 \times Hb \times SpO_2 \times 0.01) + (0.023 \times PaO_2)$$

Where:

- SO_2 = percentage saturation of Hb with oxygen.
- Hb = haemoglobin concentration in grammes per 100 mL blood.
- PO_2 = partial pressure of oxygen (0.0225 = mL of O_2 dissolved per 100 mL plasma per KPa, or 0.003 mL per mmHg).

In order to maintain good oxygen supply to major organs in a relatively low perfusion pressure, the haemoglobin level should be ensured by high inspiratory oxygen. Avoiding hypothermia also significantly improves the oxygen-carrying capacity. A careful study of the oxygen dissociation curve in Fig. 12.3 helps the reader to understand more about the oxygen-carrying capacity of haemoglobin.

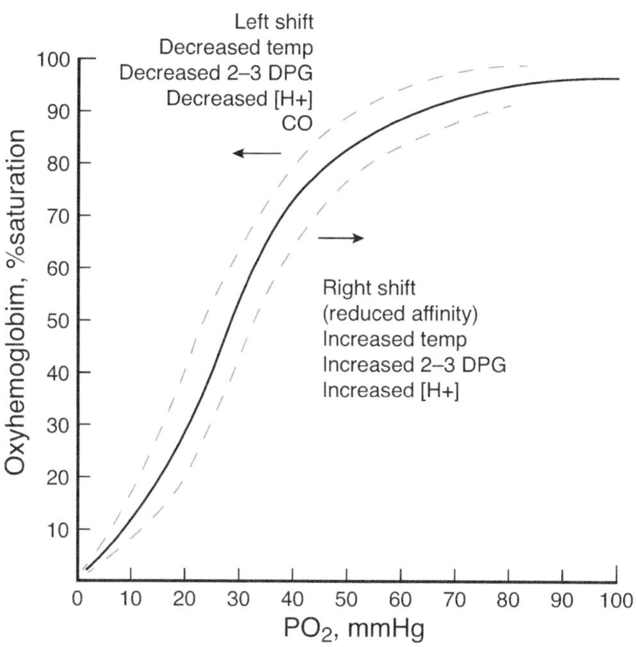

Fig. 12.3 Oxygen dissociation curve

12.5 Low CVP: How Low Is Low?

The CVP in a normal person in the supine position is 5–12 mmHg (or 7–16 cm H_2O) during liver resection. Some surgeons even request a CVP near zero.

What is the optimal CVP in liver resection? The impact of the variations in CVP on the rate of blood loss and on the perfusion of major organs has been studied. An animal study using Bama miniature pigs aimed to evaluate the haemodynamic and oxygen transport changes during liver resection at different CVP levels was carried out. The CVP was controlled from 0 to 8 cm H_2O during liver resection. The rates of blood loss and the hepatic venous pressures during liver resection were almost linearly related to the CVP. A significant drop in mean arterial pressure, cardiac output, cardiac index and in oxygen delivery happens when the CVP was less than 2 cm H_2O. In addition, there was a significant drop in oxygen consumption and oxygen extraction ratio when the CVP was less than 1 cm H_2O. Oxygen delivery, oxygen consumption and oxygen extraction ratio remained relatively constant between CVP from 2 to 8 cm H_2O. Based on this animal study, the optimal CVP for liver resection is 2–3 cm H_2O. A later study showed that these findings could also be applied to human beings.

12.6 Methods to Lower CVP

There are two main ways to lower the CVP during liver resectional surgery.

1. Physiological Way
   With the reverse Trendelenburg position (around 15 degrees head-up), the blood inside the venous compartment is pooled in the lower extremities, thus lowering the CVP (Fig. 12.4).

   This method is not often used because there is always a theoretical risk of air embolism by using this position.

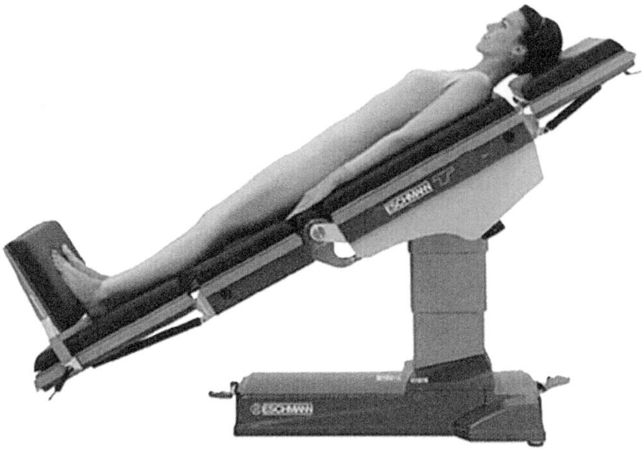

Fig. 12.4 Reverse Trendelenburg position

Many surgeons do not like to operate with the patient in this position because the instruments tend to fall off the operating table.

2. Pharmacological Ways

   Both diuretics and/or venodilators have been frequently used.

   (a) Diuretics

   The intravascular volume, especially the venous compartment, can be depleted with an intravenous bolus of loop diuretics (e.g. Frusemide 5–40 mg i.v.), which lowers the CVP. The major disadvantage is the activation of the renin-angiotensin-aldosterone system and finally raises the plasma anti-diuretic hormone, thus resulting in oliguria in the post-operative period, which may persist up to 24–48 h.

   (b) Venodilators

   The use of venodilators (e.g. glycerol trinitrate 0.5–5 mg/h) seems to be a more practicable and titratable method in achieving a low CVP. It relaxes the smooth muscle in the vessel wall and reduces the pressure inside the vessel provided that the intravascular volume remains the same. Glycerol trinitrate is the drug of choice. At a low plasma level, it has more vasodilatation effect in the venous system than the arterial system. The effect is immediate, and the half-life is short. Therefore, the CVP can be titrated to the desired level gradually by adjusting the rate of infusion.

12.7 Measures to Take When Using Low CVP for Liver Resectional Surgery

1. Preoperative Assessment

   Adequate assessment of the cardiopulmonary function to make sure that the patient can tolerate a period of low blood pressure. A history on whether the patient has been put on long-term glycerol trinitrate is important to rule out possible drug tolerance.

2. Intraoperative Measures

   After the patient is put under general anaesthesia, an intravenous drip should be set up to give normal saline at a rate of 100 mL/h, aiming to maintain urine output of >30 mL/h. A double lumen catheter should be inserted into the right internal jugular vein for CVP monitoring. Except in patients who are under 50 years of age or ASAI, an arterial line should be set up to monitor the blood pressure.

   At the start of liver parenchymal transection, an intravenous infusion of glycerol trinitrate 0.5–6 ng/h should be given to maintain a CVP at 2–3 cm H_2O. At a low CVP of 2–3 cm H_2O, there are concerns about hypoperfusion and tissue hypoxia of the major organs. This problem can be readily tackled by increasing the inspiratory oxygen concentration (FiO_2) to 80% and with the use of a vasopressor (e.g. phenylephrine 0.05 mg intravenously) to maintain a mean blood pressure of 50 mmHg, which would ensure adequate oxygen delivery to the vital organ. If excessive bleeding occurs, further fluid load can be given to replace the volume loss.

   After completion of hepatic parenchymal transection, the Pringle's manoeuvre is unclamped and adequate haemostasis is done by the surgeon on the raw liver surface. The CVP can then be brought up to more than 100 mmHg by stopping the intravenous nitrate infusion and by increasing the intravascular volume by infusion of 1–2 L of colloid solution (e.g. Gelofusine). Haemostasis is then secured one more time by the surgeon.

12.8 Conclusion

Although there is no randomized controlled trial to evaluate the relative risks and benefits of maintaining a low CVP in liver resectional surgery, it would be logical to avoid using a high CVP in liver surgery. With meticulous use, the low CVP technique is definitely helpful in liver resectional surgery to significantly reduce intraoperative blood loss.

Further Reading

Guo Y, Lin CX, Lau WY, et al. Hemodynamics and oxygen transport dynamics during hepatic resection at different central venous pressures in a pig model. Hepatobiliary Pancreat Dis Int. 2011;10:516–20.

Huntington JT, Royall NA, Schmidt CR. Minimizing blood loss during hepatectomy: a literature review. J Surg Oncol. 2014;109:81–8.

Lin CX, Guo Y, Lau WY, et al. Optimal central venous pressure during partial hepatectomy for hepatocellular carcinoma. Hepatobiliary Pancreat Dis Int. 2013;12:520–4.

13 Hepatic Vascular Inflow and Outflow Occlusions

The impact of blood loss during liver resection and the need for perioperative blood transfusions have been shown to negatively impact perioperative morbidity and mortality, as well as long-term survival outcomes. Limiting blood loss while performing safe and expeditious resection are the primary goals of liver surgeons, which can be achieved by many means.

13.1 The Vascular Inflow Occlusion of the Portal Triad (Pringle's Manoeuvre)

In 1903, von Haberer reported the use of hepatic artery ligation to treat liver injuries. We now know that the majority of the blood flow to the normal liver is through the portal venous system (75–85%), while the hepatic artery only provides a supportive role (20–25%). It is not surprising that hepatic artery ligation is ineffective in decreasing blood loss from liver injuries. Furthermore, ligation of the hepatic artery results in relative ischaemia to the liver parenchyma, thus adding more insult to the initial injury.

In 1908, Pringle reported the use of porta hepatis clamping to control bleeding in liver trauma. This manoeuvre can be done by applying a vascular clamp across or by tightening a sling around the hepatoduodenal ligament, thus occluding both the hepatic arterial and the portal venous inflow to the liver. All his patients died, and in today's modern era of evidence-based medicine, this Pringle's manoeuvre would have been banned in clinical practice. However, Pringle's experiments on rabbits and the experience by him and by others surgeons showed the Pringle's manoeuvre is of benefit not only in emergency situations to decrease bleeding, but in elective liver surgery to prevent bleeding. Pringle's manoeuvre becomes popular and is one of the most commonly adopted techniques in elective liver resection (Fig. 13.1).

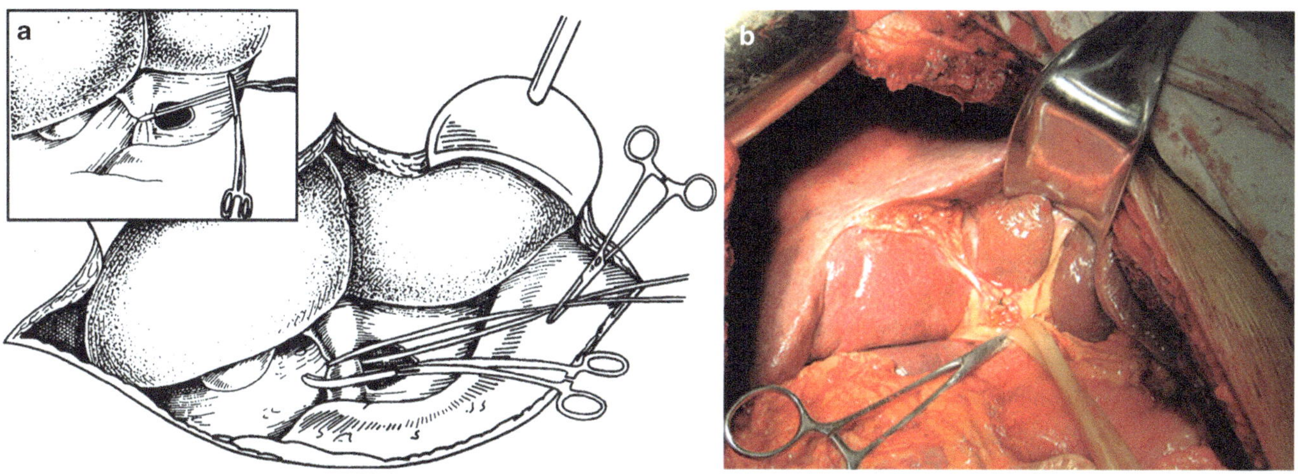

Fig. 13.1 Pringle's manoeuvre. (**a**) Using a vascular clamp, or in the insert, using a vascular sling. (**b**) Operative photograph

W. Y. Lau, *Applied Anatomy in Liver Resection and Liver Transplantation*, https://doi.org/10.1007/978-981-16-0800-1_13

13.2 Selective Vascular Inflow Occlusion

Selective vascular inflow occlusion refers to the clamping of only those arterial and portal branches which supply the part of the liver that is planned to be removed. This technique was first described as hemihepatic vascular occlusion by Makuuchi et al. in 1987. The advantages of this technique are no ischaemic insult to the remnant liver, prevention of splanchnic congestion, and better haemodynamic tolerability. On the other hand, the disadvantage is persistent bleeding from the raw surfaces of the non-clamped parts of the liver.

Selective Vascular Inflow Occlusion can be done at three levels:

1. At the level of division of the main portal triad into the right and left trunks supplying the right and the left hemilivers, respectively:
   (a) This can be achieved extrahepatically by lowering the hilar plate as the bifurcation of the right and the left trunks is extrahepatic (see Chaps. 4 and 5).
   (b) This can also be achieved intrahepatically by:
      (i) Launois' Intrahepatic Transfissural Approach (see Sect. 4.3)
      (ii) Launois' Posterior Intrahepatic Approach (see Sects. 4.3, and 15.3)
      (iii) Takasaki's Glissonean Pedicle Approach (see Sect. 15.3.1)

   At these levels, the hepatic arterial, the portal venous and the bile duct branches can be transected within the sheath (extrafascial approach), or the fascial sheath opened and the arterial, portal and biliary branches ligated individually (intrafascial approach). It is acceptable to use the intrafascial approach at the hepatic hilus. However, it is advisable that only the extrafascial but not the intrafascial approach be used at the sectoral and the segmental levels (see Sect. 4.1). In using the extrafascial approach, it is advisable that a trial clamping with a vascular clamp be used first on the pedicle to make sure the ischaemic part of the liver corresponds to the intended part of the liver to be resected.
2. At the sectoral branch level.
   Intrahepatically, the branches of the hepatic artery, portal vein and bile duct are wrapped within the Glissonian sheath. On the right side, further dissection of the right pedicle leads to the anterior sectoral and posterior sectoral pedicles, and on the left side, the medial and lateral sectoral pedicles. We have previously discussed in detail the Glissonian sheaths to segment 4 (see Sect. 6.2). Particular attention should be paid to the possibility of damage to the right posterior sectoral duct during resection of the right anterior sector due to the Hjortsjo Crook (see Sect. 6.3.1) and the possibility of damage to the right posterior sectoral duct in the left hemihepatectomy in the anomaly that the right posterior sectoral duct joins the left hepatic duct (see Sect. 6.3.2.3).
3. At the segmental level.
   Further dissection of the sectoral pedicles leads to the segmental Glissonian pedicles (see Sect. 15.4).

13.3 Selective Total Main Portal Vein Occlusion

Instead of using Pringle's manoeuvre to occlude the hepatic vascular inflow completely, there has been an attempt to occlude only the main portal vein (Fig. 13.2). The idea is to leave behind the hepatic arterial blood flow to the liver remnant, which is going to be left behind, thus causing less ischaemic-reperfusion injury during liver resection. Although animal studies have shown that in an ischaemic liver, micrometastatic foci of cancer grow more rapidly, there is no clinical data to support the use of selective total main portal vein occlusion. The main disadvantages of this method are (1) more bleeding at the liver raw area during hepatic transection; and (2) time spent to dissect out the main portal vein before occlusion can be done.

13.4 Low Central Venous Pressure (Please Also See Chap. 12)

The liver receives abundant blood supply from both the systemic arterial and portovenous systems. Intraoperative blood loss is inevitable during hepatectomy and it significantly affects post-operative morbidity and patient survival. Reducing intraoperative blood loss can significantly improve patient outcomes. One important technique used during hepatectomy to minimise blood loss is the low central venous pressure (CVP) technique. This technique has been shown to reduce intraoperative blood loss during hepatectomy. It can also reduce the amount of blood transfusion, shorten the operative time, reduce the length of hospital stay, and reduce post-operative liver and renal function impairment and morbidity.

During hepatectomy, bleeding from the hepatic artery and portal vein can be controlled readily with porta hepatis clamping (Pringle's Manoeuver). However, the significant additional amount of blood loss occurs from the ruptured branches of the hepatic veins and short hepatic veins. This blood loss is directly proportionate to the pressure gradient across the hepatic vascular wall and the diameter of the injured vessels. Since hepatic sinusoidal pressure directly correlates with central venous pressure (CVP), by maintain-

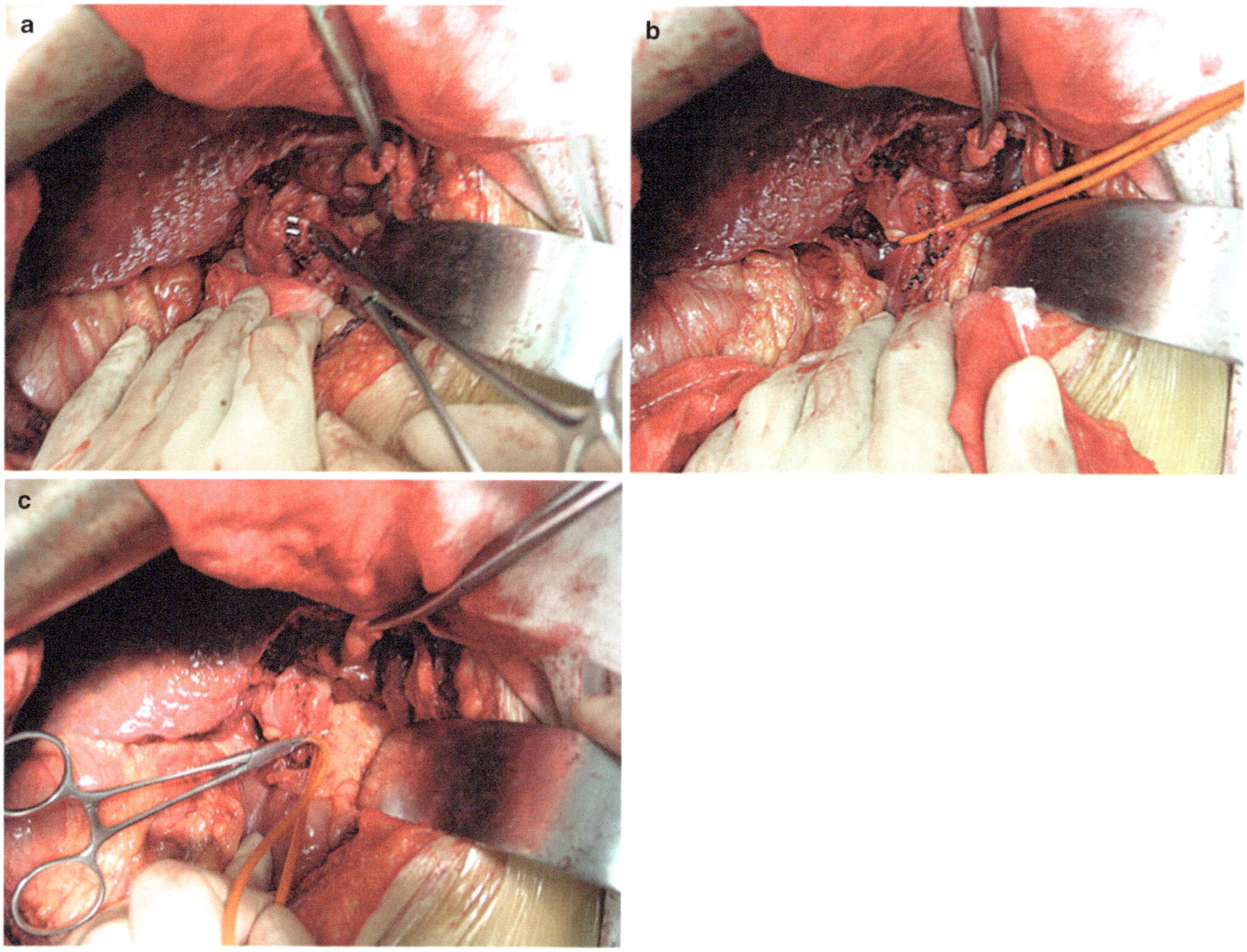

Fig. 13.2 Selective total occlusion of the main portal vein. (**a**) Isolation of common hepatic artery; (**b**) Sling around porta hepatis sparing common hepatic artery. (**c**) Occlusion of the main portal vein and common bile duct sparing common hepatic artery

ing a low CVP can reduce the hepatic sinusoidal pressure and thus the amount of blood loss intraoperatively.

The normal CVP is 5–12 mmHg (which is equivalent to 7–16 cm H_2O). The definition of low CVP is when the CVP ≤6 mmHg (or 8 cm H_2O). Most surgeons request a CVP ≤3.6 mmHg (or 5 cm H_2O). Some even request a CVP near zero.

There are two main ways to lower central venous pressure intraoperatively. The first way is to use the positioning method. With the reverse Trendelenburg position (~15-degree head up), the blood inside the vascular compartment is pooled in the lower extremities, thus lowering the CVP. There is, however, always a theoretical risk of air embolism by using this manoeuver. Some surgeons also do not like operating on a patient in this position. Thus this method is not very often used. The second way is to use pharmacological means. Two types of drugs have been frequently used: diuretics and venodilators. Intravascular volume can be depleted with a bolus of intravenous loop diuretic (e.g., frusemide 5–40 mg i.v.), which lowers the CVP. However, intravascular volume depletion activates the renin-angiotensin-aldosterone system and raises the ADH level, which may result in post-operative oliguria that can persist for 1–2 days before diuresis happens. The other means, the use of low concentration intravenous nitrate infusion (e.g. glycerol trinitrate 0.5–3 mg/h) seems to be a more practical and titratable method in achieving low CVP. At low concentration, it has more vasodilatation effect on the venous system than the arterial system. The CVP can be titrated gradually to the desired level. If excessive bleeding occurs, further fluid load can be given to replace the volume loss.

There are also concerns about hypoperfusion to major organs using the technique, but the problem can be readily tackled by increasing the inspiratory oxygen concentration (FiO_2 80%) and with the use of a vasopressor (phenylephrine 0.05 mg i.v.) to maintain a mean blood pressure of 50 mmHg which would then ensure adequate oxygen delivery to vital organs.

After completion of hepatic parenchymal transection, the Pringle's manoeuvre is unclamped and adequate haemostasis is carried out on the raw liver surface. The CVP is then brought up to >10 mmHg by stopping the intravenous nitrate infusion and increasing the intravascular volume by infusion of 1–2 L of normal saline. Haemostasis is then secured one more time on the raw surface of the liver.

With meticulous use, the low CVP technique is definitely helpful during hepatectomy to significantly reduce blood loss.

13.5 Total Vascular Exclusion

Total vascular exclusion combines total vascular inflow occlusion with Pringle's manoeuvre, and outflow occlusion of the liver by suprahepatically cross-clamping the inferior vena cava above the hepatic veins and infrahepatically cross-clamping the inferior vena cava above the renal veins (Fig. 13.3). Occasionally the inferior phrenic vessels need to be divided to facilitate dissection of the suprahepatic inferior vena cava. After ligation of the right adrenal vein, the liver is completely isolated from the systemic circulation.

The aims of total vascular exclusion are to decrease intraoperative blood loss, avoid air embolism, and allow precise dissection in a bloodless field. It is recommended for invasive tumours in the paracaval portion of the liver, and is particularly useful when a tumour thrombus is present in the inferior vena cava as total vascular outflow exclusion prevents inferior vena cava thrombus migration intraoperatively. Major hepatic veins or inferior vena cava reconstruction can safely be performed under total vascular exclusion (Fig. 13.4).

Total vascular exclusion is associated with significant haemodynamic changes and requires close monitoring and

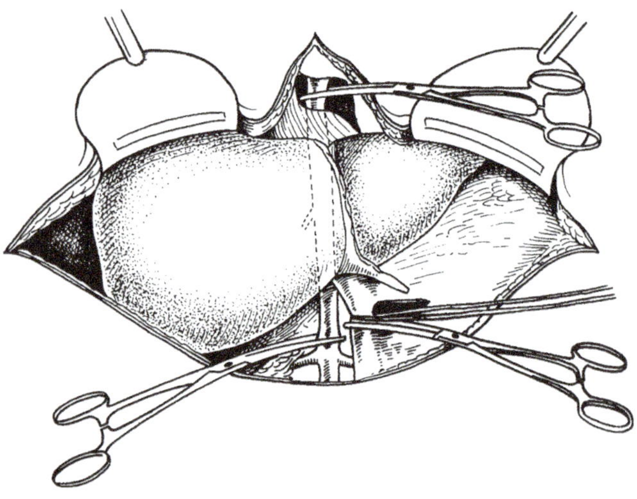

Fig. 13.3 Total vascular exclusion

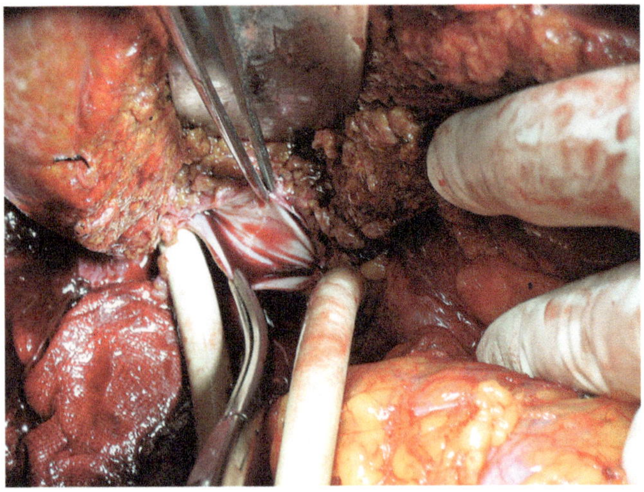

Fig. 13.4 The inferior vena cava can be opened, or part of it resected and reconstructed safely with total vascular inflow and outflow exclusion

anaesthetic expert management intraoperatively. Total vascular exclusion requires a high central venous pressure of 12–15 mmHg to maintain the cardiac preload in order to tolerate the total clamping. The high volume of fluid infused before and during total vascular exclusion leads to increased risks of post-operative liver, renal and pulmonary dysfunction, as well as abdominal fluid collection. A fall in cardiac output exceeding 50% or a decrease in mean arterial blood pressure exceeding 30% (i.e. less than 80 mmHg) in an euvolaemic patient is defined as haemodynamic intolerance to total vascular exclusion. It has been reported in 10–20% of patients. There are, in addition to the haemodynamic effects during inferior vena cava clamping, the haemodynamic and biochemical effects after unclamping the inferior vena cava and the porta hepatis.

The operative steps in total vascular exclusion are first the portal triad is occluded, followed by occlusion of the infrahepatic inferior vena cava. If this trial of dual occlusion is tolerated, the suprahepatic inferior vena cava clamp is then applied (Fig. 13.5). A trial of total clamping for 5 min should be used before liver resection is started. After liver resection, the vascular occlusions are released in a reversed order. First, the suprahepatic inferior vena cava clamp is released. If there is no active bleeding from the raw surface of the liver, the infrahepatic inferior and the porta hepatic clamps are removed.

A venovenous bypass in combination with total vascular inflow and outflow exclusion should be used if the patient cannot tolerate trial clamping or if the exclusion is expected to be long (>1 h) to prevent the side-effects, including mesenteric congestion and cardiovascular instability (Fig. 13.6).

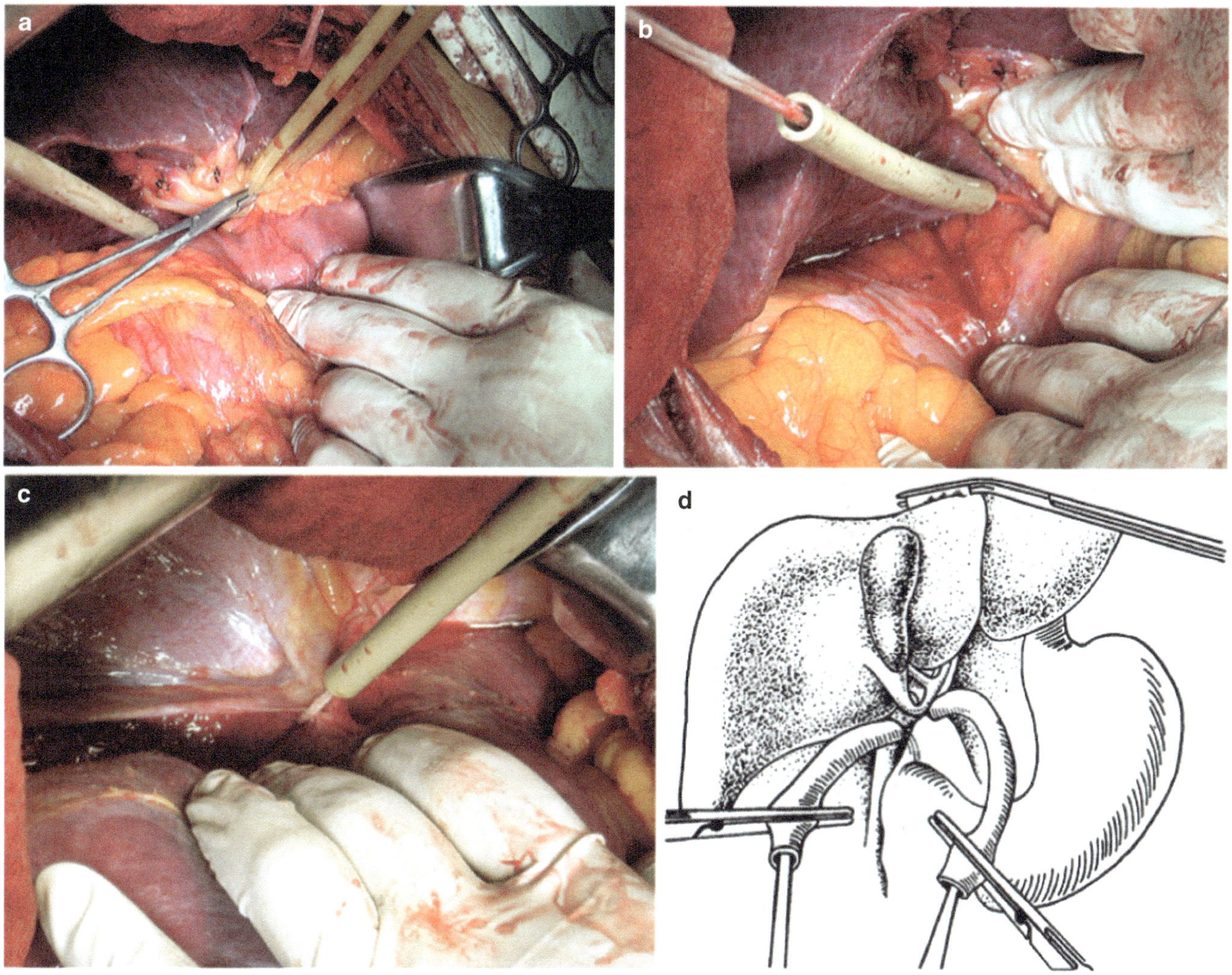

Fig. 13.5 Operative steps in total vascular exclusion. (**a**) First, the portal triad is occluded. (**b**) Then, infrahepatic vena cava is occluded. If this trial of dual occlusion is tolerated. (**c**) The suprahepatic vena cava is then occluded. (**d**) Shows the complete picture of total vascular exclusion

13.6 Selective Vascular Exclusion

13.6.1 Selective Total Vascular Exclusion

Selective total vascular exclusion entails full mobilisation of the liver from the inferior vena cava, occlusion of the hepatic inflow with Pringle's manoeuvre, and extrahepatic clamping of the right and the trunk of the middle/left hepatic veins separately. Total vascular isolation of the liver from the systemic circulation can be achieved without interruption of the caval blood flow, thus avoiding the haemodynamic and biochemical drawbacks associated with total vascular exclusion.

In the dissection and isolation of the right hepatic vein and middle/left hepatic vein trunk, the following anatomical structures are important: the hepatocaval ligament on the right side (see Sect. 3.5), the fossa between the right hepatic vein and the trunk of the middle/left hepatic veins (see Sect. 8.2.1), and the ligamentum venosum (see Sect. 6.10). To isolate the right hepatic vein, the first step is to dissect the fossa between the right hepatic vein and the trunk of the middle/left hepatic veins (Fig. 13.7a), then dissect the bare area, isolate and divide the hepatocaval ligament (Fig. 13.7b), isolate the right hepatic vein (Fig. 13.7c), and clamp the right hepatic vein (Fig. 13.7d). To isolate the middle/left hepatic vein trunk, the operative steps are to dissect the fossa between the right hepatic vein and the trunk (Fig. 13.8a), isolate and divide the ligamentum venosum near to the trunk, then the back of the trunk is dissected after turning the left liver anteriorly (Fig. 13.8b), further dissection of the fossa, then (Fig. 13.8c), a sling is passed around the trunk.

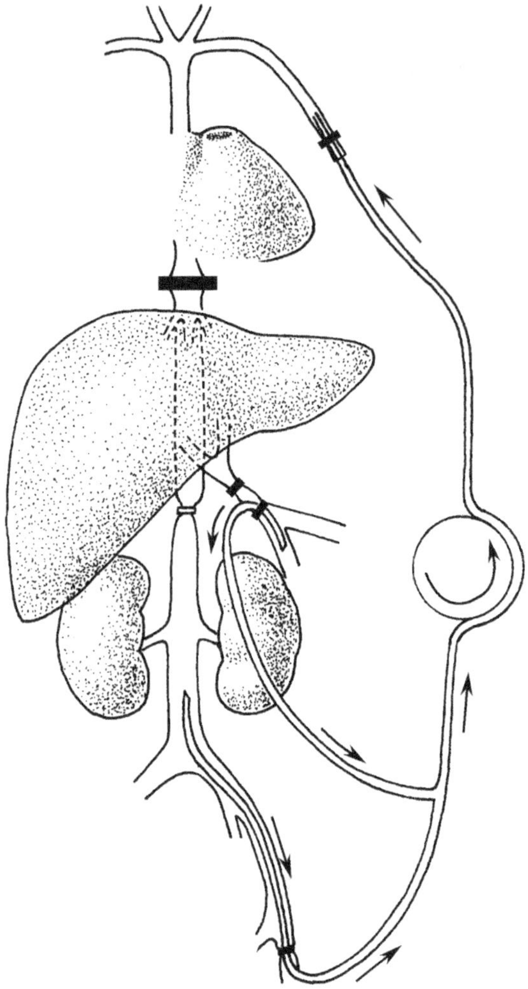

Fig. 13.6 A venovenous bypass combined with total vascular exclusion

It is usually not advisable to isolate the middle and the left hepatic veins extrahepatically as the two veins can share a common middle wall in a significant proportion of patients. Occasionally, the middle and left hepatic vein has already separated outside of the liver, and it is possible to put separate slings around them (Fig. 13.9).

An alternative way to do selective vascular exclusion is not to isolate the right and the middle/left hepatic vein trunk entirely, but to isolate them from above downward by dissecting into the fossa between them, and then to apply vascular clamps to occlude the lumens of the veins (Fig. 13.10).

13.6.2 Selective Partial Vascular Exclusion

Selective vascular exclusion can also be partial and is just for the right or the left hemiliver. It has been infrequently used, probably because dissection of the right and the middle/left hepatic veins has been considered hazardous. For selective vascular exclusion of the right hemiliver, the hepatic hilar plate is lowered (see Sect. 4.3) to gain access and control to the right portal pedicle. The right hepatic vein is controlled after division and ligation of the hepatocaval ligament (see Sect. 3.5 and above). The advantages of selective vascular exclusion on a hemiliver are the avoidance of ischaemic-reperfusion injury to the hemiliver that is left behind and the preservation of the portal flow during clamping, thus avoiding portal congestion. The main disadvantage is bleeding can still happen at the plane of transection because of the persistence of blood flow in half of the liver. For selective vascular exclusion of the left hemiliver, the left portal pedicle is controlled after lowering of the hepatic hilar plate (see Sect. 3.5) and the trunk of the middle/left hepatic veins is controlled after the division of the ligamentum venosum (see Sect. 6.10).

A modification of selective partial vascular exclusion has been developed by Chen et al. Instead of isolating the right or the left branches of the porta hepatis and the right or the main trunk of the middle/left hepatic veins or the left hepatic vein extrahepatically, these structures are controlled intrahepatically. Thus, for a right hepatectomy, a ligature is first passed and tied around the right branch of the porta hepatis (Fig. 13.11a), followed by another ligature tying around the right hepatic vein (Fig. 13.11b). After the inflow and the outflow to the right hemiliver have been occluded, hepatic parenchymal transection is then carried out along the line of demarcation. A similar technique can be used on the left liver to carry out a left hepatectomy by ligating the left branch of the porta hepatis (Fig. 13.11c) and the common trunk of the middle/left hepatic veins (Fig. 13.11d). Similarly, a left lateral sectionectomy is carrying out by ligating the porta hepatis supplying the left lateral section (Fig. 13.11e) and the left hepatic vein (Fig. 13.11f). This technique, although is quick to carry out, is not as refined, precise and as reliable as extrahepatic dissection of the individual structures.

13.6.3 Pringle's Manoeuvre + Infrahepatic Inferior Vena Cava Clamping

Chen et al. reported the use of Pringle's manoeuvre and infrahepatic inferior vena cava clamping to decrease blood loss during hepatic parenchymal transection. This is similar to total vascular occlusion with the exception that this procedure does not use a clamp to control the suprahepatic inferior vena cava. Personally, I do not see much of the advantage of this procedure over total vascular occlusion as air embolism

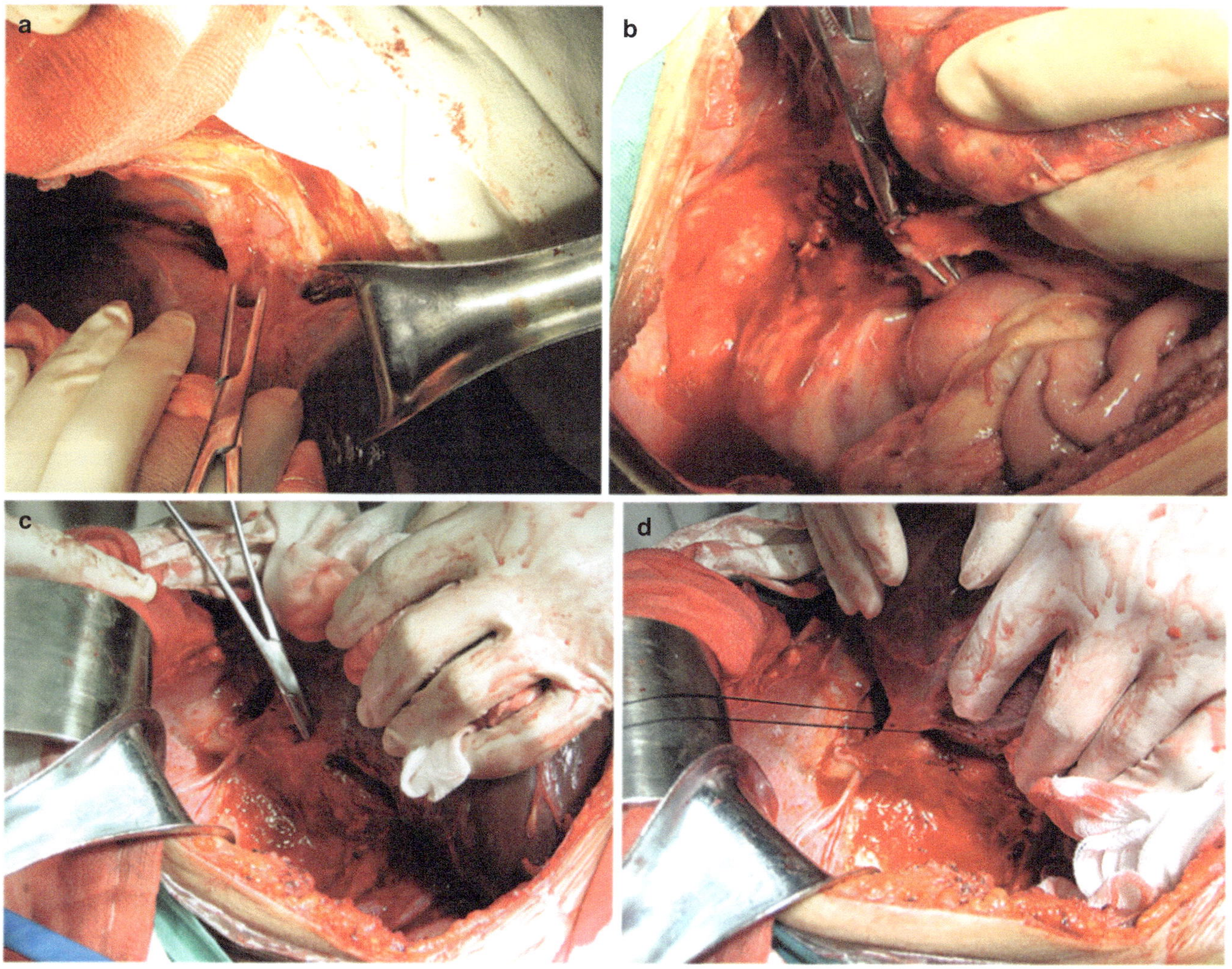

Fig. 13.7 Isolation of the right hepatic vein. (**a**) Dissect the fossa between the right and the middle/left vein trunk. (**b**) Isolate and divide the hepatocaval ligament. (**c**) Isolate the right hepatic vein. (**d**) Sling the right hepatic vein

cannot be prevented in this procedure, and it has the same drawbacks of total vascular exclusion.

13.7 Protective Strategies Against Prolonged Ischaemia During Vascular Clamping

Inflow occlusion with Pringle's manoeuvre and total/selective vascular exclusion techniques minimise blood loss during liver resection and the need for perioperative blood transfusion. Furthermore, these procedures facilitate parenchymal transection by providing a better view of the transection plane. While resection with clamping periods below 30 min is usually well tolerated, the prolonged period of continuous inflow occlusion may cause severe ischaemic/reperfusion injury in the remnant liver. Cirrhotic livers are more vulnerable to this injury than normal livers. Prolonged ischaemia can have deleterious effects on post-operative liver function, liver regeneration and survival. There are several surgical protective strategies that can be used to prolong the vascular inflow clamping time:

13.7.1 Intermitted Portal Triad Clamping

This is a technique that alternates portal triad clamping with short intervals of unclamping and portal triad reperfusion. It is most commonly carried out using 15 min of occlusion/5 min of reperfusion during liver resection. The drawback of this technique is the significantly higher blood loss during the periods of unclamping when compared to the continuous porta hepatis clamping.

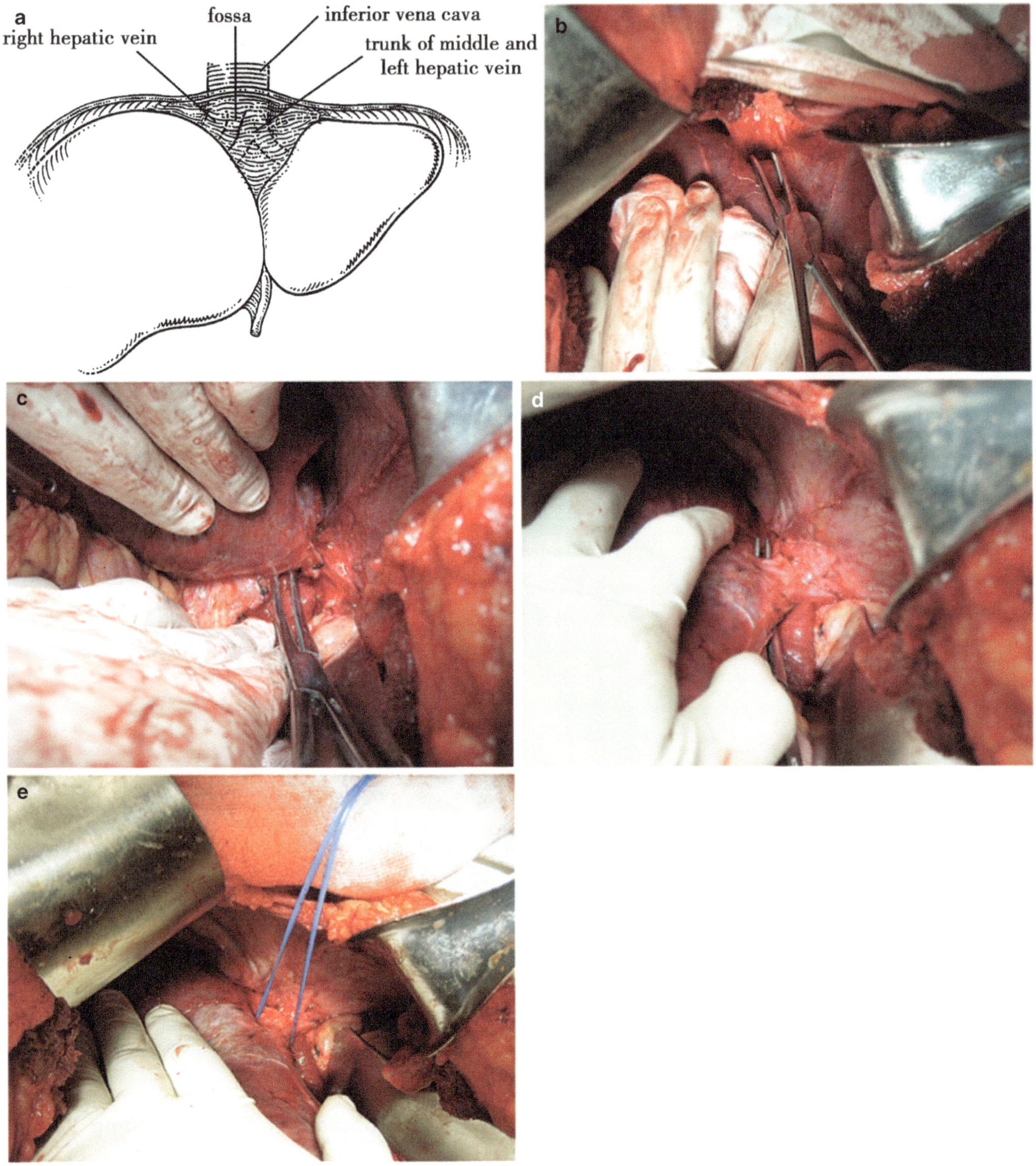

Fig. 13.8 To isolate the trunk of the middle/left hepatic veins, first identify the fossa between the right hepatic vein and the trunk of the middle/left hepatic veins (**a**). After dissecting this fossa (**b**), the left liver is lifted upwards, the ligamentum venosum is divided, and the back of the trunk is dissected (**c**). Further dissection of the fossa isolate the trunk (**d**), the trunk is slung (**e**)

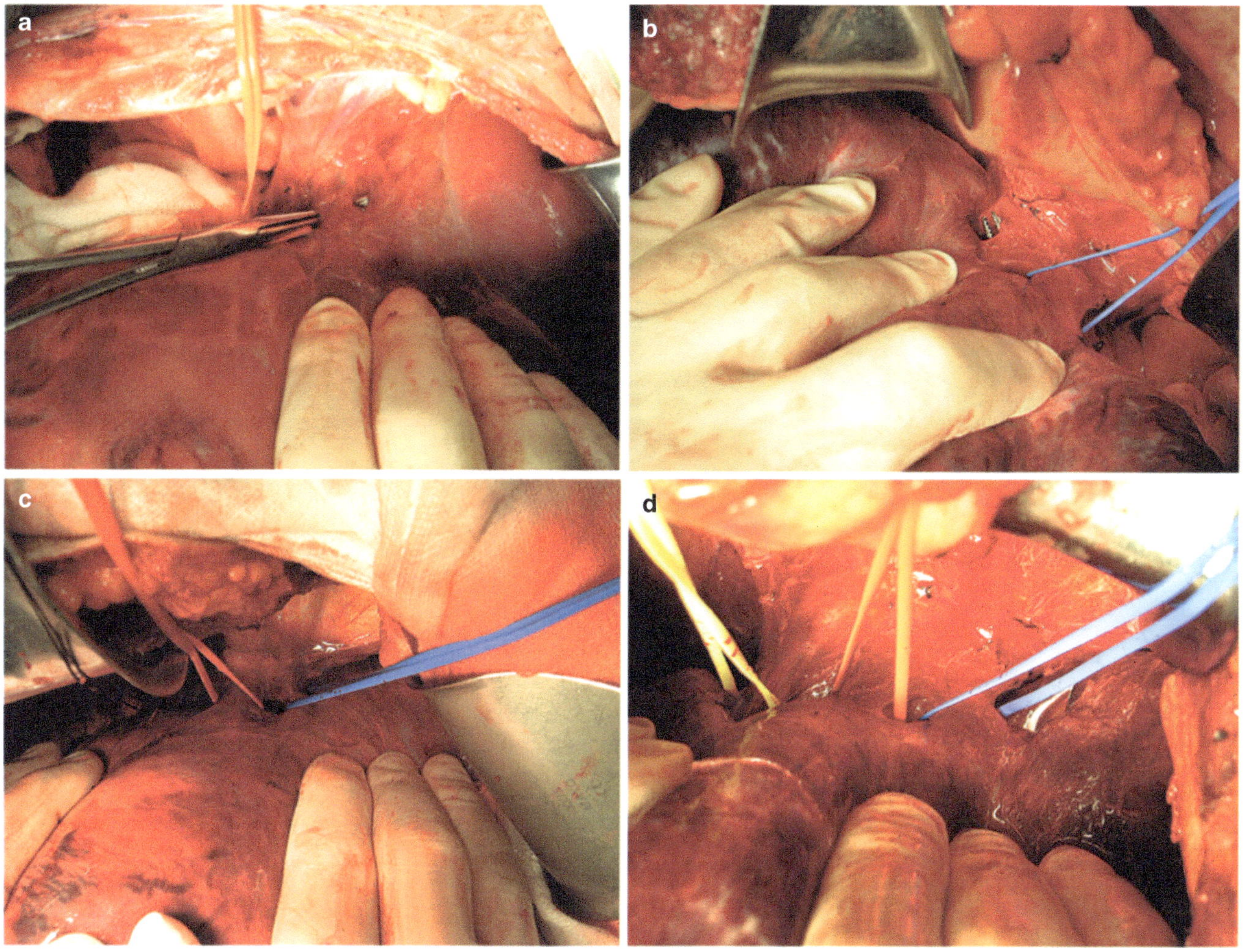

Fig. 13.9 To isolate the middle hepatic vein from the left hepatic vein (dangerous as some patients share a common wall between the two veins): the fossa between the right hepatic vein and the trunk of the middle/left hepatic veins is dissected (**a**). The trunk is isolated and slung (**b**). Further dissection to isolate the middle hepatic and the left hepatic veins (**c**). Right, middle and left hepatic veins are slung separately (**d**)

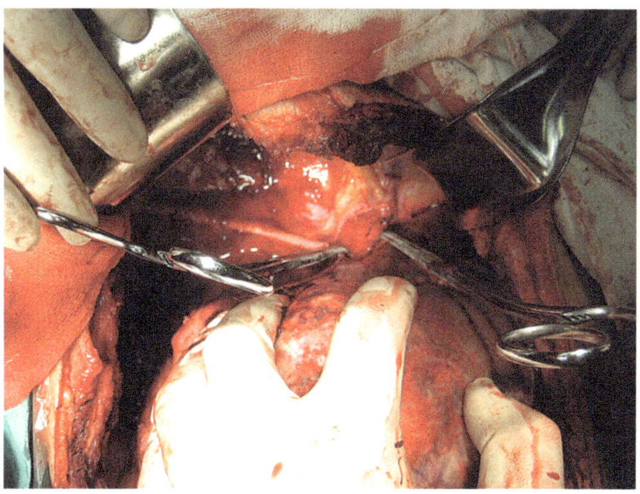

Fig. 13.10 Vascular clamps applied to the right hepatic vein and to the trunk of middle/left hepatic veins (Pringle's Manoeuvre not shown) for selective vascular exclusion

13.7.2 Ischaemic Preconditioning with Continuous Inflow Occlusion

Ischaemic preconditioning consists of a brief period of ischaemia by porta hepatis clamping (10–15 min) followed by a short interval of reperfusion (10–15 min) before transection under continuous inflow occlusion.

13.7.3 Continuous Inflow Occlusion Under In Situ Hypothermia

The advantages of continuous clamping under in situ hypothermia are prolongation of safe ischaemic times and prevention of bleeding since the method does not require cyclic unclamping and reperfusion. In situ hypothermia during liver transection can be achieved by simple topical (surface)

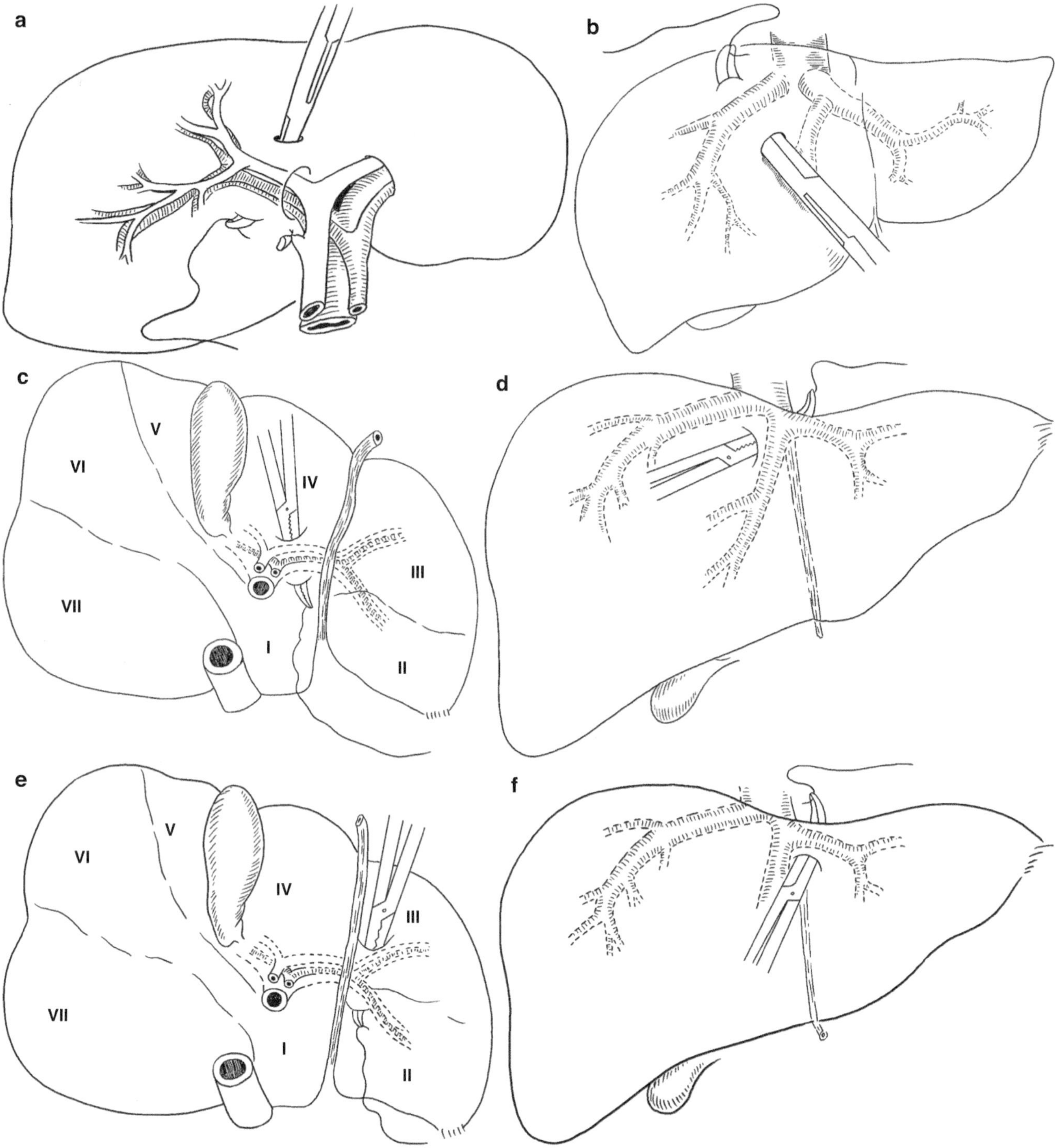

Fig. 13.11 (**a**) Ligature around the right branch of the porta hepatis. (**b**) Ligature around the right hepatic vein. (**c**) Ligature around the left branch of the porta hepatis. (**d**) ligature around the common trunk of the middle/left hepatic veins. (**e**) Ligature around the porta hepatis branch to the left lateral section. (**f**) Ligature around the left hepatic vein

cooling. Simple topical cooling can reduce the hepatic core temperature to 20–25 °C.

Represented diagrammatically, inflow occlusion ischaemia and unclamping (reperfusion) are illustrated in black and white boxes, respectively (Fig. 13.12).

13.8 Safe Clamp Times

The maximum safe duration of vascular inflow occlusion using the different surgical protective strategies are shown in Table 13.1, and they differ in normal and cirrhotic livers.

13.9 Hypothermic Infusion Techniques

In very complicated liver resection combined with vascular reconstruction of the remaining vessels, a long duration of vascular inflow and outflow control is required. To overcome the problem of the limited duration of normothermic ischaemia or hypothermic cooling by topical cooling, continuous cooling of the liver with chilled preservation fluid, e.g. University of Wisconsin (UW) solution would enhance the tolerance of the remnant liver to ischaemia. Liver perfusion with 4 °C preservation solution can be used by employing one of the following three procedures:

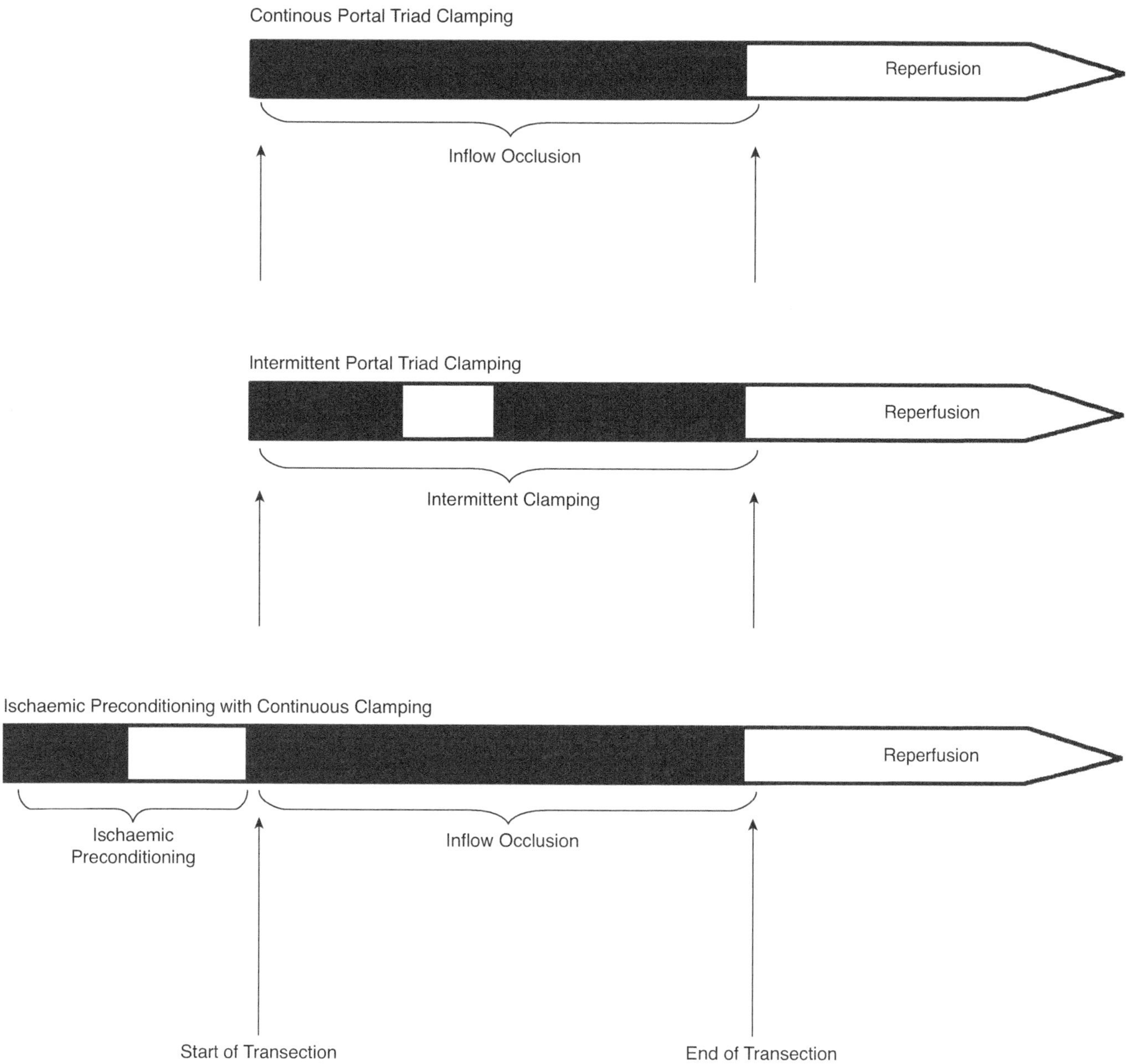

Fig. 13.12 Inflow occlusion (black) and unclamping with reperfusion (white)

13.9.1 In Situ Procedure

In the in situ procedure, hilar dissection is minimal and only the hepatic artery or the portal vein used for perfusion is dissected. In Fig. 13.13, perfusion is through the right hepatic artery to the left liver in order to perform a right hepatectomy. The hilar clamp is applied below the arterial bifurcation. The right adrenal vein is ligated and divided. Suprahepatic and infrahepatic clamps are applied to the inferior vena cava. Once total hepatic vascular occlusion is tolerated, perfusion with 4 °C UW solution starts, and a draining cavotomy done either at the infrahepatic inferior vena cava or by transecting one of the hepatic veins at its termination with the inferior vena cava. Usually, the perfusion proceeds until the caval effluent becomes clear. After completion of the liver resection, the preservation fluid is flushed from the remnant liver by perfusing 4 °C lactated Ringer. Haemostasis is completed by plicating leaking points from the raw surface with non-absorbable sutures. The cavotomy is closed with a 4/0 monofilament suture. The perfusion catheter is removed, and the orifice of the artery is closed with a ligature. After checking for haemostasis with unclamping and immediate clamping of the infrahepatic inferior vena cava clamp, obvious bleeding sites are controlled. Unclamping follows with the removal of suprahepatic clamp, then infrahepatic clamp and portal pedicle clamp.

Table 13.1 Safe clamp time

Porta hepatis clamping	Normal liver (min)	Cirrhotic liver (min)
Continuous clamping	60	30
Intermittent clamping	>120	60
Preconditioning	75	?
Hypothermia + continuous clamp	90	60

P. Clavien

In the rare instance when there is poor tolerance to total hepatic vascular exclusion, a venovenous extracorporeal bypass can be established (Fig. 13.6). Systemic heparinization is usually initiated before the extracorporeal perfusion to minimise the risk of intra-shunt clotting. Heparinization should be mild and may be neutralised after 1 h to avoid dramatic and uncontrollable haemorrhage on reperfusion of the liver.

13.9.2 Ex Situ In Vivo Procedure

The liver is mobilised by the division of the falciform, coronary and triangular ligaments. The use of venovenous bypass is essential and is performed through the left long saphenous vein, inferior mesenteric and left axillary cannulae. Portal triad clamp is applied, followed by infrahepatic clamping of the inferior vena cava. The suprahepatic clamp is applied. After resection of the right and the common trunk of the middle/left hepatic veins and ligation and division of all the short hepatic veins, the liver is placed ex situ but remains attached to the porta hepatis. Perfusion of 4 °C UW solution is through the right hepatic artery to the left liver to perform a right hepatectomy as shown in Fig. 13.14. Drainage of the perfusion fluid is through the hepatic veins. Tumour resection, vascular and biliary reconstruction are done. After completion of liver resection, the cooling system is removed. The hepatic vein of the remnant liver, sometimes after having undergone reconstruction, is reimplanted onto its original inferior vena caval orifice if possible and if not, on a cavotomy orifice. UW solution is rinsed off by 4 °C lactated Ringer solution. The perfusion catheter is removed, and the cavotomy is closed. Clamps are removed. After a few minutes, extracorporeal venovenous circulation is interrupted. Cautious systemic heparinization is used.

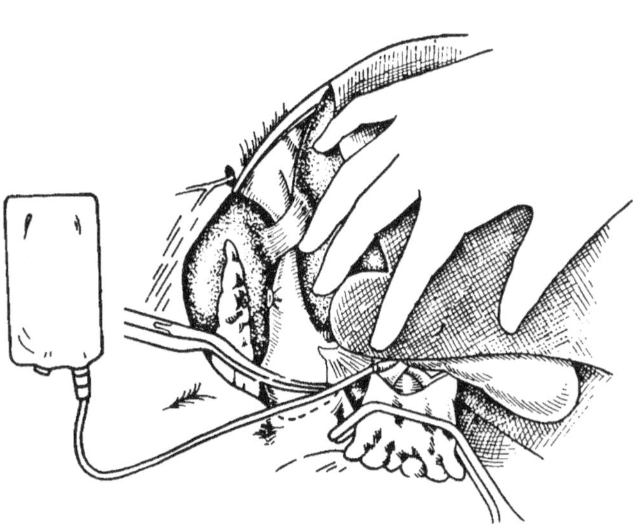

Fig. 13.13 Hypothermic Infusion using the in situ technique

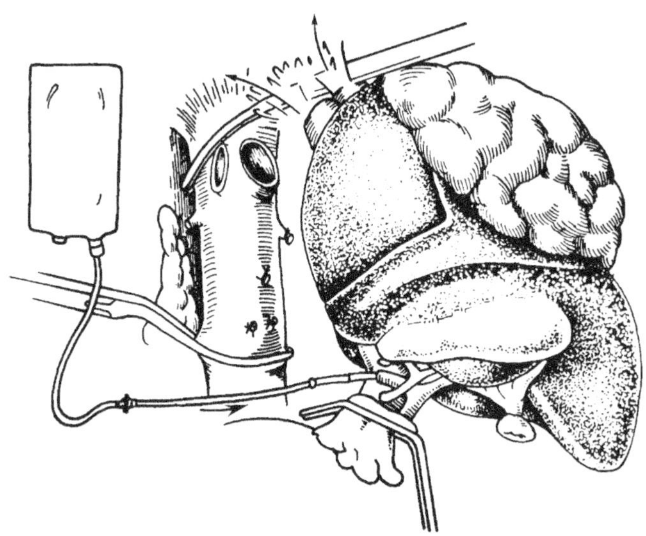

Fig. 13.14 Hypothermic Infusion using the ex situ in vivo technique

13.9.3 Ex Vivo

In the ex vivo technique described by Pichlmayr, the principles are those of auto liver transplantation. Meticulous dissection and re-anastomosis are mandatory as long vessels and patches are not available. The use of venovenous bypass is mandatory. Hypothermic liver perfusion is started in situ after caval clamping. The liver is removed within 1–2 min given the previous vascular isolation. Extracorporeal perfusion is continued on the back-bench after explantation. After completion of the liver resection, the remnant liver is brought back into the abdomen, and vascular and biliary re-anastomoses are carried out.

Further Reading

Chen XP, Qiu FZ. A simple technique ligating the corresponding inflow and outflow vessels during anatomical left hepatectomy. Langenbeck's Arch Surg. 2008;393:227–30.

Chen XP, Zhang ZW, Zhang BX, et al. Modified technique of hepatic vascular exclusion: effect on blood loss during complex mesohepatectomy in hepatocellular carcinoma patients with cirrhosis. Langenbeck's Arch Surg. 2006;391:209–15.

Fu SY, Lai EC, Li AJ, et al. Liver resection with selective hepatic vascular exclusion: a cohort study. Ann Surg. 2009;249:624–7.

Fu SY, Lau WY, Li AJ, et al. Liver resection under total vascular exclusion with or without preceding Pringle manoeuvre. Br J Surg. 2010;97:50–5.

Fu SY, Lau WY, Li GG, et al. A prospective randomized controlled trial to compare Pringle maneuver, hemihepatic vascular inflow occlusion, and main portal vein inflow occlusion in partial hepatectomy. Am J Surg. 2011;201:62–9.

Hannoun L, Borie DC. Chapter 25: Major liver resections using hypothermic perfusion. In: Mazziotti A, Cavallari A, editors. Techniques in liver surgery. London: Greenwich Medical Media; 1997. p. 239–47.

Lau WY. The history of liver surgery. J R Coll Surg Edinb. 1997;42:303–9.

Müller M, Petrowsky H, Clavien P. Chapter 23: Techniques of vascular control and protective strategies for parenchymal transection. In: Lau WY, editor. Hepatocellular carcinoma. Singapore: World Scientific; 2008. p. 507–27.

Ni JS, Lau WY, Yang Y, et al. A prospective randomized controlled trial to compare Pringle manoeuvre with hemi-hepatic vascular inflow occlusion in liver resection for hepatocellular carcinoma with cirrhosis. J Gastrointest Sug. 2013;17:1414021.

Pichlmayr R, Grosse H, Hauss J, Gubematis G, Lamesh P, Bretschneider HK. Technique and preliminary results of extracorporeal liver surgery (bench procedure) and of surgery of the in situ perfused liver. Br J Surg. 1990;77:21–6.

Takasaki K. Glissonian pedicle transection method for hepatic resection. Tokyo: Springer; 2007. p. 1–143.

Xia F, Lau WY, Qian C, et al. Surgical treatment of giant liver hemangiomas: enucleation with continuous occlusion of hepatic artery proper and intermittent Pringle maneuver. World J Surg. 2010;34:2162–7.

Xia F, Lau WY, Qian C, et al. Continuous occlusion of hepatic artery proper for prevention of blood loss in partial hepatectomy for ruptured hepatocellular carcinoma: a case-matched comparative study. Ann Surg Oncol. 2011;18:1638–43.

Yang Y, Fu SY, Lau WY, et al. Selective main portal vein clamping to minimize the risk of recurrence after curative liver resection of hepatocellular carcinoma. Hepato-Gastroenterology. 2012;59:1560–5.

Yang Y, Lai EC, Fu SY, et al. A prospective randomized controlled trial to compare two methods of selective hepatic vascular exclusion in partial hepatectomy. Eur J Surg Oncol. 2013;39:125–30.

Zhang J, Lai EC, Zhou WP, et al. Selective hepatic vascular exclusion versus Pringle manoeuvre in liver resection for tumours encroaching on major hepatic veins. Br J Surg. 2012;99:973–7.

14 Liver Transection Techniques and Haemostasis on the Raw Surface of the Remnant Liver

Most bleeding occurs at the time of liver parenchymal transection during liver resection. There are many techniques that have evolved to reduce blood loss during liver transection.

14.1 Finger Fracture Technique

The finger fracture technique, or one of its modifications, is the most commonly used technique in liver parenchymal transection. It is based on the principle that liver tissue breaks easily when crushed between the fingers, leaving behind the tougher vasculo-biliary branches within the Glissonian sheaths and the venous branches from the hepatic veins to remain intact. The larger branches can then be controlled by ligation and division while the smaller branches coagulated and divided, thus reducing blood loss. As a consequence, the operative field becomes clearer which results in a less chance of inadvertently dividing the vasculo-biliary or venous branches, thus, entering a virtuous circle of less blood loss → clearer operative field → less inadvertent damage to vessels → less blood loss.

14.1.1 History of Development of Finger Fracture Technique

In 1899, Keen noticed that the liver capsule could be stripped away from the liver parenchyma easily. Anshutz reported in 1903 that liver tissue broke easily with fingers. Such an observation was made use of clinically in 1953 when Quattlebaum broke liver tissue with the handle of the knife, thus facilitated liver parenchymal transection. A similar technique of crushing liver tissue with the fingers or breaking liver tissue with the handle of a knife was reported by Fineburg in 1953. In 1958, Tien-yu Lin reported the finger fracture technique in liver resection and he popularised the technique. This technique was adopted by Ton That Thung in 1963 and he combined the finger fracture technique with Pringle's manoeuvre, thus further reducing blood loss during liver parenchymal transection. Since then, the finger fracture technique becomes widely adopted by liver surgeons all around the world, and refinements of the finger fracture technique develop.

14.1.2 Modifications of Finger Fracture Technique

The principle used in the finger fracture technique is that the softer liver tissue is crushed between the thumb and the index finger, leaving behind the tougher blood vessels and bile ducts for diathermy coagulation, or ligation and division.

My thumb measures more than 2 cm in width (Fig. 14.1). By using the finger fracture technique, the width of transection through the liver plane has to be at least 2–3 cm wide, thus resulting in two rugged, raw surfaces at the plane of transection.

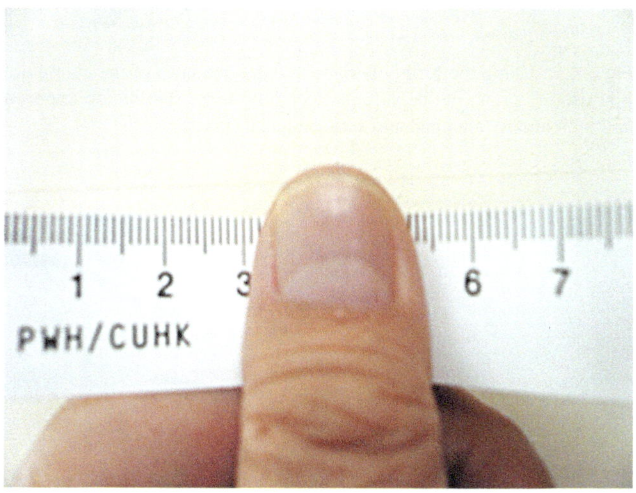

Fig. 14.1 Width of my thumb = width of liver transection plane when using the finger fracture technique

W. Y. Lau, *Applied Anatomy in Liver Resection and Liver Transplantation*, https://doi.org/10.1007/978-981-16-0800-1_14

To refine this finger fracture technique, there are a number of modifications

1. Kelly Clamp Technique
   This is the simplest of all parenchymal transection techniques. It is also known as the crush-clamp technique. It begins with the scoring of the liver capsule with electrocautery along the plane of transection. The surgeon then uses a Kelly clamp. By opening and closing the jaws of the clamp, the liver tissue is crushed. Electrocautery is used to control minor blood vessels. When larger blood vessels and bile ducts are encountered, they are individually isolated and controlled with either surgical clips, ligation or suture ligation and then divided (Fig. 14.2). These steps are repeated until the liver is transected along the predetermined plane.
   This technique has the advantages of being quick, simple and cheap. It requires the use of Pringle's manoeuvre. As the Kelly clamp has a much narrower jaw than the finger, the transection plane is narrower, with less blood loss and a more refined transection plane.

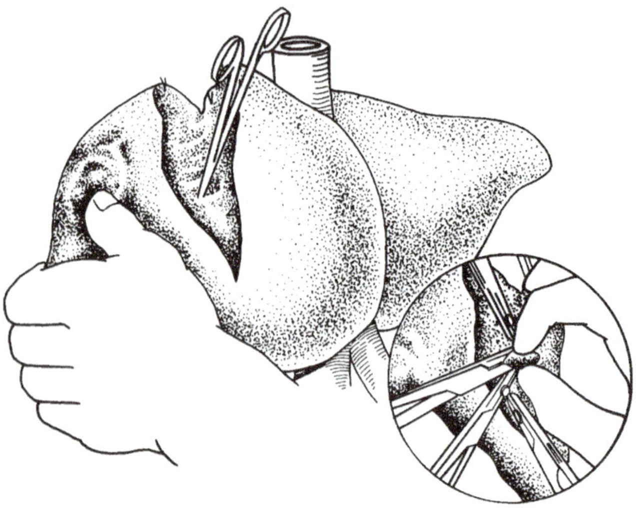

Fig. 14.2 Using the Kelly clamp technique, the jaws of the clamp are first used to crush the hepatic parenchyma; then, vessels are exposed and individually clamped and suture-ligated (inset)

2. Cavitational Ultrasonic Surgical Aspirator (SonaStar, Misonix, Cicel (Beijing) Science & Technology Co., Ltd.)
   This instrument consists of a console to which is connected a handpiece (Figs. 14.3 and 14.4). Within the handpiece, a transducer activates a hollow titanium tip along its longitudinal axis such that when the tip is brought into contact with tissue, mechanical energy is transferred, creating high- and low-pressure areas. When the pressure is below the vapour pressure of tissue fluid, vacuoles form within the cells that expand and collapse, generating forces that fragment the cell. Liver parenchymal cells have a much higher water content compared with vascular and biliary structures, which have a higher content of elastin and collagen. As a result, the ultrasonic vibrations of the handpiece tip selectively destroy liver parenchymal cells. At the same time, the tip of the handpiece is constantly irrigated. A suction line is connected to the hollow tip and aspirates irrigant, blood and tissue fragments. The negative suction pressure, the supply of irrigant, and the vibration setting can be adjusted individually. If properly adjusted, even the smallest vessels and bile ducts remain intact when cutting through the liver parenchyma (Fig. 14.5a). The ultrasonic surgical aspirator possesses no haemostatic properties, serving merely to remove or divide parenchyma, and to expose vascular structures and bile ducts. These structures need to be controlled with diathermy coagulation when small, or clipped/ligated/suture-ligated and divided when large (Fig. 14.5b). Some ultrasonic surgical aspirators could be combined with electrosurgical system.
   The main advantage of the CUSA is that it preserves biliary and vascular structures. Extensive exposure and dissection along anatomical vasculo-biliary structures are possible. Even segmental and subsegmental portal branches can be successfully searched for intraparenchy-

a

b

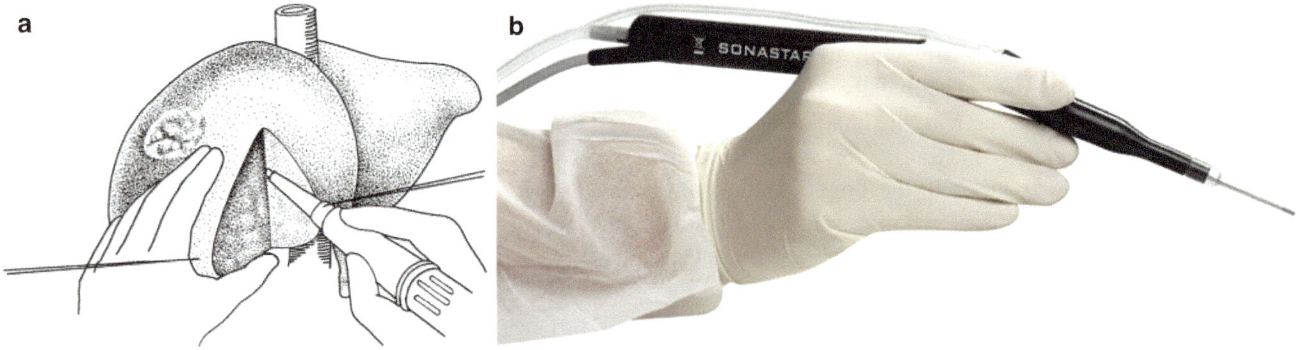

Fig. 14.3 (**a**) Hepatic resection using the cavitational ultrasonic surgical aspirator (CUSA). (**b**) The surgeon holding the handpiece

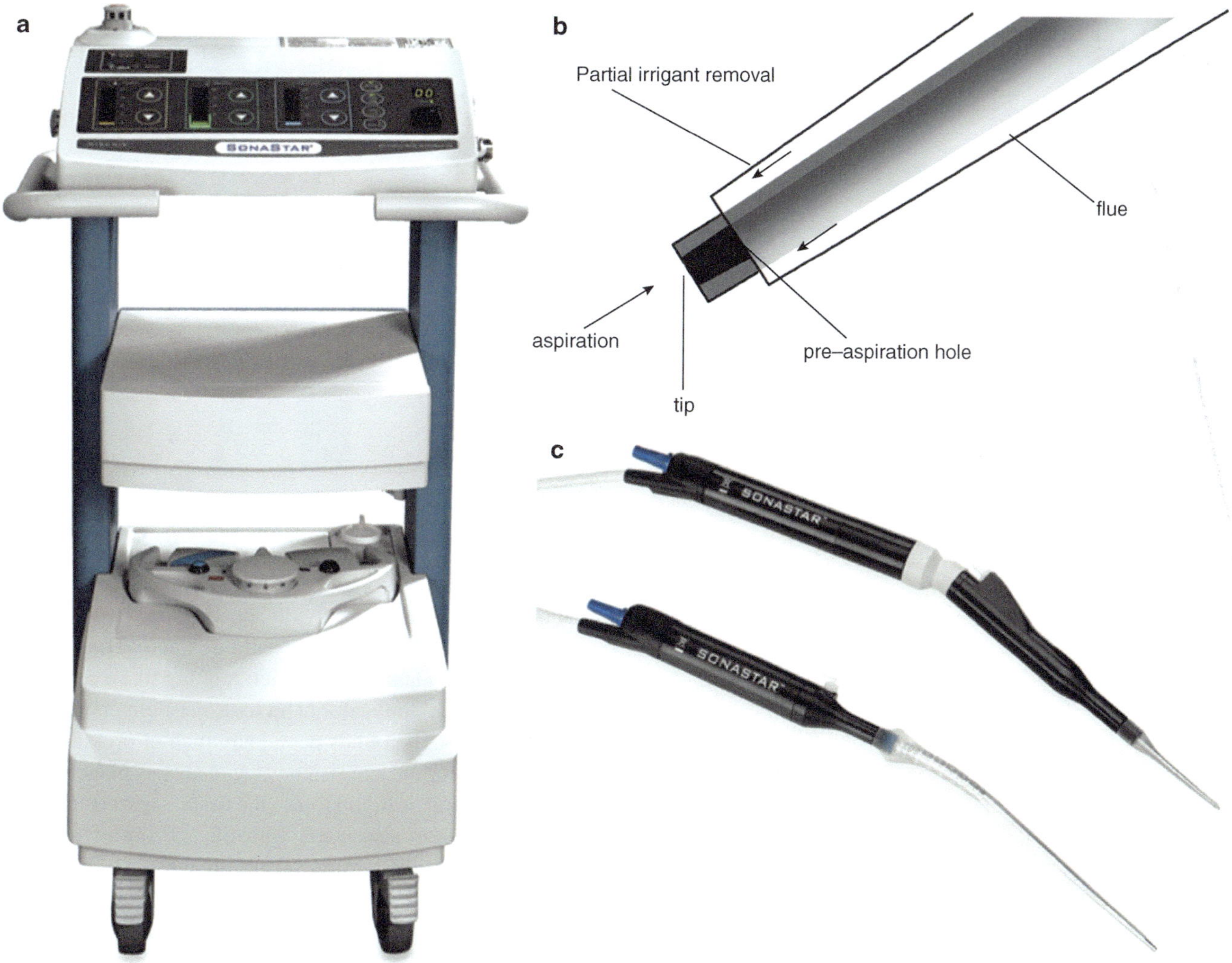

Fig. 14.4 The CUSA. (**a**) The consol. (**b**) Tip of the handpiece. (**c**) Different handpieces

mal. The other advantages are the operative precision because of the fine size of the tip of the handpiece and reduced blood loss.

The disadvantage of CUSA is the cost, and as a consequence, they are not available in many centres. The technique is reported to be slow by many authors. Also, some authors believe that CUSA works less well in cirrhotic livers when the liver parenchyma is more fibrotic.

Liver parenchymal transection using the CUSA can be carried out within a reasonably short period of time, even in cirrhotic livers, if the operating surgeon adopts the technique used by the author of this book:

(a) The operating surgeon holds the liver with his left hand while he uses the handpiece of the CUSA with his right hand.
(b) The technique begins with scoring the liver capsule along the plane of transection.
(c) The surgeon uses the left hand with four fingers at the back of the liver to left up the liver at the plane of liver transection to facilitate transection and to reduce blood loss, and he uses the thumb in front of the liver to help to open up the transection plane.
(d) The second assistant pulls the liver on the other side of the transection plane to help to open up the transection plane.
(e) The transection starts at the inferior border of the liver.
(f) Any large isolated blood vessels and bile ducts are clipped on the specimen side by the surgeon and ligated and divided on the patient's side by the first assistant.
(g) Small isolated blood vessels and bile duct are coagulated with a monopolar diathermy forceps and divided by the first assistant.
(h) The transection plane is gradually deepened from anterior to posterior and from caudal to cranial. The liver is opened up as the transection proceeds like opening up a book.

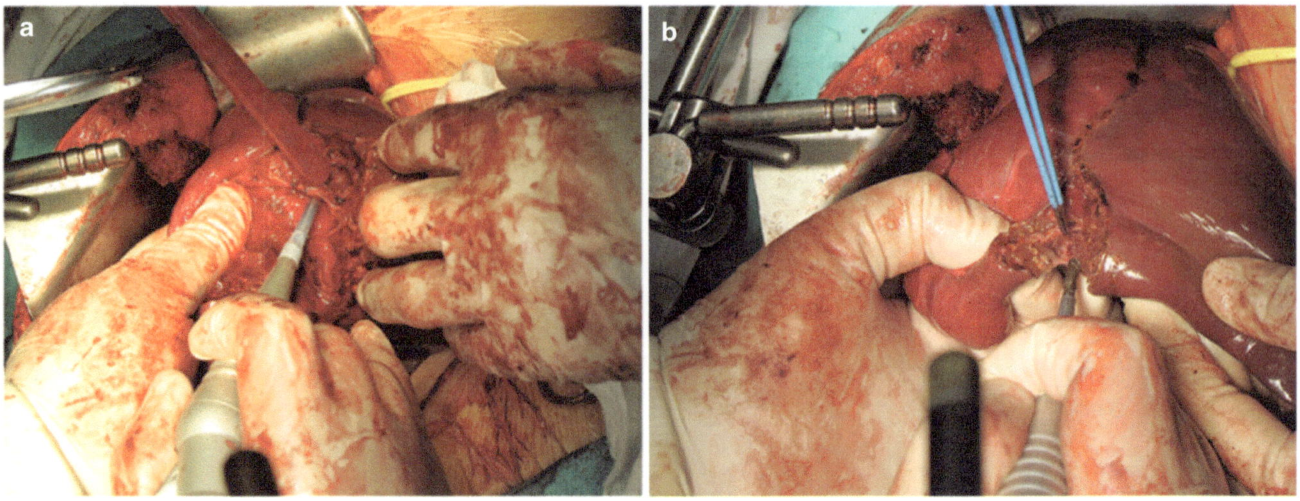

Fig. 14.5 Use of CUSA during liver parenchymal transection to enucleate a haemangioma. (**a**) Note that small vessels and bile ducts remain intact when cutting through the liver parenchyma. (**b**) Diathermy coagulation of isolated blood vessels

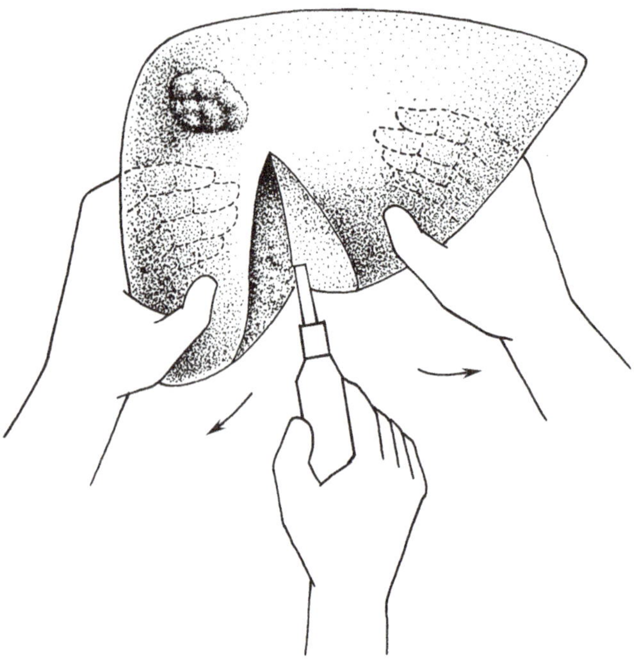

Fig. 14.6 Technique in Liver Transection using the CUSA: opening up the transection plane like opening a book

(i) Thus the surgeons work as a team with the operating surgeon transecting and clipping, and the first assistant securing the isolated vessels and biliary ducts. This cuts down the transection time tremendously.

(j) The job of the second assistant is, together with the surgeon's left hand, to open up the plane of the liver transaction (Fig. 14.6).

3. Hydrojet

It is also called water jet. The device (HydroJet, ERBE, Tubingen, Germany) uses a highly pressurised, extremely thin stream of saline to divide hepatic parenchyma, while preserving the larger vessels and bile ducts. The action is by the use of a laminar liquid jet, rotating like a drill at the surface of the applicator. The hydrojet delivers approximately 550–650 pounds per square inch of pressure for a liver with normal consistency. A cleavage plane is then created where the liquid forces the tissues apart (Fig. 14.7).

This method also starts with the scoring of the liver capsule with diathermy in the plane of transection. Pringle's manoeuvre is at the discretion of the surgeon. The water jet is used in a back-and-forth motion to divide the parenchyma until vessels and bile ducts are encountered. These structures are dealt with in the same way as in Kelly clamp or CUSA methods. The liver is gradually opened up at the transection plane. Intermittent stopping and suction are applied to clear the operative field of water build up and for assessment of bleeding. This continues until the transection is complete.

The advantages are its precision and ability to preserve blood vessels and bile ducts and decrease blood loss. Disadvantages are the cost, and the technique is slower than the other techniques.

14.2 Techniques that Divide the Liver Parenchyma and Seal Off the Vascular and Biliary Branches at the Same Time

14.2.1 Staplers

Surgical staplers work by dividing hepatic parenchyma between two staple lines. Surgical staplers are typically 30, 45 or 60 mm long. A stapler with a vascular load should be used.

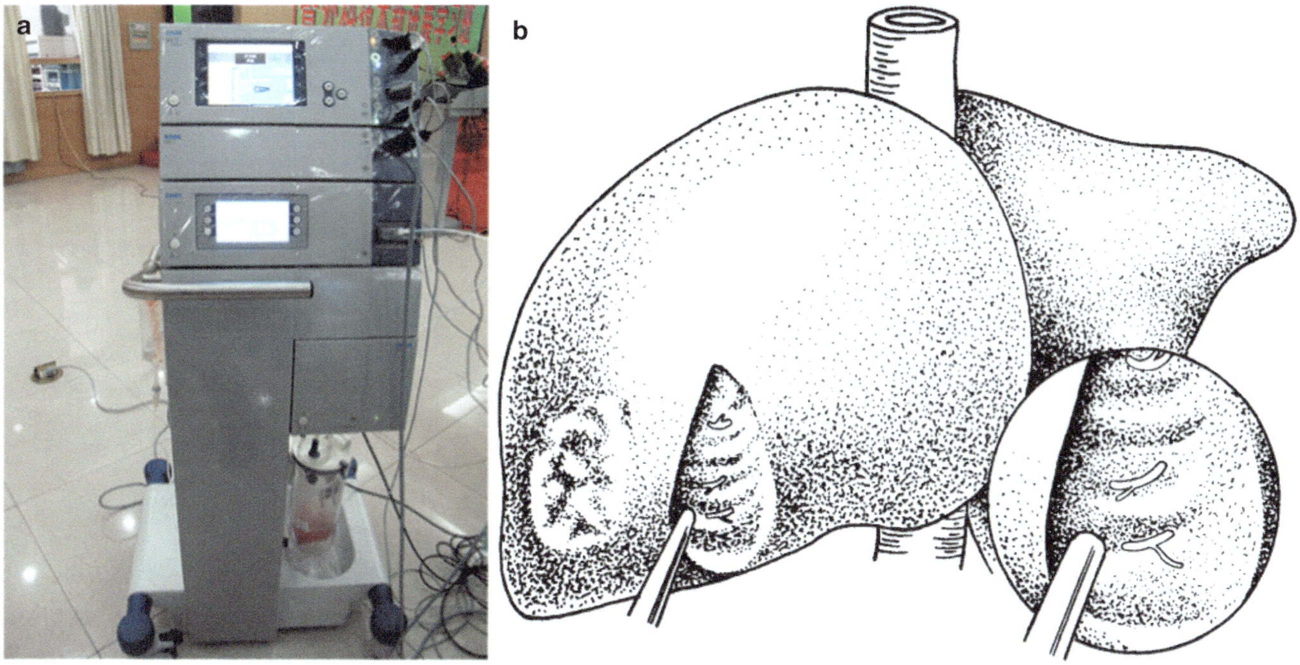

Fig. 14.7 Hydrojet. (**a**) The instrument. (**b**) Using hydrojet to dissect the hepatic parenchyma, exposing underlying blood vessels (inset), which are then isolated and divided

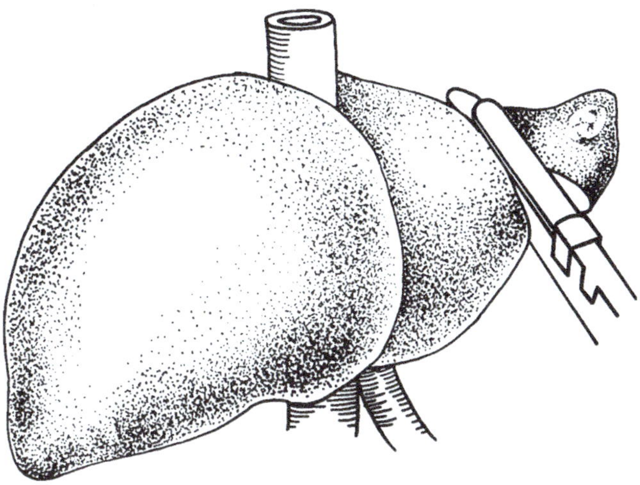

Fig. 14.8 Left lateral sectionectomy using surgical staplers with a vascular load to divide the hepatic parenchyma

The technique starts with scoring the liver capsule along the plane of live transection. The jaws of the stapler are then opened and closed to crush the liver parenchyma. The remaining biliary and vascular structures are then closed by deploying the staplers. This is repeated until the transection is complete (Fig. 14.8). The advantage is the speed. The disadvantages are the cost and an increased risk of bile leakage.

14.2.2 Harmonic Scalpel or Scissors

Harmonic scalpel or scissors, also called ultrasonic scalpel (harmonic scalpel, Ultracision, Ethicon Endo-Surgery), has three working parts: the generator, the energy transmission cable with integrating handpiece, and the scalpel or scissors. It works through a controlled electric current transmitted via a transducer which converts the current to mechanical ultrasonic vibration. The vibration is then transmitted through the rod of the device to the tip, the active blade. When the pistol grip of the device is activated, the active blade clamps against the opposing pad, compressing the target tissue and generating friction. The friction creates heat and subsequent coagulation of the target tissue (Fig. 14.9).

The technique begins by scoring the liver capsule using diathermy. The harmonic scalpel is then inserted in the liver parenchyma, and the tissue is coagulated and divided. Larger vessels or bile ducts have to be controlled with metallic clips or suture ligation. This is repeated until the liver is transected.

The advantages include minimisation of blood loss and its potential use in laparoscopic liver resection. The disadvantages are an increased risk of biliary fistula, the increased costs and decreased availability.

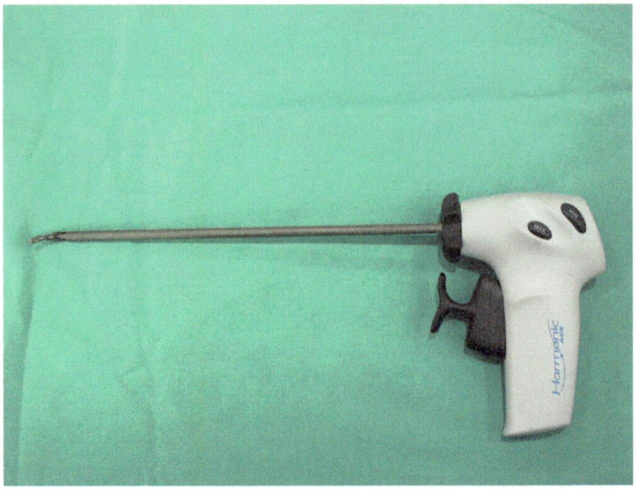

Fig. 14.9 Harmonic scalpel or scissors

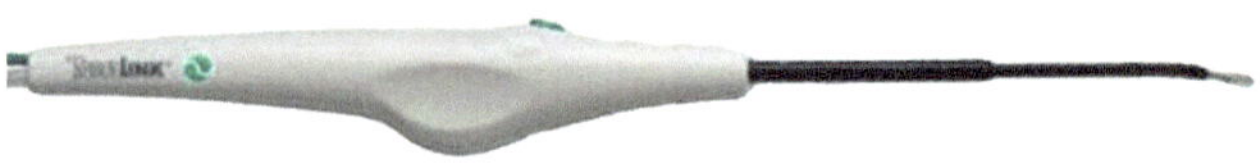

Fig. 14.10 TissuelLink

14.2.3 TissuelLink

The dissecting sealer uses radiofrequency energy from a standard electrosurgical generator delivered to tissue through a conductive fluid, e.g. saline. The saline becomes the electrode and couples the radiofrequency energy at the top of the device to the tissue, increasing the contact area and keeping the cut surface cool (below 100 °C). This shrinks cellular collagen and seals small vessels and bile ducts (Fig. 14.10). This procedure can be done with or without the use of Pringle's manoeuvre. However, larger vessels and biliary structures need to be clipped or suture-ligated and divided.

The main advantage is decrease in blood loss. There are several disadvantages: the cost, the decreased availability and the low speed in liver transection.

14.2.4 LigaSure

This device (Valleylab, Boulder, CO) is a bipolar vessel-sealing device connected to a unique power generator with a feedback control response system. A combination of pressure and energy delivered to tissue through the jaws of the device create a seal by melting the collagen and elastin in vessel walls, reforming it to a permanent seal (Fig. 14.11). It can be used for vessels up to 7 mm in diameter.

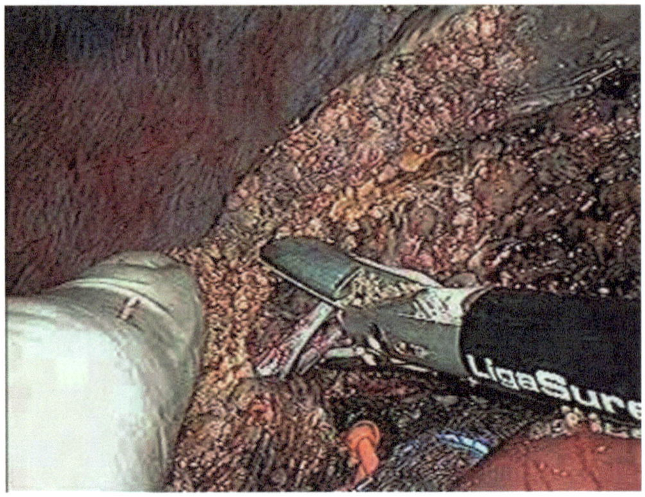

Fig. 14.11 LigaSure

The technique starts by using an electrocautery to score the liver capsule. The blades of the device are inserted into the liver substances, and the enclosed tissue is crushed between them several times, leaving vessels and bile ducts behind. The structures are then grasped, and power is applied. The jaws are released, and the coagulated blood vessels and bile ducts are divided with scissors. These steps are repeated until liver transection is complete.

The advantages are decreased blood loss, less suture ties and faster operation. The disadvantages are increased cost and difficulty in cirrhotic liver resection.

14.3 Techniques that Coagulates the Liver Parenchymal Tissues Before Liver Transection

14.3.1 Microwave Tissue Coagulator

Tabuse in 1979 first reported the use of microwave tissue coagulator to coagulate the liver parenchymal transection plane before the liver was transected with a scalpel. This resulted in decreased bleeding during liver transection. The technique requires multiple punctures of the microwave tissue coagulator into the transection plane. After the transection plane has been coagulated, the liver is divided using a scalpel with very little blood loss from the raw surfaces of the liver (Fig. 14.12).

Our study showed the thick area of necrosis at the transection plane resulted in a higher incidence of biliary fistula and right pleural effusion, although this technique is effective in decreasing blood loss during liver parenchymal transection.

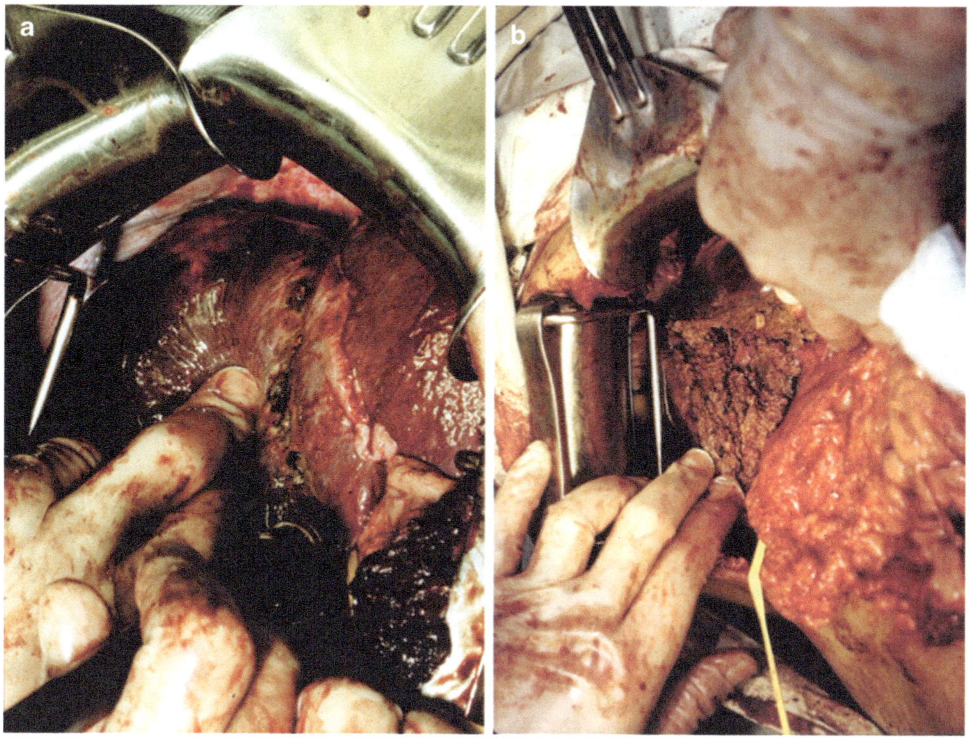

Fig. 14.12 (**a**) Liver transection plane coagulated with microwave tissue coagulation. (**b**) No bleeding from the raw surface of the liver after cutting with a scalpel and releasing the liver blood inflow/outflow

14.3.2 Radiofrequency Coagulation

Radiofrequency thermal ablation works by converting radiofrequency waves to heat. There is the cooled tip single-probe device. The new generation Habib® 4× (Emision Ltd., UK) is an instrument with an array of four electrodes in a square arrangement (Fig. 14.13a). The four needles are arranged in a 2 × 2 array with two pairs of needles (Fig. 14.13b). Two versions are available: one for open surgery and a smaller device for laparoscopic surgery. There is a long version with long electrodes of 120 mm and a short version with short electrodes of 60 mm. The electrodes are made of stainless steel covered with a nonstick coating.

Non-anatomical liver resection using the cooled tip single-probe device involves five steps (Fig. 14.14):

- Step 1: mark on liver surface the margin of the tumour (inner line).
- Step 2: mark on liver capsule 2 cm outside of inner mark.
- Step 3: use the cooled tip single radiofrequency probe and an RF generator to coagulate the plane of liver transection.
- Step 4: the number of probe application at each site that is required to obtain a zone of necrosis is related to the depth of the liver parenchyma that needs to be resected.
- Step 5: the liver parenchyma is divided with a scalpel.

Liver transection using the Habib 4× (Fig. 14.15).

- Step 1: mark on the liver surface the margin of the tumour.
- Step 2: mark on the liver capsule 2 cm away from the liver contralateral to the side of the tumour.
- Step 3: the Habib 4× is applied along the outer line.
- Step 4: at each site, the probe needs to be applied at various depths of the liver parenchyma to obtain a plane of coagulation.
- Step 5: the final division of liver parenchyma is with a surgical scalpel between the two rows of coagulated tissue.

The main advantage of this technique is the decreased blood loss and the faster liver transection. The disadvantages include the cost, decreased availability and imprecise transection line when near to vital structures.

14.4 Liver Parenchymal Transection Technique: Choice and the Basic Principle

A very often question asked is: of the many liver parenchymal transection techniques that are currently available, which is the best technique?

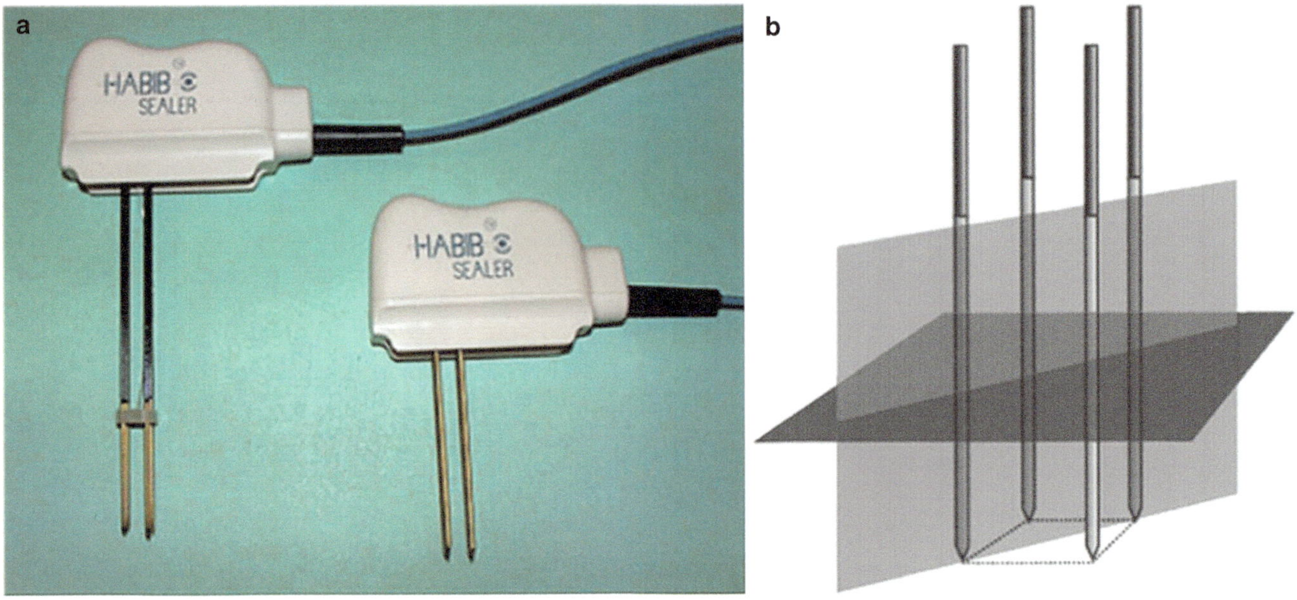

Fig. 14.13 Radiofrequency coagulator: the Habib® 4X

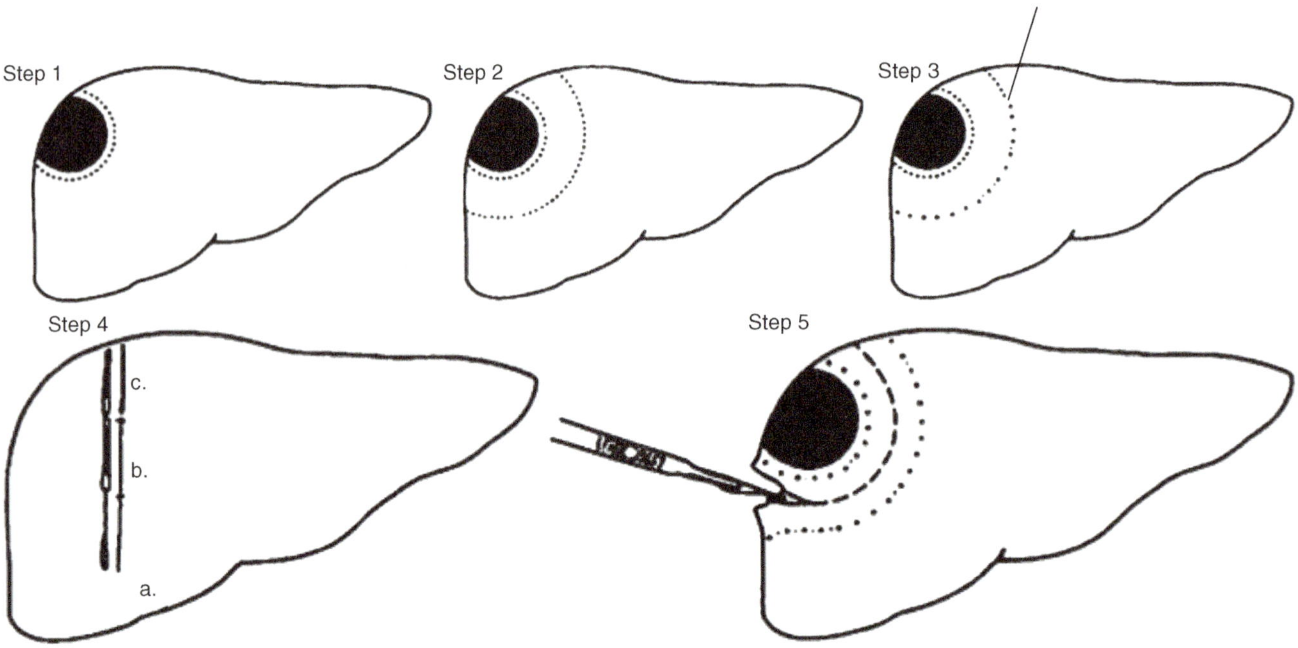

Fig. 14.14 Five steps to achieve liver resection using the radiofrequency energy single-probe device

The answer to this question is simple, the one technique that works best in your hand. Different techniques work differently under different surgeon's hands, and one should choose the technique that he is most comfortable with.

It is important to realise the proper application of the basic principle in liver transection is more important than the instrument that is required to carry it out. In 1966, Couinaud wrote, 'Finger fracture should be a gentle technique and must be respectful of anatomy; it should be carried exactly along the fissures and lead towards well-determined pedicles. It affords a quick opening of the fissures and the control of secondary vessels which usually bleed in the plane of section'. This statement is true even after 40 years since it was first published, despite all the modern technologies and new devices that are available for liver transection.

14.5 Liver Transection Techniques Using Liver Clamps and Tourniquets

14.5.1 Liver Clamps

The use of liver clamps has been reported to reduce intraoperative blood loss during liver resection. Nakayama is probably responsible for the first liver clamp designed especially for liver resection. Various clamps have been designed and applied with success in liver resection, including clamps by Stucke (1961), Storm and Longmire (1971), Lin (1973), Kanematsu et al. (1974) and de Souza (1979) (Fig. 14.16).

Liver resection varying from simple wedge resection to extended right hepatectomy has been carried out using clamps (Figs. 14.17 and 14.18), and the clamps used on most patients is that designed by Lin.

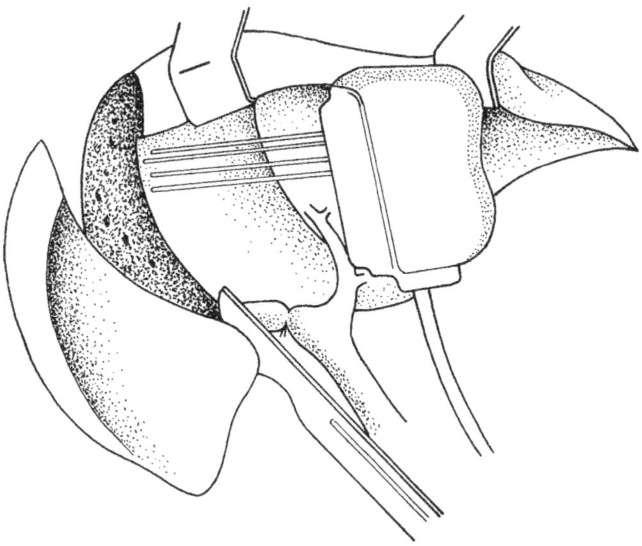

Fig. 14.15 Liver transection using the Habib® 4X radiofrequency coagulator

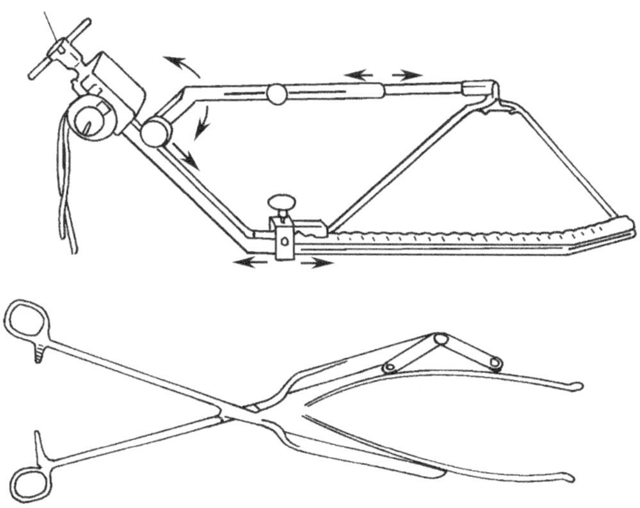

Fig. 14.16 Different types of liver clamps

14.5.2 Liver Tourniquets

Based on the same principle of the liver clamp is the use of tourniquets around the liver made with nylon Velcro reported

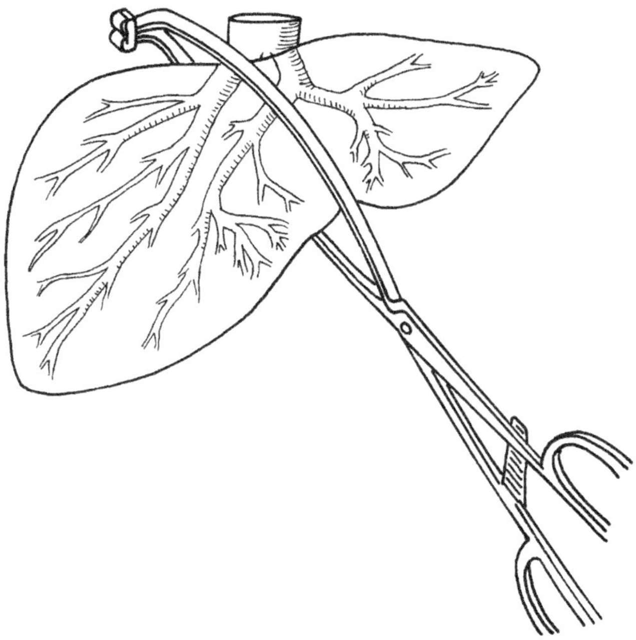

Fig. 14.17 Application of a liver clamp in major liver resection

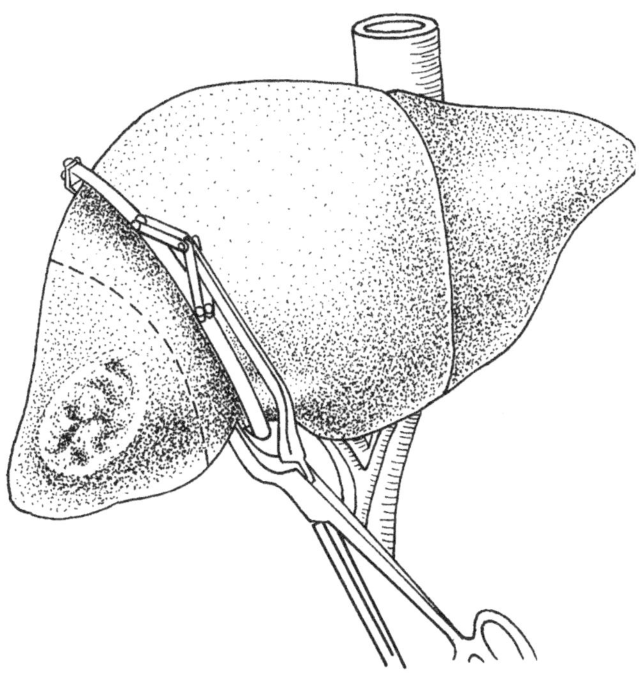

Fig. 14.18 Parenchymal compression using a Longmire clamp prior to a peripheral wedge resection

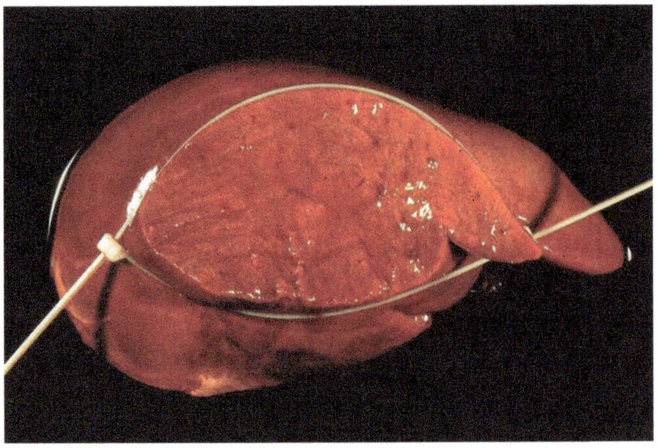

Fig. 14.19 Nylon Velcro tourniquet

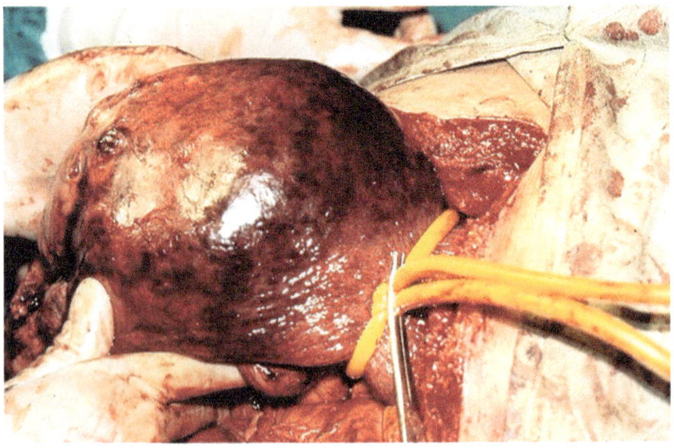

Fig. 14.20 Latex Tube Tourniquet

by Ma Xin 1981 (Fig. 14.19), or with latex tube reported by Li A in 1989 (Fig. 14.20).

All these liver clamps and tourniquets suffer from the following defects:

(a) They may slip.
(b) They may crush and lacerate the liver at the clamp/tourniquet site; and.
(c) Traction of them during liver transection may tear the small hepatic veins that drain from the liver directly into the inferior vena cava, resulting in torrential bleeding.

These liver clamps and tourniquets are now obsolete.

14.6 Haemostasis on the Raw Surface of the Remnant Liver

After the liver has been transected, adequate haemostasis on the raw surface of the remnant liver needs to be carried out.

14.6.1 History of Haemostasis of the Liver Cut Surface

Suturing of the liver to control haemorrhage was first recommended by Postemski in 1985. Von Eiseberg, Keen and Kousnetzoff and Pensky favoured individual ligation of bleeding vessels on the cut surface; however, massive bleeding in the old days made this technically difficult. The contributions of Kousnetzoff and Pensky, and of Auvray established the practice of using a blunt-ended needle with a double thread to place a through-and-through mattress suture across the entire liver above the resection line. As sutures pulled into the soft liver tissue, a series of guards was used (Fig. 14.21). Ceccherelli and Bianchi used whalebone (1894), and Payr and Martina used plates of magnesium plates (1905) to hold the liver sutures. Beck advocated plates of decalcified bone or abdominal fascia in 1902. Stem used cartilage from the scapula of a calf in 1905. Cautery was often used, but most surgeons found that a hot knife was suitable only for small vessels. A series of innovative methods was developed, but these methods have become obsolete: steam, hot air, liquid air and compressive forceps. Tumours on stalks were brought out of the belly and held out with hair pins, knitting needles and other convenient devices. Bulky removable ligatures and packs were used extensively to allow delayed sloughs and peritoneal exclusion of the tumour and its stalk.

Most of these historical techniques are obsolete.

14.6.2 Modern Techniques of Haemostasis on Raw Liver Surface

1. Raw Liver Surfaces Not Sutured Together

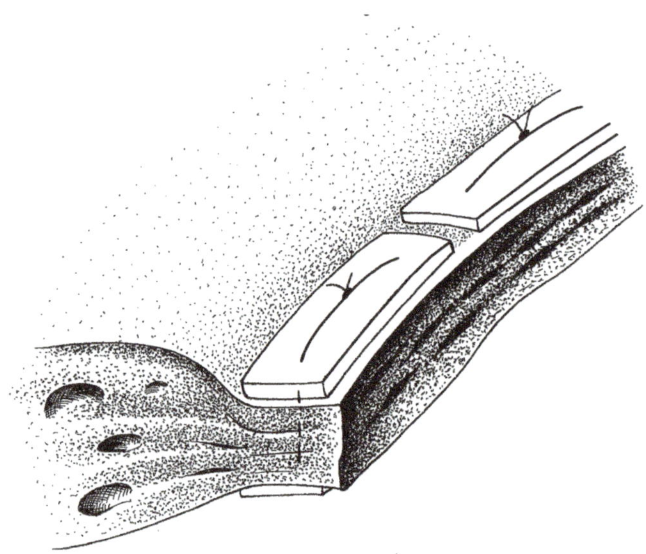

Fig. 14.21 Suturing the liver raw area using guards

Keen in 1899 advocated the plication of bleeding points on the raw surface of the liver, which is now the preferred method used for haemostasis. In modern liver surgery, most large bleeding points can be controlled using a figure-of-8 stitch around the bleeder (Fig. 14.22). In patients with a large venous opening, a deep purse-string stitch around the bleeding point may be necessary before the bleeder can be controlled (Fig. 14.23). Projecting ends of bleeding vessels can be controlled with ligaclips. The small bleeding points can be controlled with the use of diathermy. Diffuse oozing from the raw area can be controlled with either the Argon Beam Coagulator or the spraying of tissue glue. After haemostasis and checking for bile leak, the raw surface is usually left unsutured (Fig. 14.24). A pedicle omental flap may be used to cover the raw surface to decrease the chance of bile leakage or the raw area sprayed with fibrin glue (Fig. 14.25).

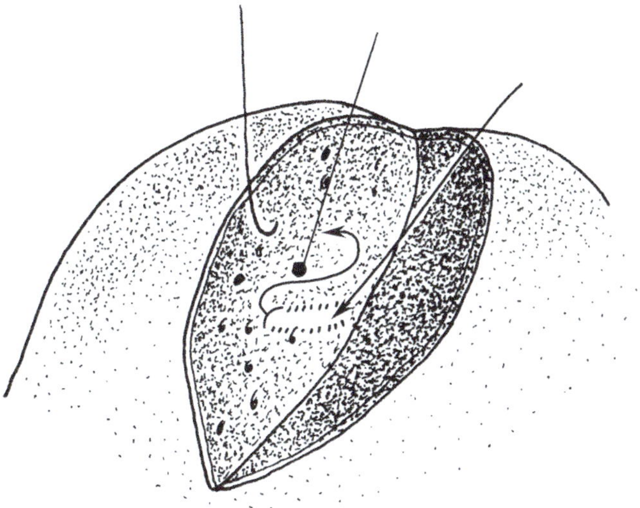

Fig. 14.22 A figure-of-8 stitch to control a bleeding vessel

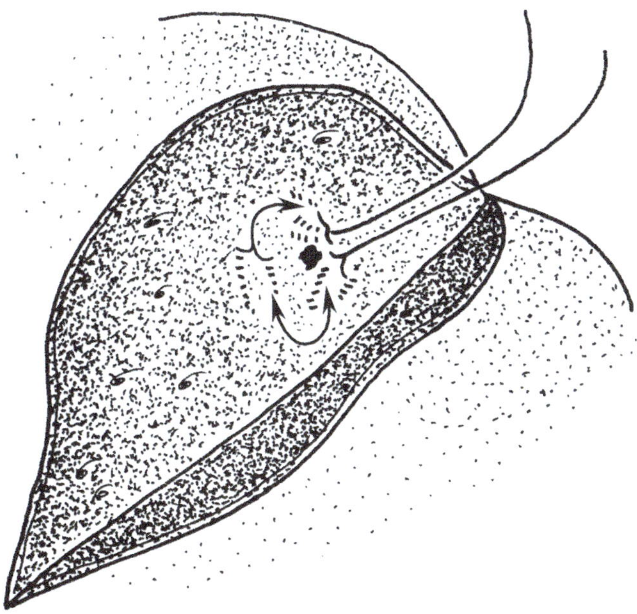

Fig. 14.23 A deep purse-string stitch to control a large venous opening

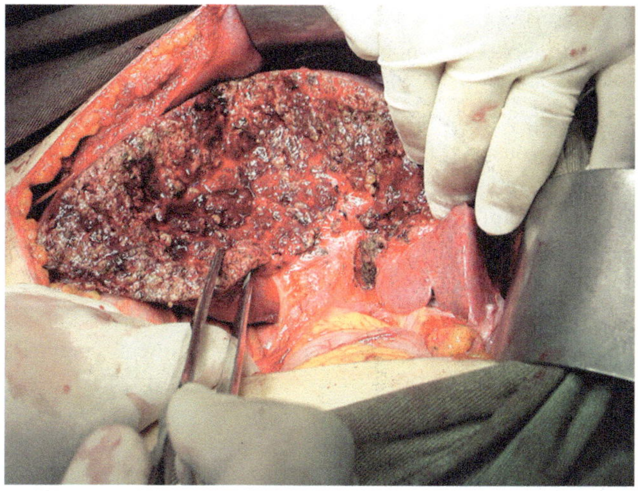

Fig. 14.24 Raw area left after resection

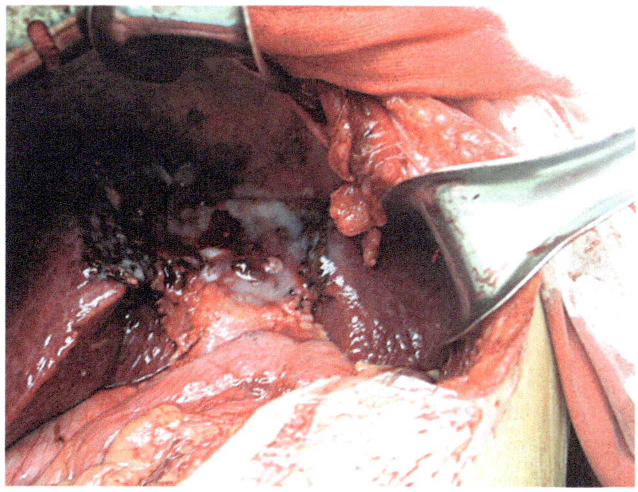

Fig. 14.25 Tissue glue used on raw surface of the liver

2. Suturing the Liver Raw Surface Together
   Under exceptional circumstances, when the patient develops bleeding tendency due to coagulopathy, the measures in (a) is not adequate to deal with the oozing from the raw surface of the liver after liver resection. If the raw area is small, it can be sutured together (Fig. 14.26). If the raw area is large, attempts to suture the liver raw surface with continuous suturing usually ends up with laceration of the suture/liver interface, resulting in more bleeding. There are two ways to deal with this scenario:

(a) Suturing the raw surface of the liver over a 'guard' of Surgicel (Ethicon, New Jersey) with horizontal mattress stitches (Fig. 14.27).

(b) Using a blunt needle with a double suture either using the Kousnetzoff and Pensky technique (Fig. 14.28) or the Wendel technique (Fig. 14.29).

If a blunt needle with a double thread is not available, the chain ligature technique of Auvray can be used (Fig. 14.30).

(c) After a small wedge liver resection, the liver can be sutured together using one of the following methods (Fig. 14.31).

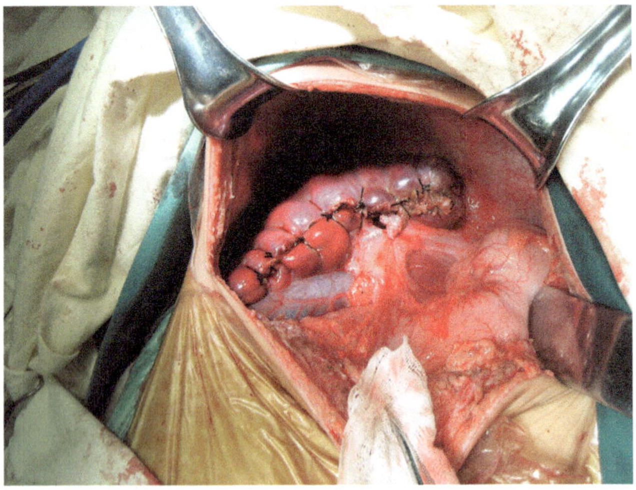

Fig. 14.26 Liver raw area is sutured together only if haemostasis is difficult and the transected raw area is small

(d) Traumatic blunt injury.

The technique in haemostasis in a rupture liver as a result of blunt injury is slightly different from that in the situation of elective liver surgery because in elective liver surgery, the condition is under control, and there should not be a lot of bleeding points in the raw area on the release of the inflow and/or outflow clamps.

The best way to identify large bleeders from the surfaces of the lacerated liver is to use two hands to compress on the liver to slow down the bleeding (Fig. 14.32). Once the large bleeding points are identified, they can be dealt with by ligation, clipping or suture ligation. One should avoid dissecting into the liver and further splitting the liver to make the laceration worse. Once the major bleeders have been dealt with, superficial liver laceration can be sutured together using either interrupted stitches or a horizontal mattress stitch to approximate the raw areas together (Fig. 14.33).

Under the situation when the laceration is deep, and the edges of the laceration are rugged and do not take stitches well, the use of a horizontal mattress suturing over a Surgicel guard on both edges of the laceration helps to prevent the sutures from cutting through the edges of the liver laceration (Fig. 14.34).

The alternative is to use interlocking stitches to reinforce the edges of the laceration before simple sutures are applied to approximate the lacerated edges together (Fig. 14.35).

Another method is to use the Robinson and Butcher suturing technique to suture the edges of the laceration together (Fig. 14.36).

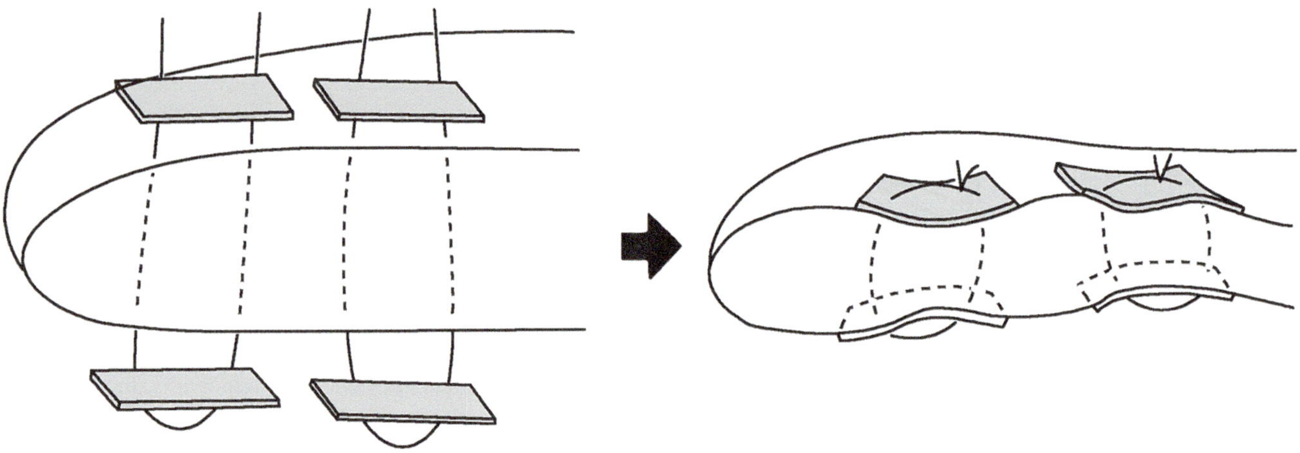

Fig. 14.27 Suturing the liver raw area over a 'guard' of Surgicel with horizontal mattress stitches

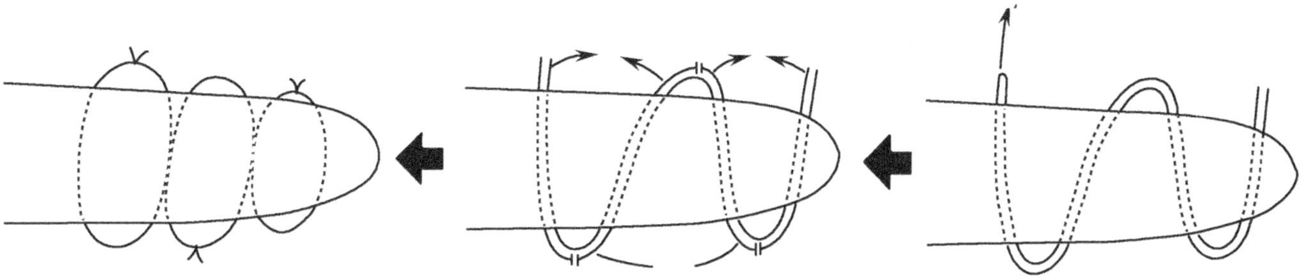

Fig. 14.28 Kousnetzoff and Pensky technique

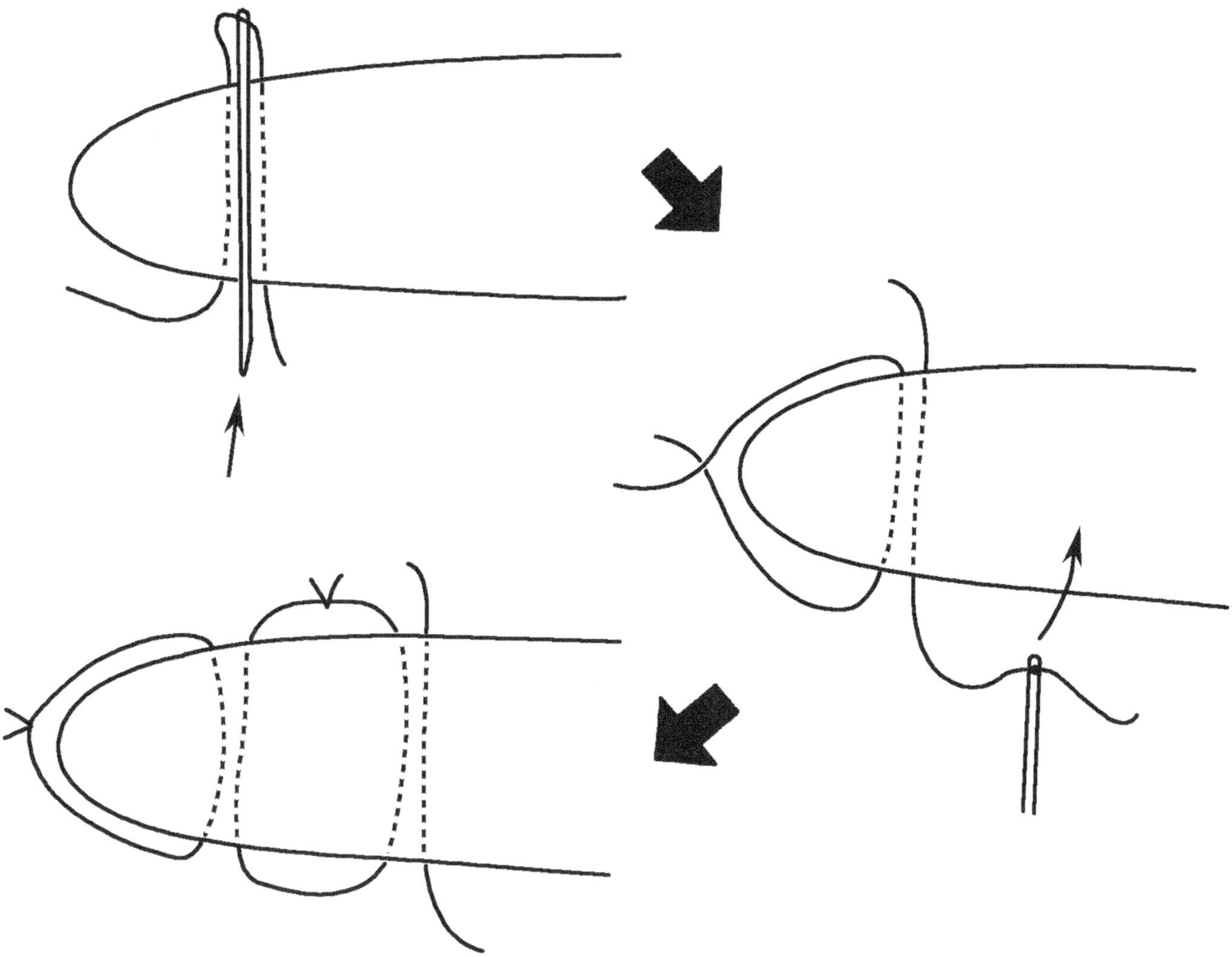

Fig. 14.29 Wendel technique

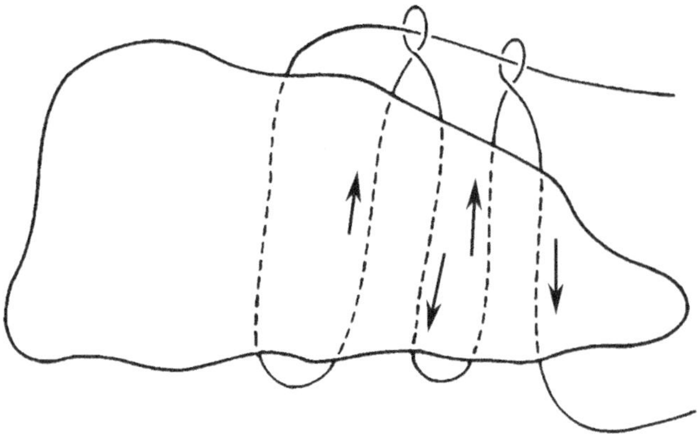

Fig. 14.30 Auvray chain ligature

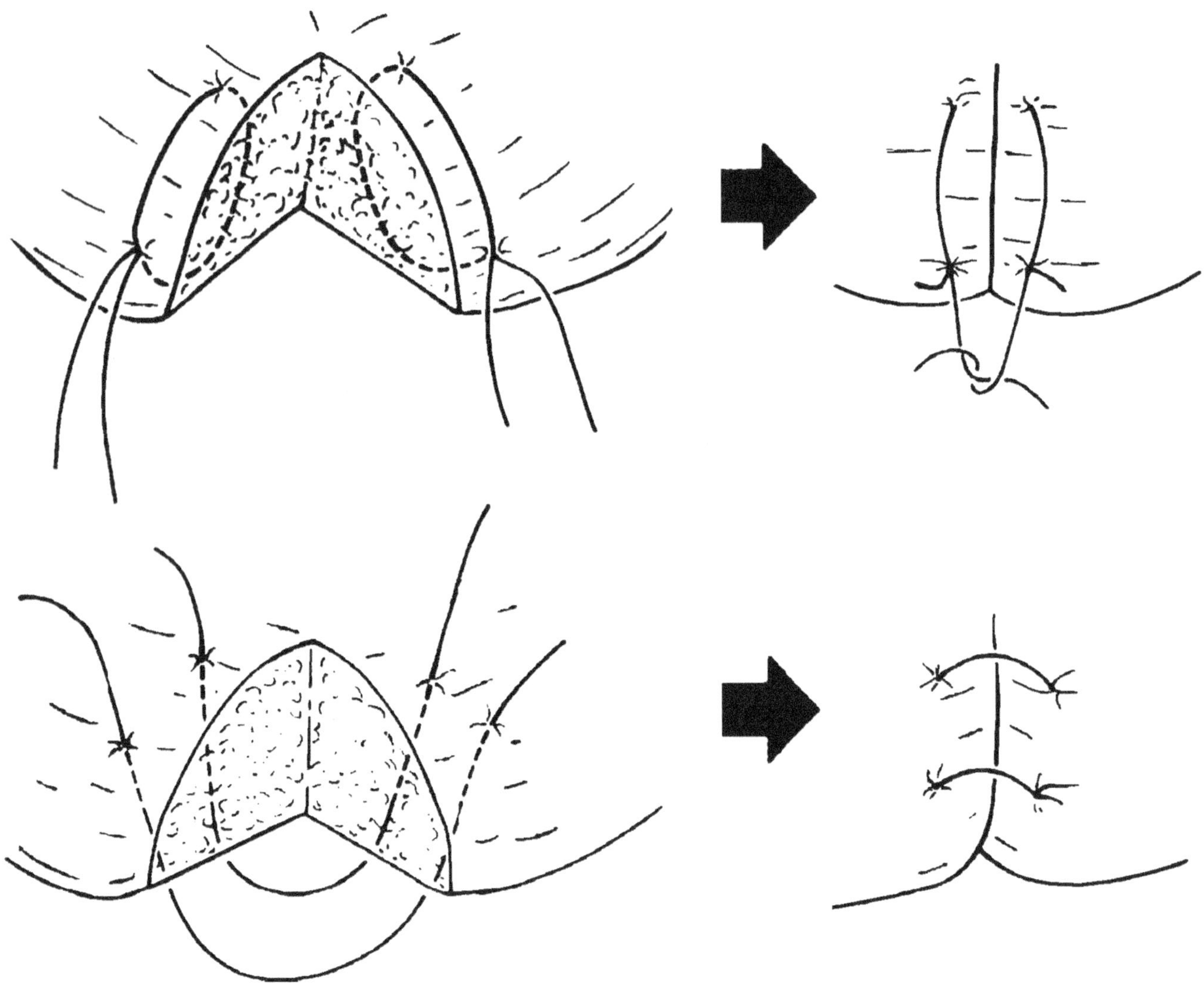

Fig. 14.31 Suturing techniques after Wedge resection of the liver

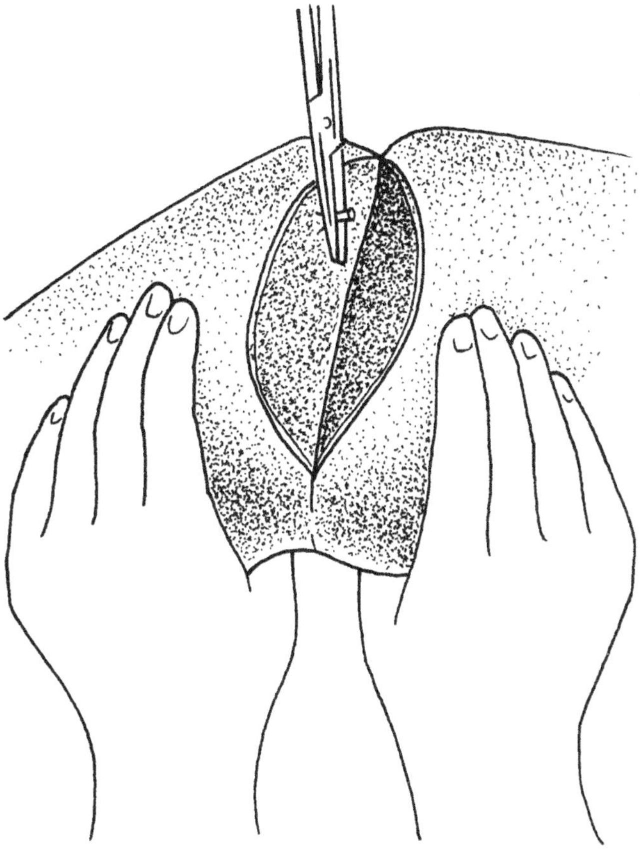

Fig. 14.32 Using both hands to compress on the edges of the lacerated liver to decrease bleeding

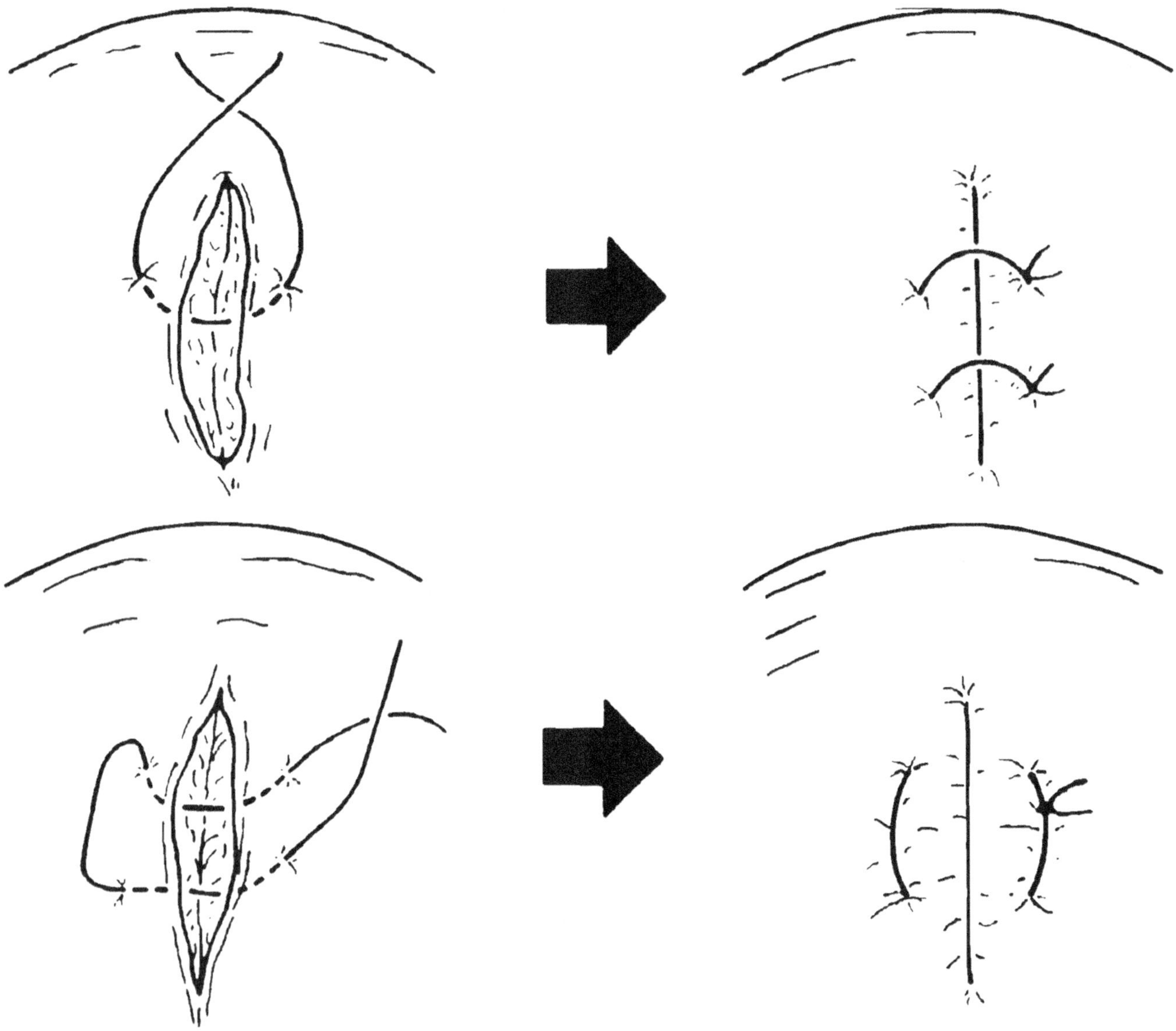

Fig. 14.33 Suturing liver lacerations using. (**a**) Simple stitches. (**b**) Horizontal mattress sutures

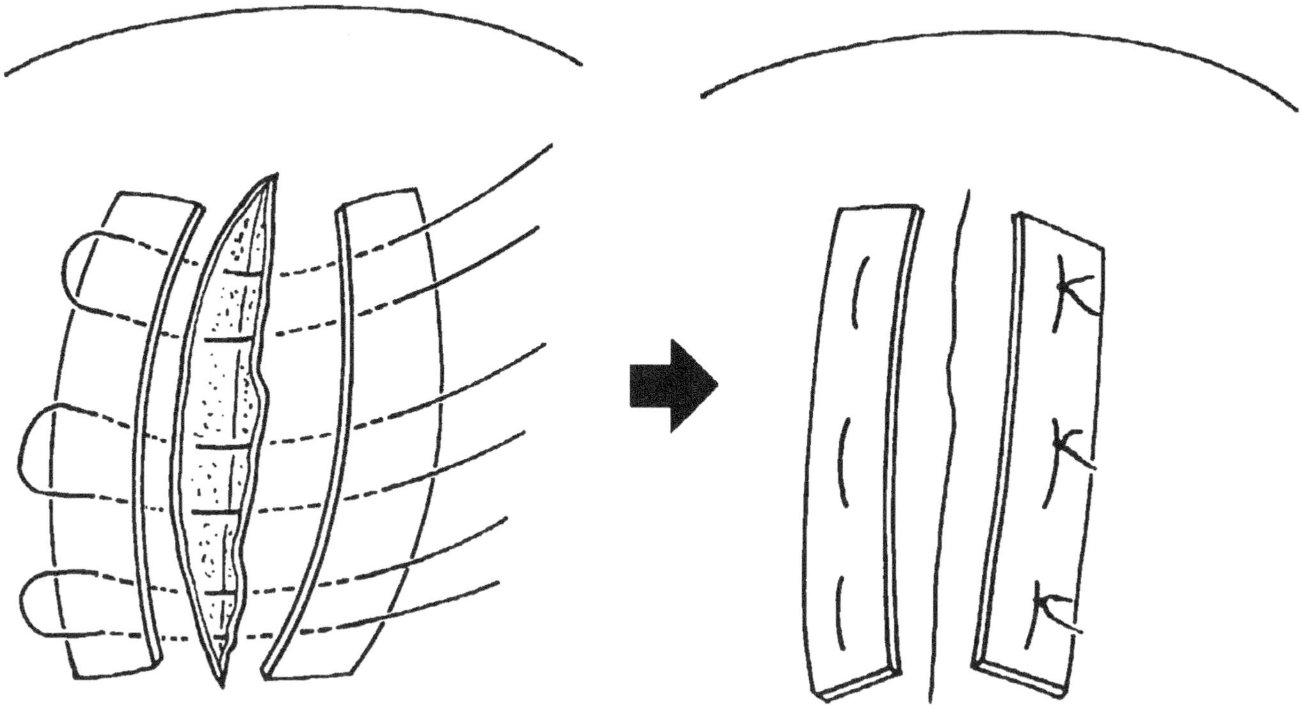

Fig. 14.34 In difficult cases of lacerated liver, suturing of laceration over a guard of Surgicel

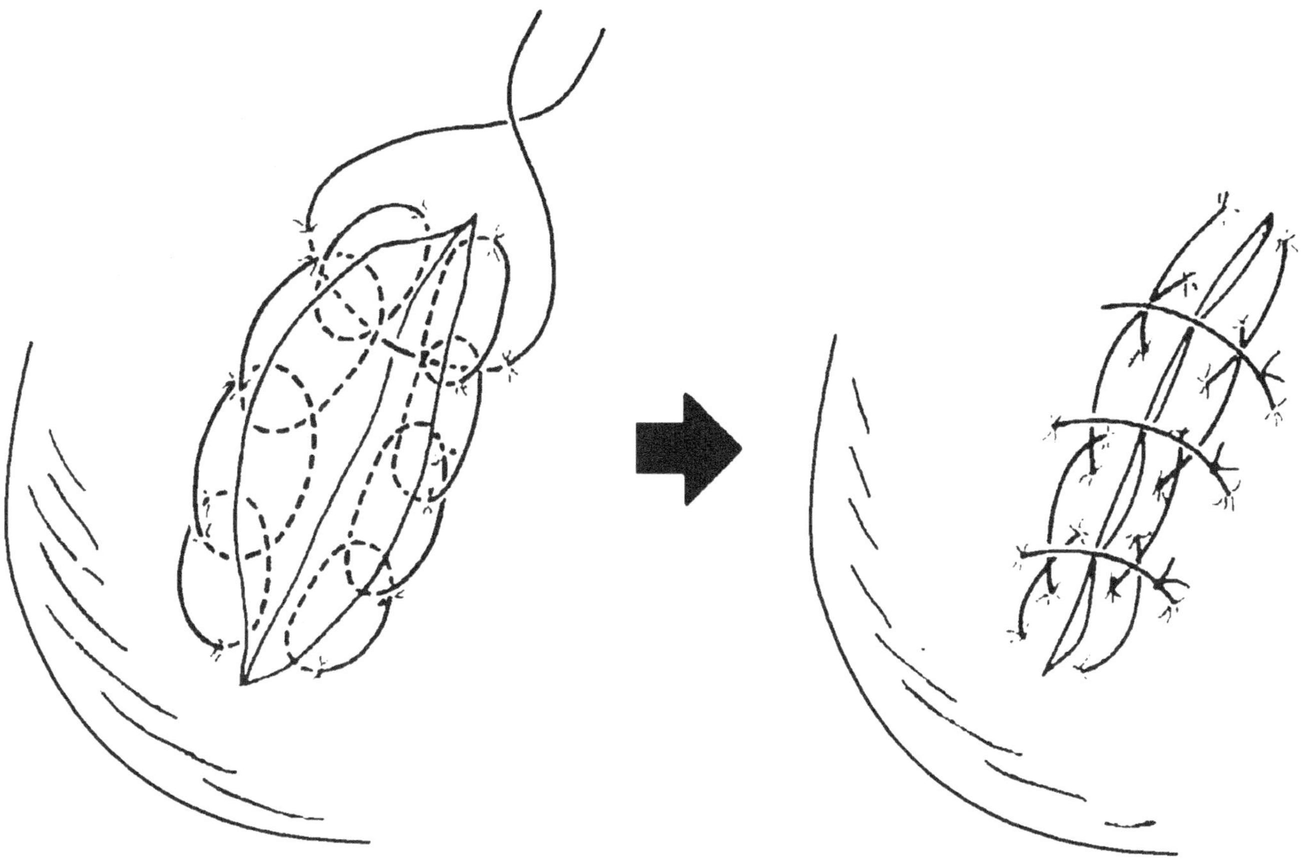

Fig. 14.35 Interlocking stitches to reinforce the laceration edge followed by simple sutures to approximate the lacerated edges

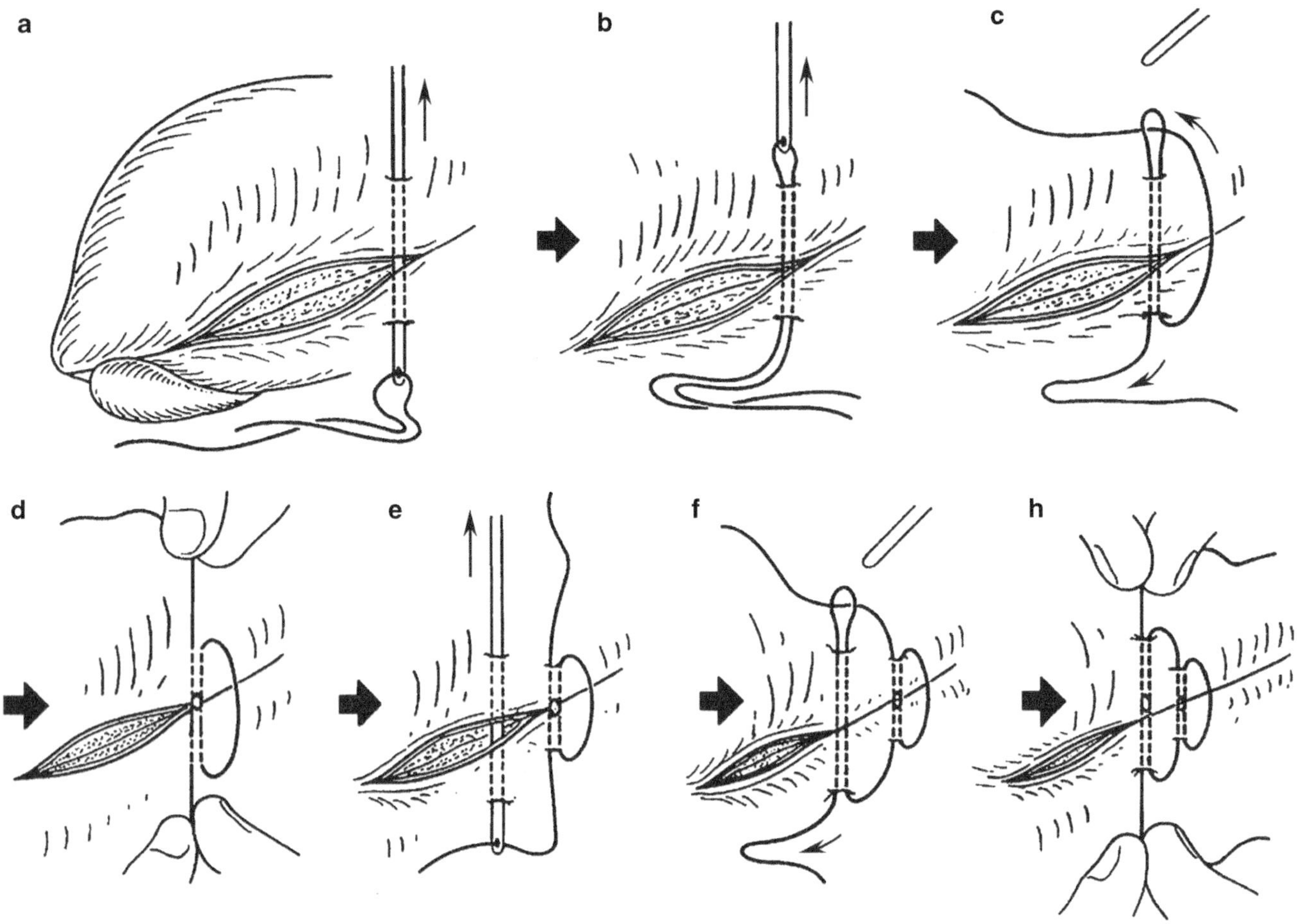

Fig. 14.36 Robinson and Butcher suturing technique

Further Reading

Castaldo ET, Pinson W. Chapter 24: Techniques of liver transection. In: Lau WY, editor. Hepatocellular carcinoma. Singapore: World Scientific; 2008. p. 529–49.

Jiao LR, Habib NA. Chapter 25: Radiofrequency-assisted liver resection. In: Lau WY, editor. Hepatocellular carcinoma. Singapore: World Scientific; 2008. p. 441–567.

Lau WY. The history of liver surgery. Roy Coll Surg Edinb. 1997;42:303–9.

Xia F, Lau WY, Mak S, Bie P. Modified conventional clamp-crushing technique in liver parenchymal transection. Hepatobiliary Pancreat Dis Int. 2012;11:442–5.

15 Different Approaches to Liver Resection

With better understanding of surgical anatomy, physiology and regeneration of the liver, and with improvements in technical developments in liver resection, alternative methods of approaches to liver resection evolved.

To make things simple, there are only five steps in liver resection before the resected specimen can be removed from the patient, and they are mobilisation of the liver by division of ligaments, interruption of the vasculo-biliary inflow, liver parenchymal transection, interruption of the venous outflow from the short hepatic veins and from the main hepatic veins. Obviously, adequate haemostasis is required before the abdomen is closed. The alternative approaches in liver resection involve the combination of these five steps in different orders. To complicate matters, the interruption of the vasculo-biliary inflow can be at a different site and level from the actual ligation and division of the vasculo-biliary inflow branches (e.g. Pringle's manoeuvre at porta hepatis with intrahepatic division of the individual branches). Similarly, the outflow control can be at a site and level different from the division and ligation of the hepatic veins (e.g. extrahepatic control of the right hepatic vein with a vascular sling coupled with intrahepatic division and ligation of the vein). Furthermore, the alternative approaches also involve the division and ligation of the Glissonian pedicle with all its portal triads together, or the individual ligation and division of the hepatic arterial branch, the portal venous branch and the biliary branch separately.

15.1 Conventional Approach

This is the approach used by Lortart-Jacob and Robert for their first successful right hepatectomy reported in 1953.

15.1.1 Right Hepatectomy

This approach starts with mobilisation of the right liver by division of the falciform, coronary and right triangular ligaments. The right hepatic artery, the right portal vein and the right hepatic duct are ligated and divided extrahepatically (Fig. 15.1). The right liver is then dissected from the inferior vena cava. Short hepatic veins that drain from the right liver into the inferior vena cava are ligated and divided (Fig. 15.2a, b). The hepato-caval ligament is then divided between clamps and ligated. The right hepatic vein is dissected extrahepatically (see Sect. 13.6.1) divided and sutured (Fig. 15.2c). A line of demarcation is seen marking the devascularised right liver from the left liver (Fig. 15.3). Liver parenchymal transection can then be done, transecting the liver on the right border of the middle hepatic vein. Occasionally, the transection line of a right hepatectomy can be along the left border of the middle hepatic vein. Under this situation, care must be taken to preserve the segment 4 branch or the liver segment 4 would become congested (see Fig. 15.4).

The segment 8 and segment 5 branches of the middle hepatic vein need to be ligated intrahepatically and divided.

Blood loss during parenchymal transection can be decreased by the use of Pringle's manoeuvre and a low central venous pressure (see Chap. 12). Instead of using low central venous pressure, the trunk of the middle and the left hepatic veins can also be controlled with a clamp extrahepati-

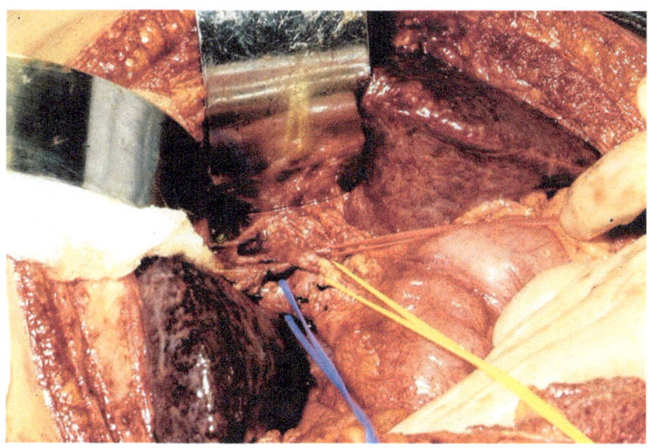

Fig. 15.1 Dissection and slinging of the right hepatic artery (red sling), right portal vein (blue sling) and right hepatic duct (yellow sling)

W. Y. Lau, *Applied Anatomy in Liver Resection and Liver Transplantation*, https://doi.org/10.1007/978-981-16-0800-1_15

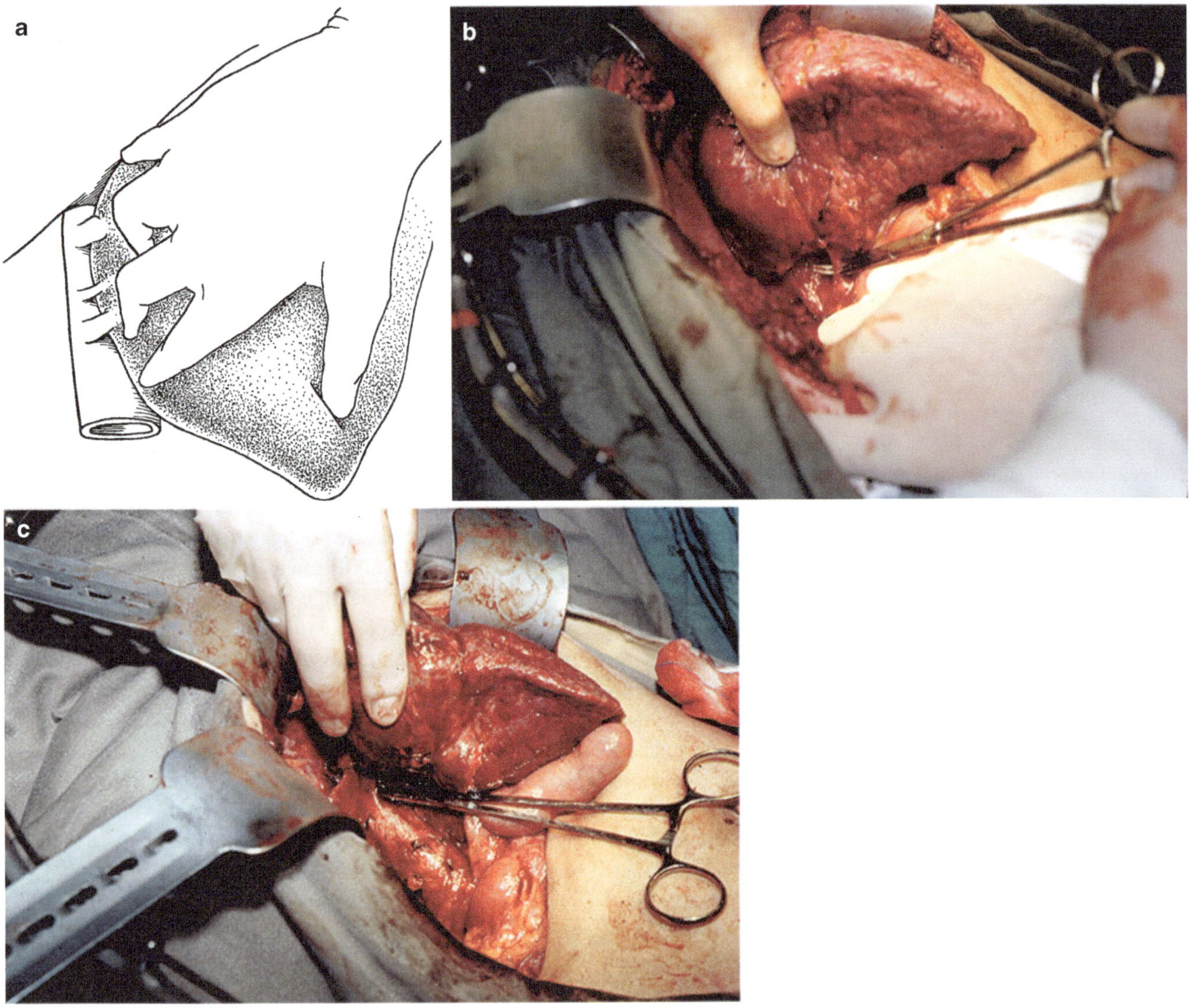

Fig. 15.2 (a) Short hepatic veins and right hepatic vein from the right liver draining into the inferior vena cava. (b) Isolation of a short hepatic vein prior to ligation and division. (c) Isolation of the right hepatic vein before ligation and division

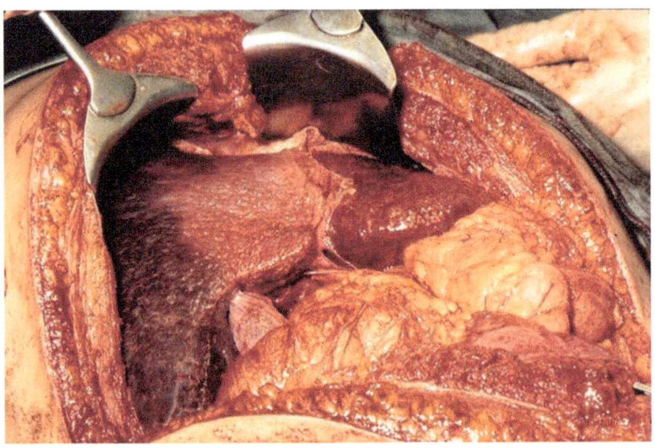

Fig. 15.3 Line of demarcation between right and left livers after the division of blood inflow and outflow to right liver

cally to prevent blood loss during parenchymal transection (selective hepatic vascular exclusion). The dissection of the trunk of the middle and left hepatic veins can be facilitated by the division of the ligamentum venosum near to its termination at the trunk of the hepatic veins (see Sect. 13.6.1). Once the ligamentum venosum is divided, the trunk can be dissected and looped with a vascular sling (see Sect. 6.10).

15.1.2 Left Hepatectomy

The procedure is similar to right hepatectomy with mobilisation of the left liver by division of the falciform and left triangular ligaments. Extrahepatic branch of the hepatic artery, left portal vein and left hepatic duct are divided and ligated extrahepatically. The trunk of the middle and left hepatic veins are isolated and slung. Parenchymal transection is along the plane demarcated by the ischaemic left liver along a plane on the left side of the middle hepatic vein. The left hepatic vein is ligated intrahepatically. Again, blood loss can be reduced by using Pringle's manoeuvre plus either low central venous pressure or selective hepatic vascular exclusion by clamping the right hepatic vein as well. Please note that there is no short hepatic vein that drains from the left liver (segments 2, 3, 4) to the inferior vena cava.

15.1.3 Extended Right Hepatectomy (Right Trisectionectomy)

This is similar to carrying out right hepatectomy with the exception that the hepatic arterial, portal venous and bile ducts branches to segment 4 are also divided extrahepatically, and the right and middle hepatic veins are divided, leaving behind the portal triad supplying the left lateral section and the left hepatic vein.

15.1.4 Extended Left Hepatectomy (Left Trisectionectomy)

This is similar to carrying out left hepatectomy, with the exception that the hepatic arterial, portal venous and bile duct to the right posterior section, and the right hepatic vein are preserved, and the additional resection of the right anterior section. If the right inferior hepatic vein is large (see Sect. 7.4) this vein should be preserved, or the venous drainage to segment 6 can be affected.

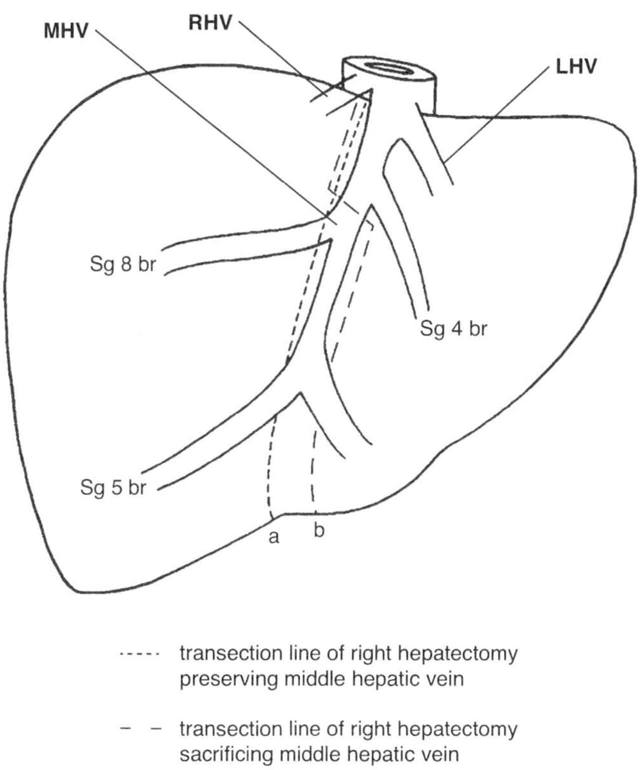

Fig. 15.4 Line of parenchymal transection for right hepatectomy

15.2 Parenchymal Transection with Early Intrahepatic Control of Glissonian Sheath

15.2.1 Anterior Approach

This approach starts off with liver parenchymal transection of the liver. This procedure is commonly accompanied by temporary control of the portal triad using Pringle's manoeuvre and low central venous pressure. Occasionally, liver parenchymal transection can proceed with minimal blood loss using some of the technical devices described in Chap. 14 without Pringle's manoeuvre. Vascular inflow control can soon follow parenchymal transection by early intrahepatic approach to, and control of, the hepatic pedicle structures which are contained within the Glissonian sheath (Fig. 15.5). Further parenchymal transection and intrahepatic control of vessels follow until the parenchymal transection is complete. This is then followed by ligation and division of the hepatic venous branches, which include one or two of the major hepatic veins and the short hepatic veins (Fig. 15.6). Mobilisation of the falciform, coronary and triangular ligaments completes the resection and removal of the specimen.

The beneficial effects of the anterior approach are there is less blood loss compared with the conventional approach, there is less chance of disseminating the tumour as a result of less manipulation on the tumour, there is less compromised blood flow to the remnant liver as a result of less kinking of the remnant liver during mobilisation, and there is less chance of rupturing the tumour when the tumour is large.

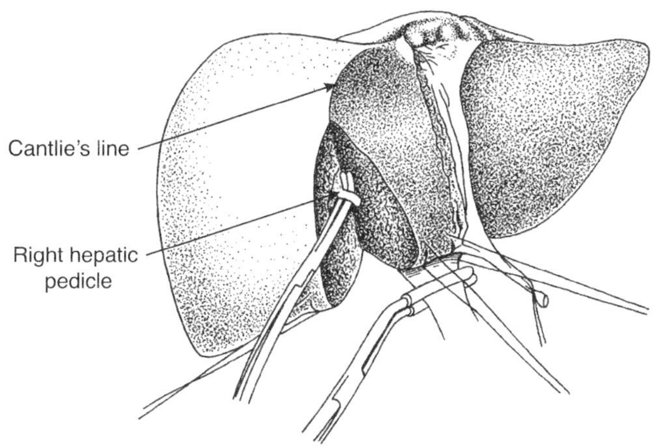

Fig. 15.5 Right hepatectomy using the anterior approach

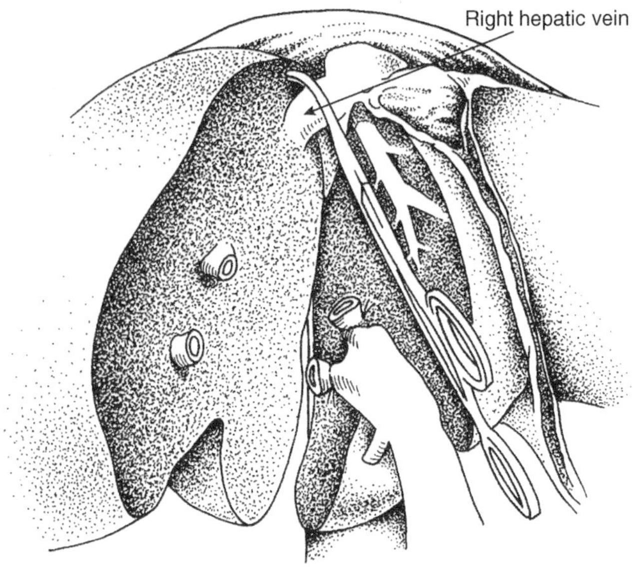

Fig. 15.6 Right hepatectomy using the anterior approach. Please note the completion of parenchymal transection before control of the venous structures (the right hepatic vein and the short hepatic veins) at the black of the liver

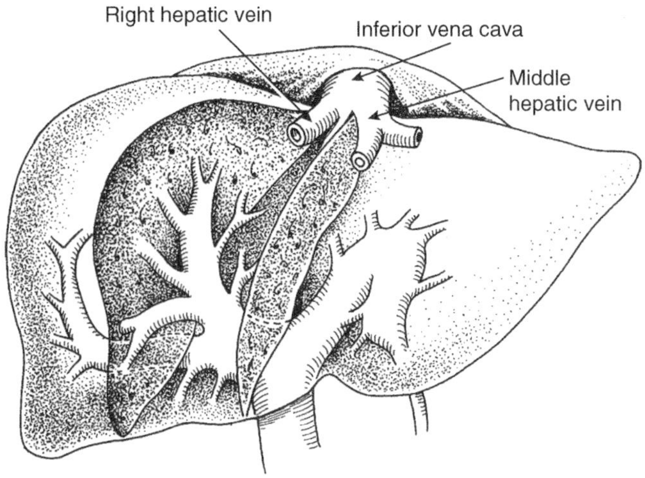

Fig. 15.7 Belghiti's liver hanging technique transecting the liver at the mid plane

There are some limitations to the anterior approach. Some patients with a large tumour compressing a major hepatic vein can develop venous collaterals. In these patients, parenchymal transection without prior mobilisation of the liver is associated with massive bleeding from the venous collaterals. In such situation, after inflow clamping, the outflow should be resumed by mobilisation of the liver with an upper traction on the liver to facilitate venous flow into the inferior vena cava.

Nagasue in 1985 advocated the additional clamping of the hepatic vein draining the part of the liver being resected, claiming there to be less blood loss during resection than when the Pringle's manoeuvre is used in isolation. However, this procedure requires early mobilisation of the liver, thus defeats the original purpose of the anterior approach: parenchymal transection to get early inflow occlusion, then further parenchymal transection, followed by outflow occlusion and mobilisation of the ligaments.

15.2.1.1 Belghiti's Liver Hanging Technique

Belghiti proposed a liver hanging technique to facilitate the anterior approach.

The idea is to put a sling between the back of the liver and the front of the inferior vena cava at the 'avascular tunnel' (see Sect. 3.3) right at the back of the midplane of the liver (Fig. 15.7).

Intraoperative ultrasound is used to confirm the absence of tumour infiltration and abnormal short hepatic veins at the 10–11 O'clock position of the anterior surface of the retrohepatic inferior vena cava. After opening the anterior leaf of the coronary ligament and the anterior part of the right triangular ligament (to expose the anterior and left sides of the right hepatic vein), the fossa located between the right hepatic vein and the middle hepatic vein is dissected 3–4 cm downwards with a right-angled dissector (Fig. 15.7). For caudal retrohepatic dissection, the caudal edge of the caudate lobe is lifted from the inferior vena cava (Fig. 15.8a), and small short hepatic veins are divided and ligated up to the level of the inferior right hepatic vein. A long, tight, curved aortic clamp is inserted behind the caudate lobe just to the left side of the inferior right hepatic vein and is passed cranially along the anterior surface of the inferior vena cava between the 10 and 11 O'clock positions, identifying the position of the clamp tip by ultrasonography. By successfully opening and closing the clamp, the clamp is directed towards the previously dissected space between the right hepatic vein and the middle hepatic vein, while the clamp tip reaches suprahepatically (Fig. 15.8b). A 10 mm-wide soft silicon multitubular drain is seized with the clamp and pulled down through the retrohepatic space (Fig. 15.8c). When the right hepatectomy includes the middle hepatic vein, the tap is switched from the right to the left side of the middle hepatic vein; this allows safer dissection of the middle hepatic vein near the vena cava confluence. The caudate lobe is divided to place the tap near the right portal pedicle. The tap pulls up the liver, thus facilitated the anterior approach by pulling the liver tissue away from the inferior vena cava and protecting it from injury during transection (see Fig. 15.8d).

The Belghiti's hanging technique has a success rate of 80–92%. Adhesions between the liver and the inferior vena cava and direct tumour involvement are the main causes of failure. In 4–6% of cases, major bleeding happens because of torn short hepatic veins.

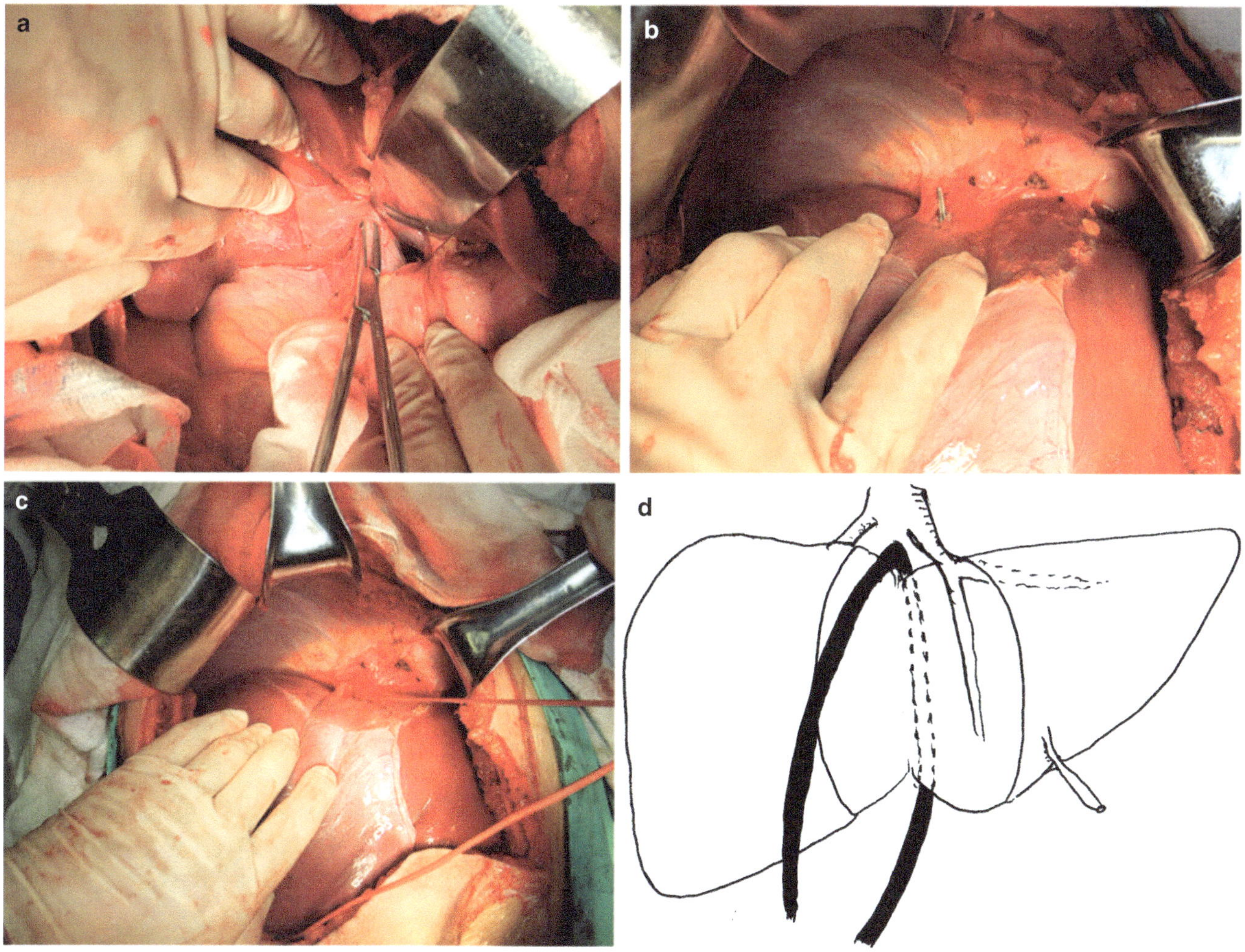

Fig. 15.8 Belghiti's liver hanging technique. (**a**) Dissecting retroperitoneal tunnel. (**b**) Forceps came out of previously dissected plane at fossa between right hepatic vein and middle hepatic vein. (**c**) Silicon sling pulled liver up. (**d**) Diaphragmatic representation

15.2.1.2 Chen's Double Liver Hanging Technique Through the Retrohepatic Avascular Tunnel

This technique involves dissecting a tunnel on the right side of the inferior vena cava. As this is the medial part of the bare area, there is no vessel that runs in this space.

The procedure starts with dividing the posterior peritoneum, 2–3 cm wide, on the right side of the inferior vena cava, just inferior to the liver. The right adrenal gland is exposed and protected. The operator uses his right index finger, dissects the space from below upwards between the hepatic parenchyma and the anterior and superior edges of the right adrenal, then along the right side of the inferior vena cava. The right coronary ligament is opened suprahepatically on the right side of the inferior vena cava for 2–3 cm. The operator uses his left index finger to dissect the retrohepatic space from above downwards along the right side of the inferior vena cava. The retrohepatic tunnel is built when the fingers touch each other. A clamp is used to place two tapes around the liver for liver suspension. One tape is pulled to the right and the other to the left. Liver transection is done along a plane to the right of the middle hepatic vein as determined by intraoperative ultrasonography for right hepatectomy (Fig. 15.9). The liver double-hanging manoeuvre provides better exposure of the operative fields and easier manipulation during liver parenchymal transection. When the liver parenchymal transection reaches the right hepatic vein, the origin of the right hepatic vein from the inferior vena cava is dissected, doubly ligated and divided. Liver transection is then carried out along the right border of the inferior vena cava, dividing the caudate process and ligating and dividing the short hepatic veins.

With this technique of right hepatectomy, the entire caudate lobe is preserved. While in Belghiti's liver hanging techniques, parts of the paracaval portion and the caudate process are resected. Thus, this technique represents a right hepatectomy which is anatomically more correct.

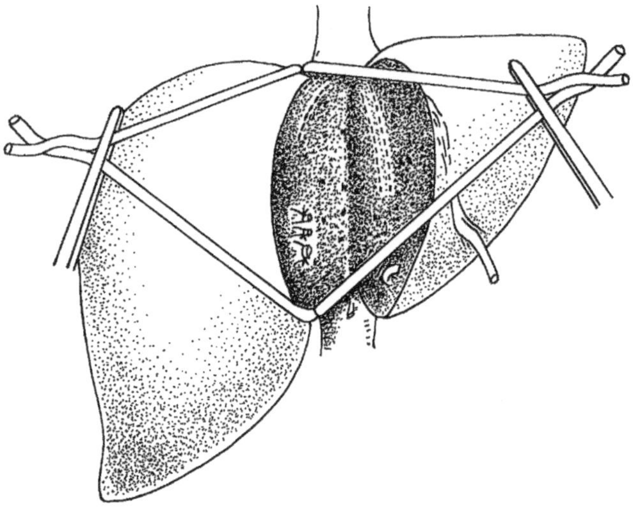

Fig. 15.9 Chen's double liver hanging manoeuvre

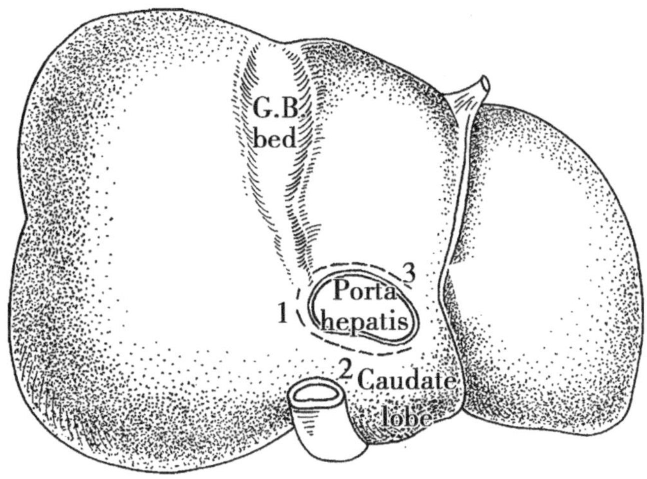

Fig. 15.10 Launios and Jameison's posterior intrahepatic approach to the hepatic pedicles: Method 1: incision 1; Method 2: incisions 2, then 3

The main advantage of this technique is the lack of bleeding in developing the tunnel as the tunnel goes through a true avascular space containing loose connective tissues only.

15.3 Glissonian Sheath Approach

This approach starts off with minimal parenchymal transection to expose the intrahepatic Glissonian sheaths followed by control of the portal triad within the sheath, then further parenchymal dissection till the liver is divided, then controlling the vascular outflow branches. Mobilisation of the liver can be done at the beginning of the operation or towards the end of the operation.

Launois described two approaches:

1. Launois Anterior Intrahepatic Approach
2. Launois and Jameison's posterior intrahepatic approach to Glissonian Sheath (also see Sect. 4.3 Intrahepatic Transfissural Approach to the Conference of the Bile Duct)

Both the two approaches to the hepatic pedicles contained within the Glissonian sheath aid early vascular inflow control. The Launois and Jameison's posterior intrahepatic approach can be accomplished in one of two ways. Firstly, the caudate process behind the gallbladder bed is divided vertically along the right side of the inferior vena cava. An extension of this incision is then made in the posterior aspect of the gallbladder bed (Incision 1 in Fig. 15.10). The surgeon's forefinger and thumb of one hand are now placed within the liver substance, with the forefinger in the caudate process incision and thumb in the gallbladder fossa incision. At a depth of 10–20 mm, a sheath is usually encountered. It is encircled by a large curved clamp, and a tape is passed around it, this is often a sheath to segment 6. Dissection can

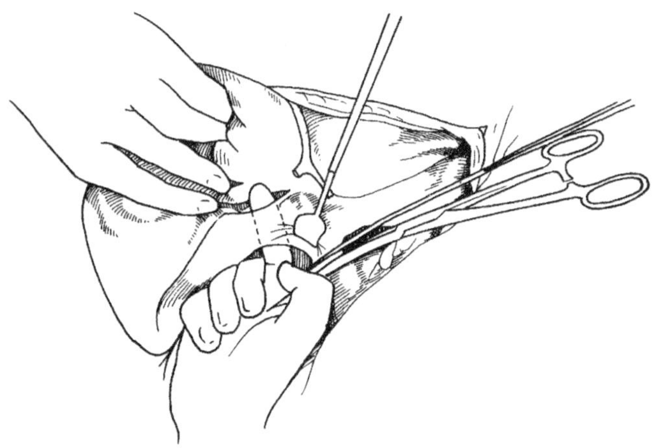

Fig. 15.11 Launios and Jameison's posterior intrahepatic approach: passing an index finger

now be undertaken further centrally or peripherally, depending on which sheath is being sought.

Secondly, they described a more central approach. First, an incision is made at the junction of the hilum with the liver substance of the caudate process, i.e. behind the hilum. This incision is approximately 20 mm in length (Incision 2 in Fig. 15.10). A second incision is then made in front of the hilum, and parallel to the first incision, extending from the gallbladder bed on the right to the umbilical fissure on the left (Incision 3 in Fig. 15.10). An index finger is passed into the incision behind the hilum and the undersurface of the sheath is kept above the finger until the superior part of the sheath is reached Fig. 15.11). A large curved clamp is then used to pass a tape around this region of the confluence. Traction on the tape tends to exteriorize both the right and left main sheaths (Fig. 15.12). Further dissection to the right exposes the anterior sectoral pedicle to segments 5 and 8, which passes upwards. The posterior sectoral pedicles to segments 6 and 7 run posterolateral. Dissection to the left exposes the sectoral branches to segment 4, and segments 2/3. Further dissection identifies the hepatic pedicle going into the individual liver segments making resection of individual or combination of sectors and individual or combination of segments possible.

15.3.1 Takasaki's Glissonian Sheath Approach

Takasaki divided the liver into three almost equal parts: the right segment (equivalent to Couinaud segments 6, 7, or the right posterior sector); the middle segment (Couinaud segments 5, 8 or the right anterior sector); and the left segment (Couinaud segments 4, 3, 2 or the left medial and lateral sectors) (Figs. 15.13 and 15.14a, b).

The Glissonian pedicles entering the liver for these three Takasaki 'segments' can be exposed by dissecting into the liver parenchyma at the hepatic hilum. Further dissection of these pedicles leads to the hepatic pedicles supplying the individual Couinaud segments. As a consequence, any Couinaud segment can be resected individually or in combination by ligation and division of the pedicle to the segment. Parenchymal transection can then follow the demarcation line of the ischaemic liver segment. The operation ends with the division of the outflow venous pedicle.

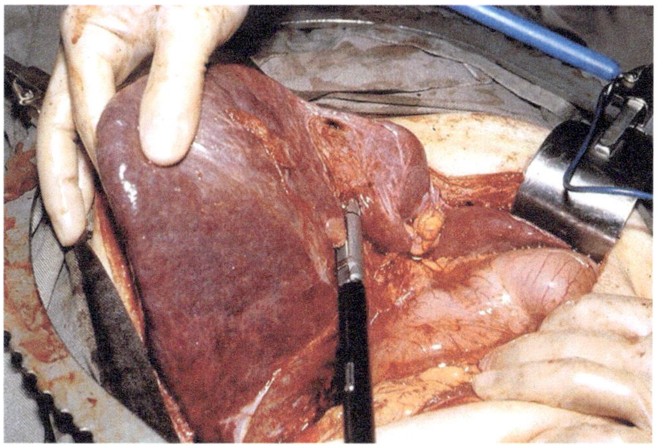

Fig. 15.12 Using Launios and Jameison's posterior intrahepatic approach, the right pedicle was isolated. A vascular stapler was applied. Note the demarcation separating the ischaemic right liver and the normally perfused left liver

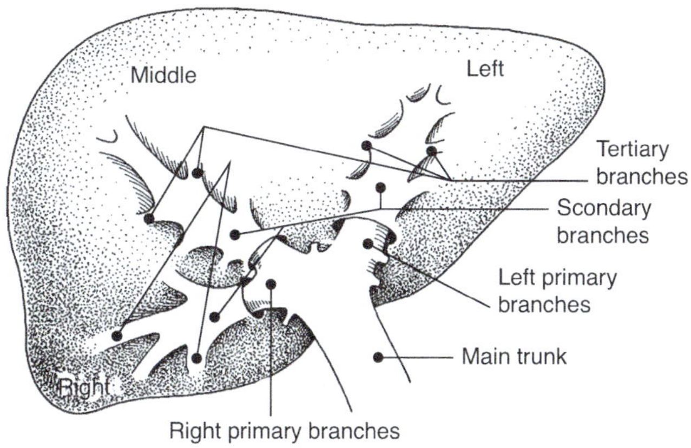

Fig. 15.13 Takasaki's liver 'segments'

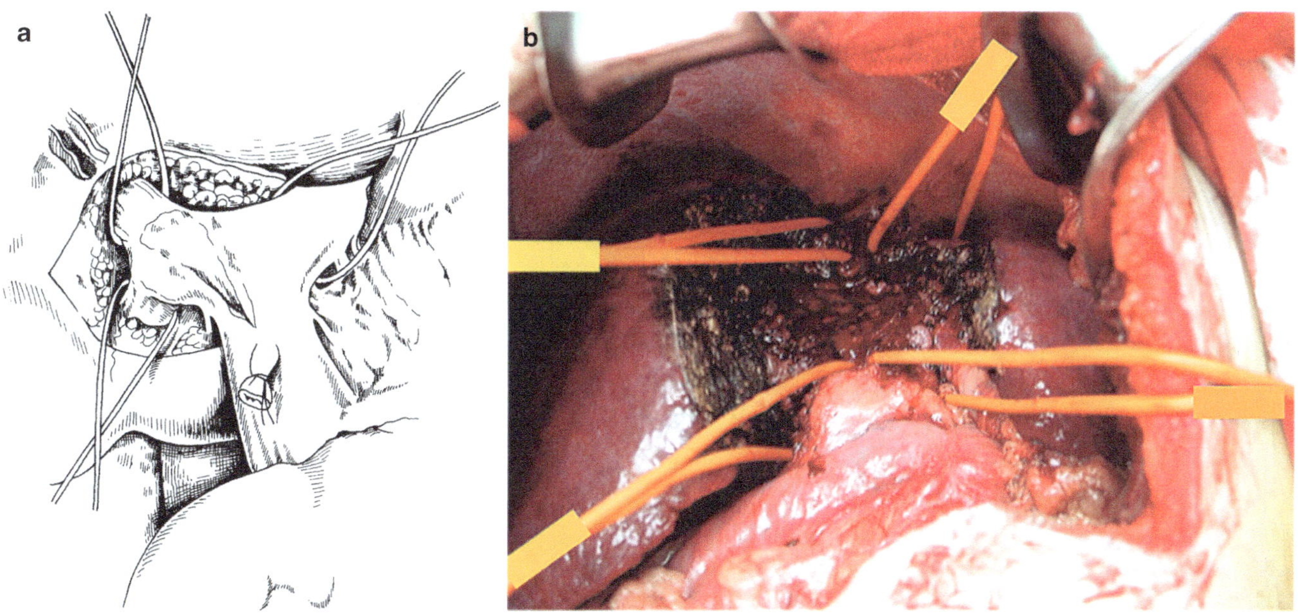

Fig. 15.14 (**a**) Dissecting of the liver parenchyma at the hepatic hilus exposed the 3 pedicles to Takasaki's 'segments'. (**b**) Operative photograph showing liver after resection of segments 4, 5, 8. Sling (**a**) around right hepatic vein. Sling (**b**) around left hepatic vein. Sling (**c**) around right posterior sectional pedicle (to segments 6, 7) and sling (**d**) around left lateral sectional pedicle (to segments 2, 3)

15.4 Segment-Based Liver Resection

Liver resection based on Couinaud liver segments is called segment-based liver resection. As each liver segment receives its own tributaries from the portal pedicles and drains independently into the tributaries of the hepatic veins, each segment is an independently functional unit and can be resected individually or in combination with other liver segments.

15.4.1 Advantages of Segment-Based Liver Resection

There are many theoretical advantages of segment-based liver resection:

1. The anatomical boundaries between the individual liver segments are not crossed by large branches of the portal pedicle (hepatic artery, portal vein and bile duct), these boundaries are relatively avascular planes that facilitate surgical resection and decrease intraoperative blood loss.
2. By avoiding damages to the portal pedicle, segment-based liver resection avoids leaving behind devitalized liver parenchyma in the liver remnant, thus avoiding the risk of infection and bile duct fistulation.
3. By predetermining the liver segments to be removed and by following the intrahepatic anatomy during parenchymal transection, an adequate resection margin can be guaranteed while at the same time preserving the largest amount of non-tumorous liver parenchyma.
4. There is a good oncological rationale for using segment-based liver resection because of liver tumour characteristics. Hepatocellular carcinoma usually grows and is confined to one liver segment in the early phase of the disease. Intrahepatic tumour spread originates from tumour invasion of the portal venous branches, giving rise first to satellite metastases within the same liver segment, followed by involvement of the corresponding part of the same sector, and ultimately a complete hemiliver or bilateral spread to the whole liver. Indeed vascular invasion and intrahepatic metastases are the risk factors that most strongly influence post-operative prognosis. Given that early satellite metastases lie in the same liver segment as the main tumour, segment-based liver resection should be used to give the best chance of oncological tumour clearance.

15.4.2 Techniques of Segment-Based Liver Resection

The application of the principles of segment-based liver resection has been facilitated by the development of modern imaging techniques. Preoperative ultrasonography, computed tomography or magnetic resonance imaging can accurately relate the location of the tumour in the liver segment(s). Thus, a preoperative planning can be made on the liver segment(s) to be resected (Sect. 10.2).

15.4.2.1 Surface Anatomy + IOUS

Intraoperative ultrasound (IOUS) together with the liver surface markings can help to define the exact location of the tumour (see Chap. 10), to determine the liver segment(s) to be resected and the amount of resection margin that can be achieved (Fig. 15.15).

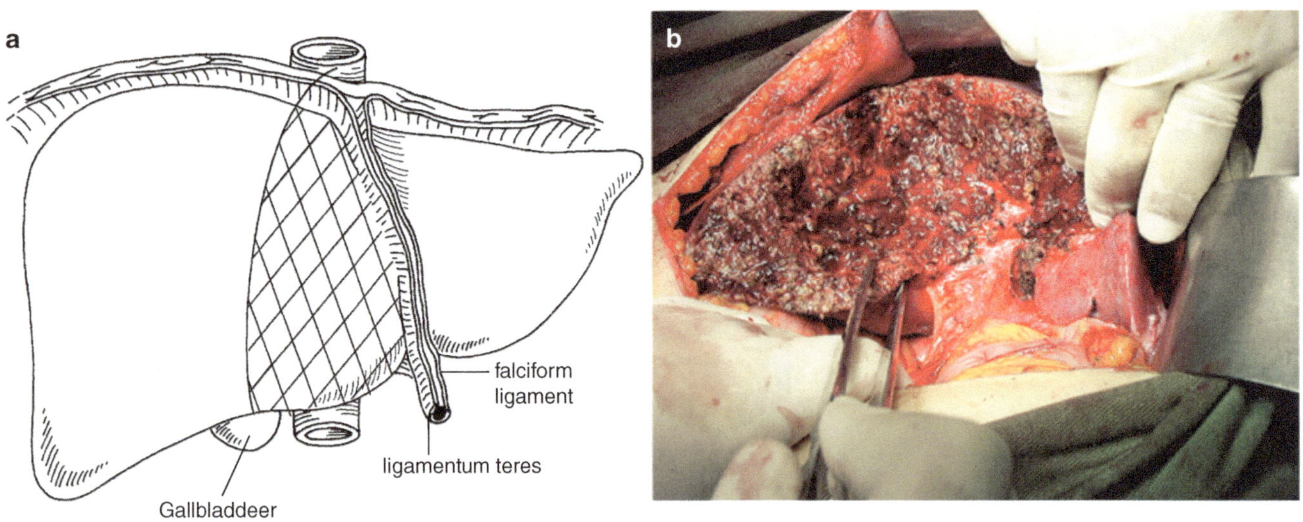

Fig. 15.15 Resection of segments 4B based on surface anatomy. (**a**) Segment 4B between midplane of liver and falciform ligament. (**b**) After segment 4B resection (note segment 4A left behind)

After marking the planes of liver parenchymal transection on the surface of the liver, the liver is transected. The pedicles of the vessels and the bile ducts of the relevant liver segments are divided at the end of the parenchymal transection (Fig. 15.16). Pringle's manoeuvre and a low central venous pressure are usually used. It is essential for surgeons to have a detailed knowledge of the intrahepatic anatomy and skill in IOUS before this technique can be used with good results.

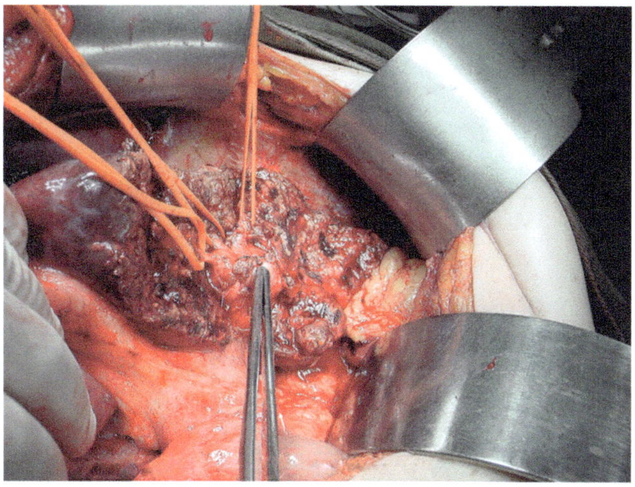

Fig. 15.16 Resection of segments 4B. Slings around segments 6 pedicle (**a**); sling around pedicle 7 (**b**); and sling around anterior right sectional pedicle (**c**) (segments 5, 8). Forceps pointing to divided pedicles of segment 4

15.4.2.2 Preliminary Control of the Glissonian Pedicle of the Liver Segment to Be Removed

The Glissonian pedicles are approached at the hepatic hilus using either the Launois or the Takasaki's technique. Lowering the liver plate helps to increase the extrahepatic length of these pedicles (see Chap. 4). By dissecting the right pedicle distally, the right anterior sectoral pedicle (segments 5, 8) and the right posterior sectoral pedicle (segments 6, 7) are found (Fig. 15.17). Similarly, by dissecting and tracing the left pedicle, the segment 4 pedicle and the segments 2/3 pedicle are found. Further dissection distally exposes the pedicles to the individual liver segments, but this dissection requires further parenchymal transection. Occlusion of the relevant pedicle by a bulldog clamp results in a change in colour of the liver segment. The arterial and portal pedicle are ligated and divided at the end of the parenchymal resection. This technique requires more tissue dissection and a longer operating time than the other techniques, and it is technically more difficult in patients with cirrhosis and portal hypertension.

An alternative technique is to isolate the pedicles to the segments or sectors and to clamp the pedicle with a vascular clamp to show the colour changes in the respective segment/sector (Fig. 15.18).

15.4.2.3 Ultrasound-Guided Puncture of Portal Vein Branch and Injection of Dye

The portal vein branch supplying the liver segment to be resected is punctured under ultrasound guidance. A few millilitres of methylene blue or Congo red is then infused into the portal branch. The dye stains the liver segment corre-

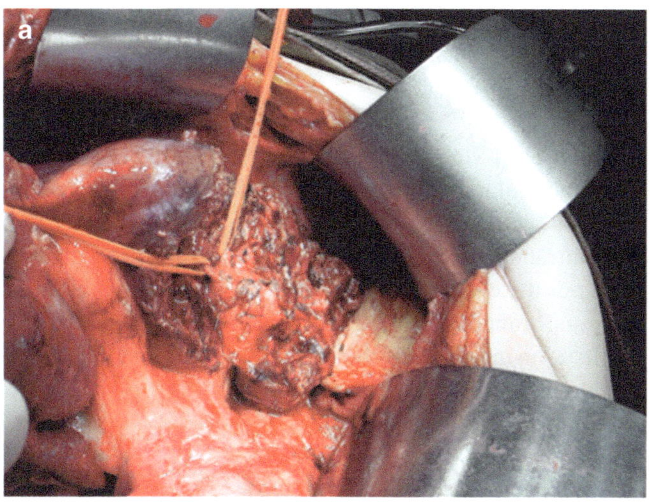

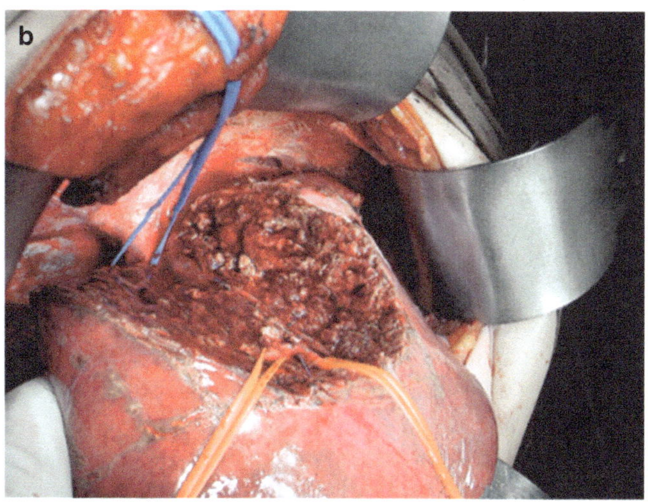

Fig. 15.17 Dissection of right hepatic pedicle distally. (**a**) The right anterior sectional pedicle to segments 5, 8 (sling, a) and the right posterior sectional pedicle to segments 6, 7 (sling b). (**b**) Further dissection distally showed segment 5 pedicle (sling c). The right posterior sectional pedicle to segments 6, 7 at sling d. Pedicle to segment 8 divided for isolated segment 8 resection (arrow). Note sling e around right hepatic vein

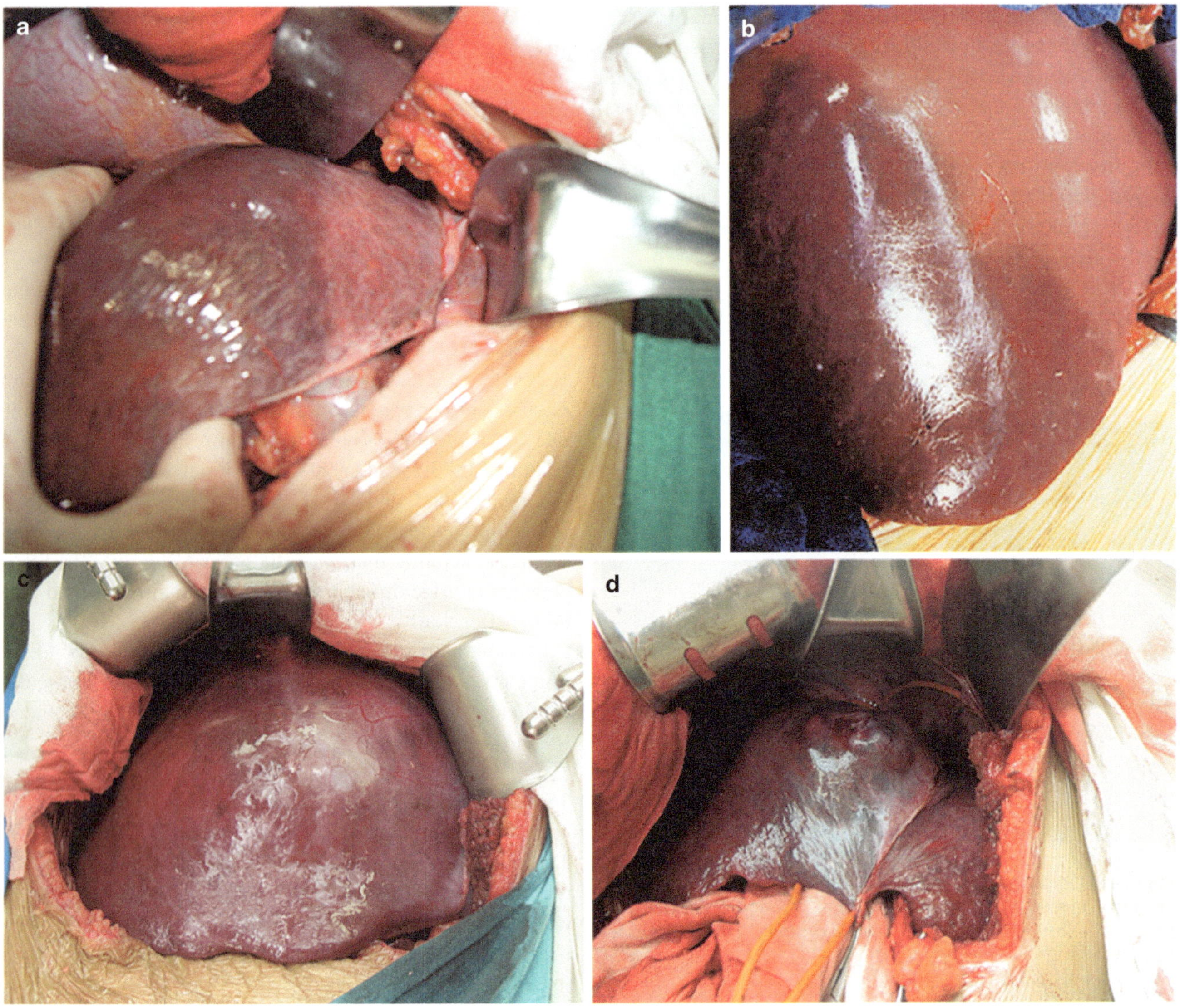

Fig. 15.18 Occlusion to hepatic pedicle showing colour changes in liver parenchyma. (**a**) To right hepatic pedicle. (**b**) To right posterior sectional pedicle supplying segments 6, 7. (**c**) Colour changes in segments 4, 5, 8. (**d**) Colour changes in segment 4

sponding to the limits of the liver transection plane. Transection is then carried out. The technique requires great expertise in interventional ultrasound, and for this reason, has not gained wide acceptance.

15.4.2.4 Selective Portal Venous Occlusion Using a Balloon Catheter Through a Branch of the Superior Mesenteric Vein

This technique is carried out during open surgery using a bilateral subcostal incision with an upward midline extension. The liver is completely mobilised by the division of the liver ligaments. A 6 French balloon catheter is inserted into the portal vein via a branch of the superior mesenteric vein. The catheter is guided to the corresponding branch of the portal vein (either the right or the left), where the hepatocellular carcinoma is situated, with the surgeon's hand in the porta hepatis. Once the tip of the catheter is in the intrahepatic portal venous system, further advancement of the catheter into the sectoral and segmental portal venous branches is done by rotating and advancing the catheter using the trial-and-error method. Guidance of the catheter tip into the desired portal venous branch is assisted with ultrasound and the surgeon's hand in the porta hepatis.

When the balloon catheter is in the right position, the balloon is inflated with 3 mL of normal saline to occlude the venous branch, a few millilitres of methylene blue is injected through the catheters to delineate the liver segment to be resected. If the balloon catheter enters into a wrong pedicle supplying a liver segment that needs to be preserved, counter-staining with methylene blue helps to identify the segment to be resected because this segment remains unstained. The line of demarcation is marked on the liver surface with a diathermy device. The procedure is repeated if more than one liver segment needs to be delineated. The time required to get the catheter in the right position is around 10 min. The hepatic parenchyma is then transected along the line of demarcation. After haemostasis on the raw liver surface, the balloon catheter is deflated and removed, and the entry hole of the catheter into the liver segment is sutured. The branch of the superior mesenteric vein is ligated.

15.5 Non-Anatomical Liver Resection

Non-anatomical resection of centrally located tumour often results in inadequate tumour resection, especially at the deep margins. The operation is often associated with bleeding and results in a higher chance of biliary fistula and infection because devitalized liver segments are left behind.

Non-anatomical resection should only be carried out (1) when the tumour is situated at the border of several segments; (2) when the tumour is small and is situated peripherally at the edge of the liver. Under the latter situation, a wedge excision made in the shape of an arch or box is a simpler operation than a segment-based liver resection. Wedge excision should not be done in a V-shape because of the higher chance of the resection margin being involved by the tumour on histological study.

15.6 Caudate Lobe Resection

15.6.1 Classification of Caudate Lobectomy

The caudate lobe consists of three parts: the Spigelian lobe, the paracaval portion and the caudate process. In the majority of cases, the three parts receive different blood supply making partial caudate lobectomy a possibility (see Chap. 3). When resection of the caudate lobe is required for tumour clearance, the operation may be an isolated caudate lobe resection or a caudate lobe resection combined with a major hepatectomy, e.g. right hepatectomy. Thus, caudate lobectomy can be classified into four types: isolated complete caudate lobectomy, combined complete caudate lobectomy, isolated partial caudate lobectomy and combined partial caudate lobectomy (Fig. 15.19).

15.6.2 Approaches to Caudate Lobectomy

There are five approaches.

15.6.2.1 Bilateral Approach

This is the most often used approach for isolated caudate lobectomy as it gives good exposure. The caudate lobe is approached from both the right and the left sides.

Technique: The liver is fully mobilised by the division of all its ligaments. The suprahepatic and the infrahepatic inferior vena cava is controlled by passing a tape around it so that total vascular outflow occlusion can be applied if necessary (see Chap. 13). The right adrenal vein is ligated, and the right adrenal gland is detached from the liver.

The peritoneal reflection between the Spigelian lobe and the inferior vena cava is incised. The right hepatic vein is controlled by a loop after the division of the hepato-caval ligament. The common trunk of the middle and left hepatic veins is controlled by a loop after the division of the ligamentum venosum. It is usual to use Pringle's manoeuvre.

The caudate lobe is detached from the inferior vena cava by dissecting along the anterior surface of the retrohepatic inferior vena cava. The short hepatic veins are identified, ligated and divided (Fig. 15.20a). On the left side, the hepatogastric ligament (lesser omentum) is detached from the under surface of the liver (Fig. 15.20b), the fibrous hepatocaval ligament needs to be divided to free the Spigelian lobe from the inferior vena cava and the diaphragm. All short hepatic veins are ligated and divided, thus freeing the caudate lobe from the inferior vena cava (Fig. 15.20c).

On the right side, the caudate process is divided before the portal triad to the caudate process is isolated and divided. The branches to the paracaval portion of the caudate lobe from the right portal vein, right hepatic artery and right hepatic duct, and the branches to the Spiegelian lobe from the left portal vein, left hepatic artery and left hepatic duct are sought, ligated and divided.

By careful dissection, the liver is detached from the surrounding liver and the right, middle and left hepatic veins. There are two important landmarks for this dissection: the top of the caudate lobe which is located at the angle between the left hepatic vein and the inferior vena cava, and the point where the caudate process meets the right liver. An imaging line joining these two points can be considered as the caudate boundary for liver transection. Transection can start from either end or from both ends. Meticulous care should be paid not to injure the major hepatic veins or torrential bleeding can result, which can be difficult to control. After removal of the specimen, adequate haemostasis should be ensured.

15.6.2.2 Left-Sided Approach

Some small tumours <3 cm can be resected using a left-sided approach, especially if an isolated partial caudate lobectomy or a left hepatectomy is combined with a complete caudate lobectomy is carried out. The procedure is very similar to the bilateral approach with the exception that the dissection is mainly from the left side of the liver.

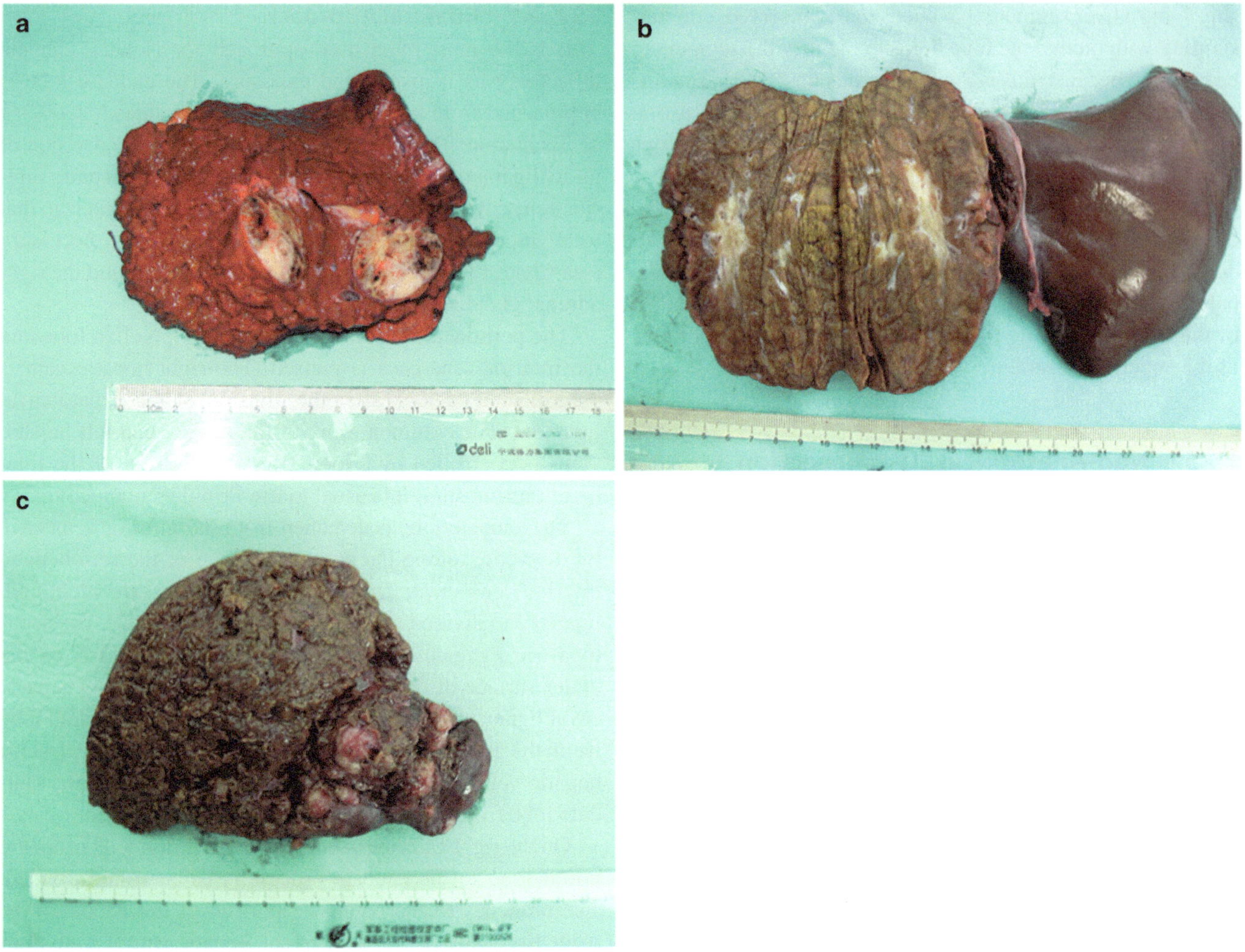

Fig. 15.19 Caudate lobectomy. (**a**) Isolated complete Caudate lobectomy. (**b**) Combined left hepatectomy and caudate lobectomy for a huge focal nodular hyperplasia. (**c**) Combined right hepatectomy and caudate lobectomy

15.6.2.3 Right-Sided Approach

In a thin patient with a right hepatectomy combined with a complete caudate lobectomy, a right-sided approach can be done. Again the procedure is very similar to the bilateral approach with the exception that the dissection is mainly from the right side of the liver.

15.6.2.4 Anterior Approach

This is the approach for a caudate tumour larger than 4 cm, especially when the tumour is located in the paracaval portion or is in close contact with the major hepatic veins. This approach provides a better operative field by opening up the midplane of the liver widely so as to expose the major hepatic veins and the hilar plate to direct vision, thus facilitating dissection of the tumour from the main vessels.

Technique: The initial steps of the operation are similar to the bilateral approach.

After mobilising the liver by division of all the ligaments and having control of the suprahepatic and infrahepatic inferior vena cava, and the right and the trunk of the middle/left hepatic veins, the caudate lobe is freed from the retrohepatic inferior vena cava. Pringle's manoeuvre is then applied. The liver is transected through the midplane, starting from the point between the roots of the right hepatic vein and the middle hepatic vein to the fossa of the gallbladder (which should have been removed). The transection is continued to the plane 1 cm from the tumour, as shown on intraoperative ultrasound (Fig. 15.21a). The transection then goes around a plane 0.5 cm from the tumour surface (Fig. 15.21b). When the transection reaches the hilar plate at the hilum, the caudate portal triads are isolated and divided (Fig. 15.21b, c).

The caudate lobe together with the tumour is then separated from the major hepatic veins by careful parenchymal transection as described in the bilateral approach (Fig. 15.21d).

After removal of the specimen, all bleeding points and bile leaks are controlled individually.

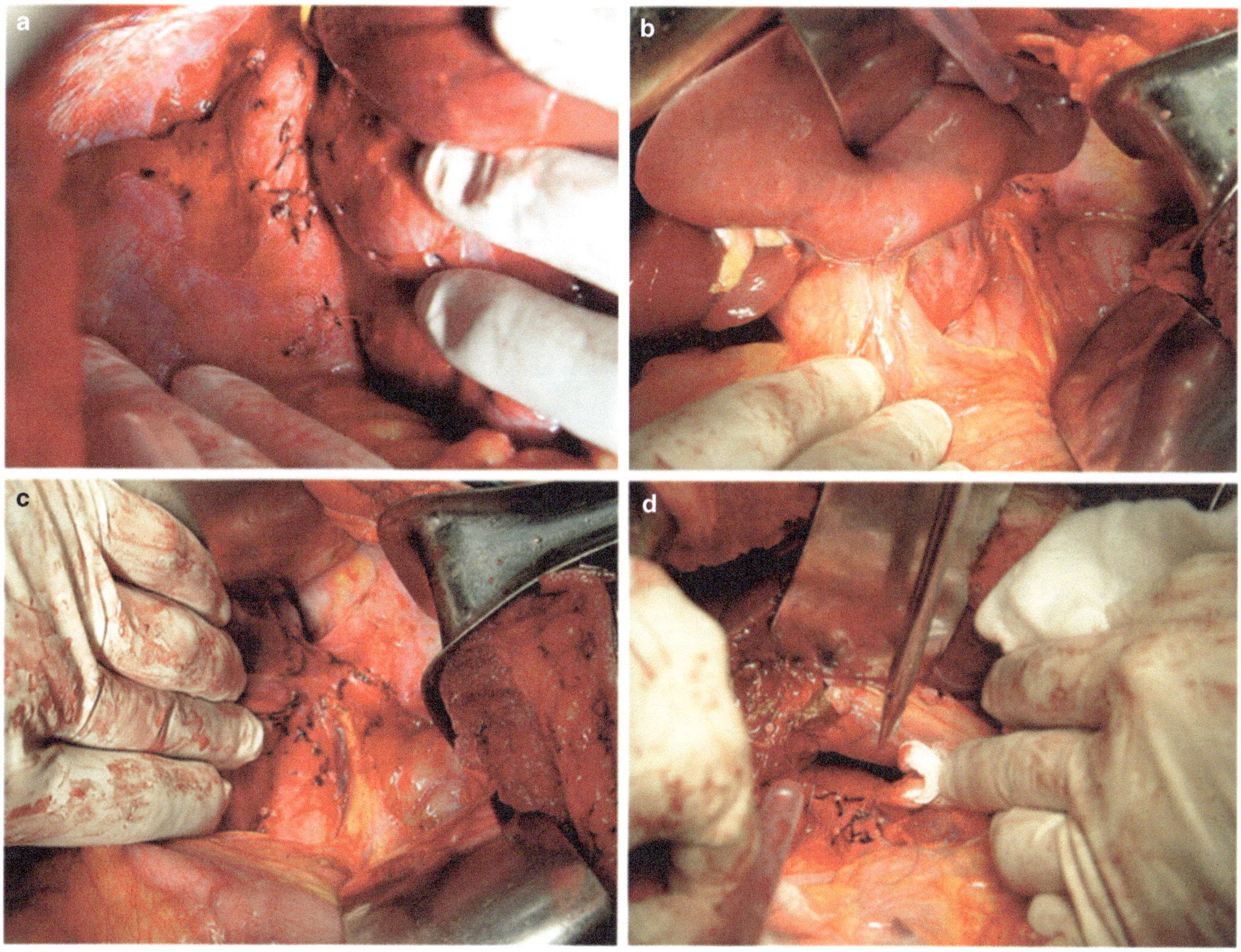

Fig. 15.20 Bilateral approach caudate lobectomy. (**a**) Approach from right side to divide short hepatic veins to free the caudate lobe from the inferior vena cava. (**b**) Approach from left side, division of gastrohepatic ligament (lesser omentum). (**c**) Caudate lobe mobilised from left side after division of short hepatic veins. (**d**) Space left behind after removal of caudate lobe

15.6.2.5 Retrograde Caudate Lobectomy

All the technical approaches described above emphasise the early detachment of the caudate lobe from the retrohepatic inferior vena by division of the short hepatic veins. However, if the caudate lobe tumour is closely adherent to or has infiltrated the inferior vena cava or if it is too large in size to be turned from one side to the other, the retrograde caudate lobectomy has to be used instead. In this operation, the division and ligation of the short hepatic veins are left to the final stage of the operation.

Technique: Mobilisation of the liver by division of all the ligaments. The hepatoduodenal ligament, the suprahepatic and the infrahepatic inferior vena cava are looped for temporary occlusion if necessary.

The anterior transhepatic approach should be used, although occasionally the bilateral approach can be used. The liver is transected along the midplane down to 1 cm from the tumour (Fig. 15.22a). The hepatic veins are exposed under direct vision and meticulously dissected from the specimen. The caudate portal triads from the right/left hepatic arteries and right/left portal veins are ligated and divided. In combined right/left hepatectomy with caudate lobectomy, the right/left hepatic pedicle can be transected (Fig. 15.22b). The specimen is attached only to the inferior vena cava.

The distal and proximal parts of the caudate lobe are dissected from the inferior vena cava (Fig. 15.22c). Short hepatic veins, when encountered, are ligated and divided. If the tumour is attached to the inferior vena cava, part of the vein can be resected with the tumour (Fig. 15.22d). The inferior vena cava is then repaired with 40 prolene or reconstructed with an artificial graft.

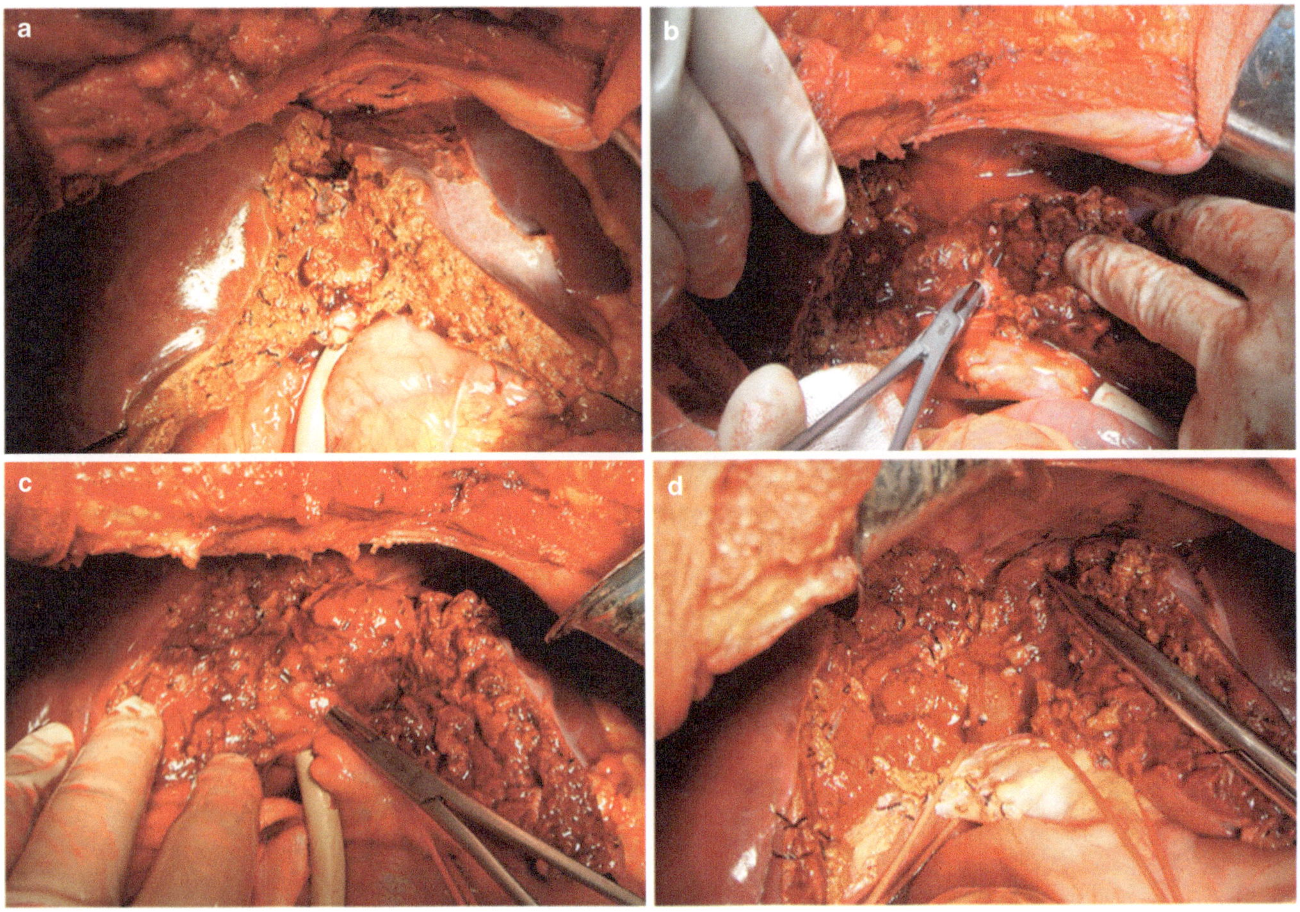

Fig. 15.21 Anterior approach for caudate lobectomy. (**a**) Liver is transected at midplane down to 1 cm form tumour. (**b**) Dissection around tumour. Dissection of caudate portal traid from left hepatic pedicle. (**c**) Dissection of caudate portal traid from right posterior hepatic pedicle. (**d**) Space left after caudate lobectomy. *IVC* inferior vena cava. *RHV* right hepatic vein, *LHV* left hepatic vein

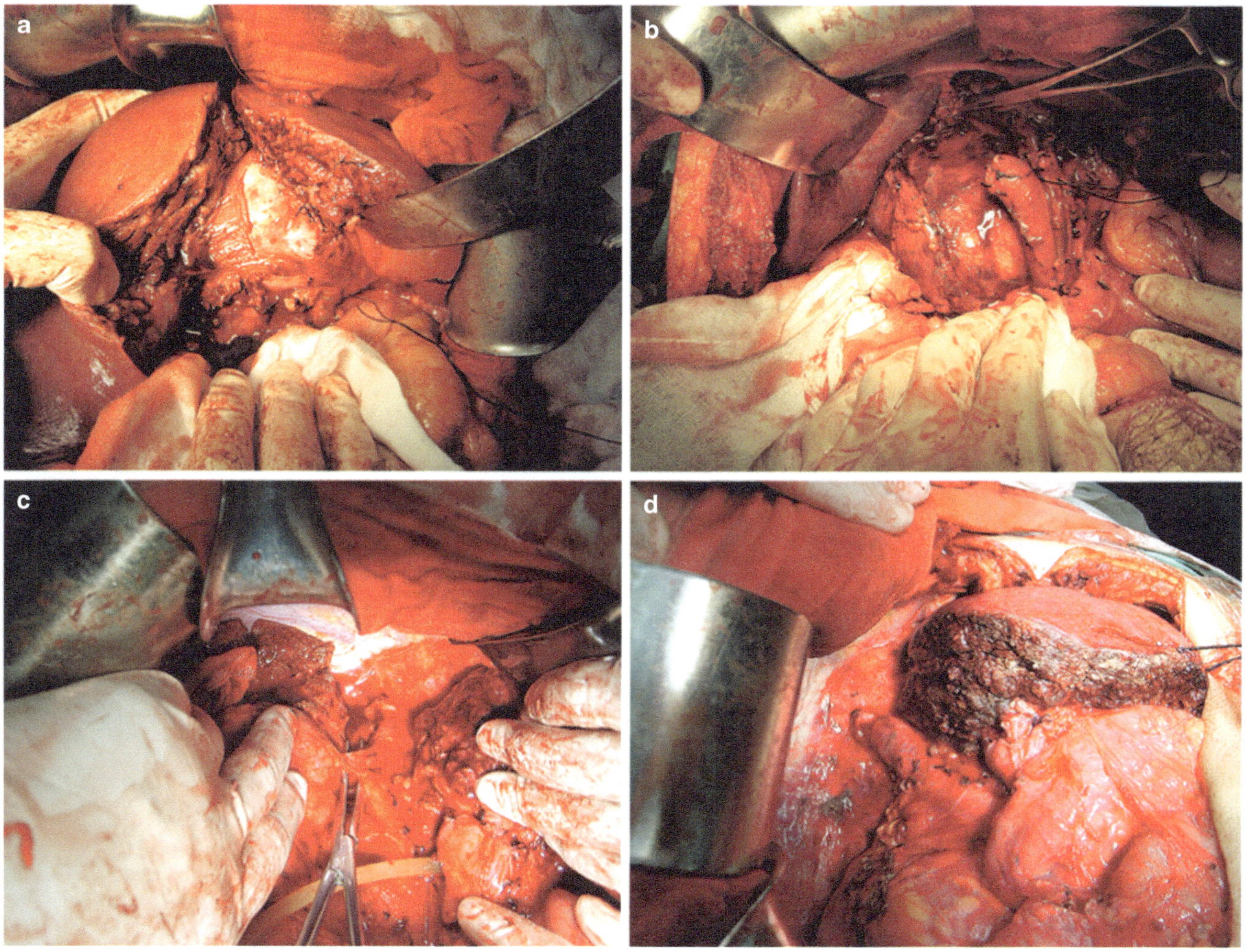

Fig. 15.22 Retrograde caudate lobectomy combined with right hepatectomy. (**a**) Liver transected at midplane down to 1 cm from tumour. (**b**) Division of right hepatic artery, portal vein and right duct. (**c**) Specimen isolated from inferior vena cava, and left attached to right hepatic vein. (**d**) Space left behind after resection

Further Reading

Belghiti J. Chapter 16: Surgical treatment. In: Lau WY, editor. Hepatocellular carcinoma. Singapore: World Scientific; 2008. p. 387–408.

Belghiti J, Alkofer B. Chapter 17: Anterior approach using the hanging technique. In: Lau WY, editor. Hepatocellular Carcinoma. Singapore: World Scientific; 2008. p. 409–18.

Chen XP, Zhang WG, Lau WY, Qiu FZ. Right hepatectomy using the liver double-hanging manoeuvre through the retrohepatic avascular tunnel on the right of the inferior vena cava. Surgery. 2008;144:830–3.

Lau WY, Lai ECH. Chapter 18: Segment-based liver resection. In: Lau WY, editor. Hepatocellular carcinoma. Singapore: World Scientific; 2008. p. 419–28.

Launois B, Tay KH. Chapter 19: Intrahepatic Glissonian approach. In: Lau WY, editor. Hepatocellular carcinoma. Singapore: World Scientific; 2008. p. 429–46.

Peng SY. Chapter 21: Isolated caudate lobe resection (Resection of Couinaud Segment 1). In: Lau WY, editor. Hepatoceullular carcinoma. Singapore: World Scientific; 2008. p. 465–89.

Takasaki K. Glissonian pedicle transection method of hepatic resection. Tokyo: Springer; 2007.

16 Laparoscopic Liver Surgery

16.1 Minimally Invasive Surgery

The introduction of minimally invasive surgery has revolutionised surgical practice in the past two decades. Minimally invasive surgery has been shown to be safe and effective for the surgical management of several gastrointestinal pathologies. It benefits patients with the better cosmetic outcome, less pain and earlier recovery, and medical institutions with lower cost and a shorter hospital stay. Traditionally, liver surgery is considered as one of the most complex surgeries among abdominal surgical procedures. Its minimally invasive surgical development has been lagging behind other gastrointestinal surgical developments. The concerns of excessive bleeding, gas embolism and compromised oncological outcomes contributed to its delayed development.

16.2 Development of Laparoscopic Liver Surgery

In 1992, Gagner et al. reported the first complex laparoscopic hepatectomy for a 6 cm focal nodular hyperplasia, using an ultrasonic dissector, monopolar cautery and clip appliers (Gagner et al. 1992). It was until the early 2000s that there was increasing interest in laparoscopic liver surgery. There were a lot of concerns of using laparoscopic surgery to treat malignant tumours because of the fear of compromising oncological resection margins. Most surgeons at that time concentrated on laparoscopic liver surgeries for benign conditions. However, as laparoscopic liver resection requires advanced techniques in surgeons who have to process laparoscopic skills, the development was very slow. It took nearly 10 years of increasing experience and refinement in technology before laparoscopic hepatectomy becomes recognised as a technique that is feasible and safe. Since then, more and more studies on laparoscopic hepatectomy for liver malignancies have been published. It has now almost 23 years since laparoscopic hepatectomy was first carried out.

16.3 Indications and Different Approaches of Laparoscopic Hepatectomy

There are currently no universally accepted indications for laparoscopic hepatectomies and selection criteria may vary among institutes. Laparoscopic hepatectomy has been most widely used for symptomatic benign tumours, or those benign tumours with a potential of malignant transformation or a potential to develop local complications in the future. Although most laparoscopic hepatectomies have been performed for benign diseases, there were more and more operations for malignant diseases, which include primary and metastatic liver tumours. Tumours in liver segments 2, 3, 4b, 5, 6 are more assessable to laparoscopic resection, whereas those in segments 1, 7, 8 and 4a are more difficult to resect laparoscopically. Tumours that are either large (>5 cm), centrally situated, multiple, bilateral or adjacent to the liver hilum, major hepatic veins or inferior vena cava are now not considered for laparoscopic resection in most centers. Laparoscopic live donor hepatectomy remains the most controversial application of laparoscopic liver surgery, and it should be carried out under the confines of a worldwide registry.

16.4 Advantages

The potential advantages of laparoscopic hepatectomy are those of minimally invasive surgery, such as better cosmetic outcome, earlier recovery, and shorter hospital stay (Table 16.1). The post-operative recovery after hepatectomy by laparoscopic approach may also be improved in patients with liver cirrhosis and portal hypertension because the abdominal wall is preserved, collateral venous drainage is preserved, and there is less post-operative ascites.

W. Y. Lau, *Applied Anatomy in Liver Resection and Liver Transplantation*, https://doi.org/10.1007/978-981-16-0800-1_16

Table 16.1 Potential advantages of laparoscopic hepatectomy

Operation	Recovery
Improved visualisation	Less post-operative pain
Reduced blood loss	Earlier ambulation
Reduced blood transfusion requirement	Improved perioperative pulmonary function
Less intra-abdominal adhesion formation	Fewer wound complications
	Reduced perioperative immune suppression
	Better cosmetic outcome
	Shorter post-operative recovery time
	Shorter post-operative hospital stay
	Reduced ascites in patients with portal hypertension

16.5 Approaches of Laparoscopic Hepatectomy

The traditional approaches of laparoscopic hepatectomy include the pure (or total) laparoscopic approach, the hand-assisted laparoscopic approach, and the hybrid technique. In the pure laparoscopic approach, the entire procedure is completed through laparoscopic ports, although a small incision may be required for specimen extraction. The hand-assisted laparoscopic approach is defined by the elective placement of a hand port through which a surgeon's hand can be put into the peritoneal cavity to facilitate the procedure. The hybrid technique is a procedure that starts off as a pure laparoscopic or a hand-assisted procedure, but liver resection is performed through a mini-laparotomy incision. The hand-assisted approach has been developed to bridge the gap between the open and the pure laparoscopic approaches. The advantages of the hand-assisted approach in hepatectomy include: (1) facilitate retraction; (2) improve safety in liver parenchymal transection; (3) use of manual compression if a hepatic vein tear occurs for immediate haemostasis and prevention of air embolism; and (4) use of tactile sensation to facilitate the assessment of surgical resection margins. The pure laparoscopic approach is superior to the hand-assisted approach in terms of pain and cosmetic outcome because the hand-assisted approach requires a 6–8 cm length of incision for placement of a hand-port device. Another disadvantage of the hand-assisted approach is obstruction of the visual field by the surgeon's own hand.

16.6 Expert Consensus on Laparoscopic Hepatectomy

To summarise the current role and status of laparoscopic hepatectomy, the first international consensus meeting on laparoscopic hepatectomy was held in Louisville, USA, in November 2008, to incorporate the opinions of the world's experts in laparoscopic and open liver surgery (Buell et al. 2009). The organising committee selected forty-five international experts with extensive experience in both laparoscopic and open liver surgery to attend the meeting. The conclusions of this meeting were (1) the currently accepted indications for laparoscopic hepatectomy are patients with a solitary lesion, ≤ 5 cm, which is located in the antero-peripheral liver segments 2–6. (2) Laparoscopic left lateral sectionectomy should be considered as a standard practice. (3) Laparoscopic hepatectomy is a safe and effective approach provided it is performed by trained surgeons with experience in hepatobiliary and laparoscopic surgery. (4) Although most types of hepatectomy can be performed laparoscopically, including major hepatectomy, the more complex resections should be reserved for experienced liver surgeons skilled at complex laparoscopic resections. Since then, the National Hepatic Surgery Group of the Society of Surgery, a professional society of the Chinese Medical Association, also gathered 60 expert hepatic surgeons in Wuhan, China, in December 2012, and an 'Expert Consensus on Laparoscopic Hepatectomy' was developed (National Hepatic Surgery Group Society of Surgery, Chinese Medical Association 2013; Lai 2013). The types of hepatectomy, indications and contraindications, preoperative preparation, anesthesia, patient position, insufflation pressure, port positions, control of hepatic inflow and outflow, indications for conversion to open surgery, and surgical devices and equipment were reviewed. The techniques and procedures of the various laparoscopic hepatectomies were also discussed. The Second International Consensus Conference on Laparoscopic Liver Resections was held in Morioka, Japan, from October 4 to 6, 2014, to evaluate the current status of laparoscopic liver surgery and to provide recommendations to aid its future development (Wakabayashi et al. 2015). The organising committee invited 43 renowned surgeons from 18 countries, with 34 as expert panel members with experience in laparoscopic hepatectomies, and 9 as jury members, to provide evidence and draw recommendations. The jury concluded that laparoscopic minor hepatectomy is a standard practice in many surgical centres, but it is still in an assessment phase because it is gradually adopted by an increasing proportion of surgeons, and laparoscopic major hepatectomy is an innovative procedure which is still in an exploration or learning phase with incompletely defined risks. It should be introduced cautiously.

16.7 Laparoscopic Major Hepatectomies

The technical feasibility and safety of conventional laparoscopic major hepatectomies in experienced hands have been reported. Hwang et al. summarised the combined data of 265 laparoscopic major hepatectomies in 12 tertiary medical centres in Korea from 2001 to 2011 (Hwang et al. 2013). The most frequently performed procedure was left hemihepatectomy (165 patients), followed by right hemihepatectomy (53 patients). A pure laparoscopic procedure was performed in 190 patients, including 19 patients using the robotic approach. The hand-assisted laparoscopic major hepatectomies were performed in three patients and the hybrid approach in 55 patients. The most frequent indications for laparoscopic major hepatectomies were primary hepatolithiasis ($n = 131$), and hepatocellular carcinoma ($n = 62$). The open conversion was required in 17 patients (6.4%). The mean operative time and the estimated blood loss in laparoscopic major hepatectomies were 399.3 min and 836.0 mL, respectively. Intraoperative transfusion was required in 65 patients (24.5%). The mean post-operative length of stay was 12.3 days. Post-operative complications were detected in 53 patients (20.0%), and in-hospital mortality occurred in two patients (0.75%). resection was achieved in 120 patients with hepatic tumours, but R_1 resection happened in eight patients. The mean distance of the resection margin was 14.6 mm. Dagher et al. summarised the combined data of 5388 laparoscopic major hepatectomies in 18 international centres carried out from 1996 to 2014 (Dagher et al. 2014). The most frequent indication for laparoscopic right hepatectomy was colorectal liver metastases (37.0%). Seven centres used the hand-assisted or the hybrid approach selectively for laparoscopic right hepatectomy, mostly at the start-up of their experience. Seven centres used routine Pringle's manoeuvre. The conversion rate for all laparoscopic major hepatectomies was 10%, and the mean operation time was 291 min. The mean estimated blood loss for all the laparoscopic major hepatectomies was 327 mL, and the R_0 margin rate was 96.5%.

Based on the current evidences, laparoscopic major hepatectomies are still limited to a few expert centres, and they cannot be considered as standard procedures. Laparoscopic major hepatectomies require a high level of technical skill and has a long learning curve. Further evaluation and technical refinements are required.

16.8 Perioperative Outcomes of Laparoscopic Hepatectomy

No randomised trials are available. All data have been reported as case series or nonrandomised comparative studies. Most data were obtained from prospectively maintained databases. Nguyen et al. in 2009 analyzed 127 published articles on laparoscopic hepatectomy with 2804 procedures (Nguyen et al. 2009a). Fifty percent were performed for malignant pathologies, 45% for benign pathologies, 1.7% for live donor hepatectomies, and the rest were indeterminate in pathologies. The most common type of hepatectomy was wedge resection or segmentectomy (45%); followed by left lateral sectionectomy (20%), right hemihepatectomy (9%), and left hemihepatectomy (7%). Conversion to open laparotomy and to the hand-assisted approach happened in 4.1% and 0.7% of cases, respectively. The overall mortality rate was 0.3%, and the morbidity rate was 10.5%, with 0% intraoperative mortality. Based on three recent meta-analyses, laparoscopic hepatectomy for benign and malignant liver pathologies is safe and feasible with significant benefits for patients because of fewer complications, less blood loss and shorter length of hospital stay than those of their open counterparts (Mirnezami et al. 2011; Mizuguchi et al. 2011; Nguyen et al. 2011). The benefits of the laparoscopic approach have also been observed in cirrhotic patients. However, there were limitations in these meta-analyses. The data from each included study in the analyses are affected by selection biases, because these patients were carefully selected on clinical grounds for the laparoscopic approach. Furthermore, significant heterogeneity is present in the included studies.

16.9 Oncological Outcomes

Controversies regarding tumour cell seeding and port-site metastases in laparoscopic hepatectomies for malignancies become less and less because of increasing evidences coming from studies using laparoscopic surgery to treat malignancies. However, the rules of oncological surgery must still be followed in minimally invasive operations, just as in their open counterparts. Laparoscopic hepatectomy for hepatocellular carcinoma (HCC) and colorectal liver metastasis (CRLM) results in surgical margins equivalent to those of open surgery. Survival is comparable in the majority of the retrospective comparative studies (Lai et al. 2009a, b, 2011). However, a very careful interpretation of the results is required in the light of significant patient selection bias, which arises from the retrospective nature of the included studies.

16.9.1 Hepatocellular Carcinoma (HCC)

At least five meta-analyses have evaluated the evidences coming from nonrandomized comparative studies of laparoscopic hepatectomy for HCC. Zhou et al. included ten studies which involved 494 patients, with 213 patients in the laparoscopic group and 281 patients in the open group (Zhou et al. 2011). Fancellu et al. included nine studies which involved 590 patients, with 227 patients in the laparoscopic group and 363 patients in the open group (Fancellu et al. 2011). Xiong et al. included nine studies that involved 550 patients, with 234 patients in the laparoscopic group and 316 patients in the open group (Xiong et al. 2012). Yin et al. included 15 studies that involved 1238 patients, with 485 patients in the laparoscopic group and 753 patients in the open group (Yin et al. 2013). Their results showed that laparoscopic hepatectomies for HCC were superior to the open approach in terms of its perioperative results and did not compromise the oncological outcomes. A recent meta-analysis by Morise et al. included more number of studies (Morise et al. 2015) (twenty-one comparative studies). The analysis also found that laparoscopic hepatectomy generally resulted in improved short-term outcomes without compromising the long-term oncological outcomes. In addition, on analysis of nine studies that reported on the incidences of post-operative ascites and six studies that reported on the incidences of post-operative liver failure, the incidences of post-operative ascites and liver failure were reduced with laparoscopic hepatectomy for HCC.

16.9.2 Colorectal Liver Metastasis (CRLM)

Nguyen et al. retrospectively reviewed 109 patients who underwent minimally invasive liver resection for CRC metastasis from February 2000 to September 2008 in multiple medical centres from the United States and Europe (Nguyen et al. 2009b). The minimally invasive approaches included totally laparoscopic (56%) and hand-assisted laparoscopic (41%) approaches. There were four conversions to open surgery (3.7%), all due to bleeding. The extents of resection included wedge/segmentectomy (34%), left lateral sectionectomy (27%), right hepatectomy (28%), left hepatectomy (9%), extended right hepatectomy (0.9%), and caudate lobectomy (0.9%). Major hepatectomies were performed in 45% of patients. The median operating time was 234 min (range, 60–555 min) and the blood loss was 200 mL (range, 20–2500 mL), with 10% of patients receiving blood transfusion. There were no perioperative deaths but a 12% complication rate. The median length of hospital stay for the entire series was 4 days (range, 1–22 days). Negative margins were achieved in 94.4% of patients. The 1-, 3- and 5-year actuarial overall survivals for the entire series were 88%, 69% and 50%, respectively. The 1-, 3- and 5-year disease-free survivals were 65%, 43% and 43%, respectively. Schiffman et al. systematically analyzed the clinical evidences in case-matched studies comparing laparoscopic hepatectomy with open hepatectomy in patients with colorectal liver metastasis for perioperative and oncologic outcomes (Schiffman et al. 2015). Eight articles were identified consisting of 610 patients (242 laparoscopic hepatectomy, 368 open hepatectomy). The two groups were well-matched for age, sex, American Society of Anesthesiologists score, tumour size, number of metastases, the extent of major hepatectomy, and use of neoadjuvant/adjuvant chemotherapy. The mean number of metastases in the laparoscopic hepatectomy and the open hepatectomy groups were 1.4 and 1.5, respectively ($P = 0.14$). The estimated blood loss was significantly less in the laparoscopic hepatectomy group (262 versus 385 mL). The transfusion rate was significantly less in the laparoscopic hepatectomy group (9.9% versus 19.8%). There was no difference in the operative time (248.7 versus 262.8 min). The length of stay was significantly less in the laparoscopic hepatectomy group (6.5 versus 8.8 days). The overall complication rate was significantly less in the laparoscopic hepatectomy group (20.3% versus 33.2%). The 1-, 3- and 5-year disease-free survival rates in the laparoscopic hepatectomy group were 73.4%, 47.1% and 31.9%, and in the open hepatectomy group were 78%, 40.4% and 25.5%, respectively. There was no significant difference between the groups. The 1-, 3- and 5-year overall survival rates in the laparoscopic hepatectomy group were 95.4%, 72.7% and 51.4% and in the open hepatectomy group were 95.4%, 67.2% and 45.9%, respectively. There was also no significant difference between the two groups. The authors concluded that in carefully selected patients with limited colorectal liver metastasis (1 or 2 tumours), laparoscopic hepatectomy provided marked perioperative benefits without compromising the oncologic outcomes or the long-term survival. Specifically, laparoscopic hepatectomy offered decreased blood loss, length of stay, and overall complication rates with comparable 5-year overall survival and disease-free survival. Recently, a multi-institutional Japanese study was reported by Beppu et al. (2015) From 2005 to 2010, 1331 patients with hepatectomies for colorectal liver metastasis were enrolled. By propensity score matching, 171 laparoscopic hepatectomy and 342 open hepatectomy patients showed similar preoperative clinical characteristics. The median estimated blood loss (163 versus 415 g) and the median post-operative hospital stay (12 versus 14 days) were significantly reduced in the laparoscopic hepatectomy group. The morbidity and mortality rates were similar. The 5-year rates of recurrence-free, overall, and disease-specific survival did not differ significantly between the two groups. The R_0 resection rates were similar also.

16.10 Conclusion

Laparoscopic hepatectomy is a complex procedure requiring expertise in both laparoscopic technique and liver surgery. Careful selection of patients and meticulous operative techniques reduce post-operative complications. The minimal invasive approach offers benefits in perioperative short-term outcomes without compromising oncological outcomes when compared to open hepatectomy.

References

Beppu T, Wakabayashi G, Hasegawa K, et al. Long-term and perioperative outcomes of laparoscopic vs open liver resection for colorectal liver metastases with propensity score matching: a multi-institutional Japanese study. J Hepatobiliary Pancreat Sci. 2015;22(10):711–20. https://doi.org/10.1002/jhbp.261.

Buell JF, Cherqui D, Geller DA, et al. World consensus conference on laparoscopic surgery. The international position on laparoscopic liver surgery: the Louisville statement, 2008. Ann Surg. 2009;250:825–30.

Dagher I, Gayet B, Tzanis D, et al. International experience for laparoscopic major liver resection. J Hepatobiliary Pancreat Sci. 2014;21:732–6.

Fancellu A, Rosman AS, Sanna V, et al. Meta-analysis of trials comparing minimally-invasive and open liver resections for hepatocellular carcinoma. J Surg Res. 2011;171:e33–45.

Gagner M, Rheault M, Dubuc J. Laparoscopic partial hepatectomy for liver tumour. Surg Endosc. 1992;6:99.

Hwang DW, Han HS, Yoon YS, et al. Laparoscopic major liver resection in Korea: a multicenter study. J Hepatobiliary Pancreat Sci. 2013;20:125–30.

Lai EC. A step forward in laparoscopic hepatectomy: comments on "Expert consensus on laparoscopic hepatectomy (2013 version) by National Hepatic Surgery Group, Society of Surgery, Chinese Medical Association". Front Med. 2013;7:520–2.

Lai EC, Tang CN, Ha JP, et al. Laparoscopic liver resection for hepatocellular carcinoma: ten-year experience in a single center. Arch Surg. 2009a;144:143–7.

Lai EC, Tang CN, Yang GP, et al. Minimally invasive surgical treatment of hepatocellular carcinoma: long-term outcome. World J Surg. 2009b;33:2150–4.

Lai EC, Tang CN, Yang GP, Li MK. Multimodality laparoscopic liver resection for hepatic malignancy--from conventional total laparoscopic approach to robot-assisted laparoscopic approach. Int J Surg. 2011;9:324–8.

Mirnezami R, Mirnezami AH, Chandrakumaran K, et al. Short- and long-term outcomes after laparoscopic and open hepatic resection: systematic review and meta-analysis. HPB (Oxford). 2011;13:295–308.

Mizuguchi T, Kawamoto M, Meguro M, et al. Laparoscopic hepatectomy: a systematic review, meta-analysis, and power analysis. Surg Today. 2011;41:39–47.

Morise Z, Ciria R, Cherqui D, Chen KH, Belli G, Wakabayashi G. Can we expand the indications for laparoscopic liver resection? A systematic review and meta-analysis of laparoscopic liver resection for patients with hepatocellular carcinoma and chronic liver disease. J Hepatobiliary Pancreat Sci. 2015;22:342–52.

National Hepatic Surgery Group Society of Surgery, Chinese Medical Association. Expert consensus on laparoscopic hepatectomy (2013 version). J Huazhong Univ Sci Technolog Med Sci. 2013;33:791–7.

Nguyen KT, Gamblin TC, Geller DA. World review of laparoscopic liver resection-2,804 patients. Ann Surg. 2009a;250:831–41.

Nguyen KT, Laurent A, Dagher I, et al. Minimally invasive liver resection for metastatic colorectal cancer: a multi-institutional, international report of safety, feasibility, and early outcomes. Ann Surg. 2009b;250(5):842–8.

Nguyen KT, Marsh JW, Tsung A, Steel JJ, Gamblin TC, Geller DA. Comparative benefits of laparoscopic vs open hepatic resection: a critical appraisal. Arch Surg. 2011;146:348–56.

Schiffman SC, Kim KH, Tsung A, Marsh JW, Geller DA. Laparoscopic versus open liver resection for metastatic colorectal cancer: a meta-analysis of 610 patients. Surgery. 2015;157:211–22.

Wakabayashi G, Cherqui D, Geller DA, et al. Recommendations for laparoscopic liver resection: a report from the second international consensus conference held in Morioka. Ann Surg. 2015;261:619–29.

Xiong JJ, Altaf K, Javed MA, et al. Meta-analysis of laparoscopic vs open liver resection for hepatocellular carcinoma. World J Gastroenterol. 2012;18:6657–68.

Yin Z, Fan X, Ye H, Yin D, Wang J. Short- and long-term outcomes after laparoscopic and open hepatectomy for hepatocellular carcinoma: a global systematic review and meta-analysis. Ann Surg Oncol. 2013;20:1203–15.

Zhou YM, Shao WY, Zhao YF, Xu DH, Li B. Meta-analysis of laparoscopic versus open resection for hepatocellular carcinoma. Dig Dis Sci. 2011;56:1937–43.

17 Laparoscopic Liver Resection Using the Lowering of Hilar Plate Approach

17.1 Design of the Operation

This operation was designed based on the following surgical concepts.

1. The Hemihepatic Vascular Inflow Control by Makuuchi

   In 1987, Makuuchi (Fig. 17.1) first reported the use in open surgery the hemihepatic vascular inflow control in liver surgery. He used the conventional technique to dissect out the right or the left hepatic pedicle. Then carried out either the right or the left hemiheptaic vascular inflow control to facilitate right or left partial/total hemihepatectomy (Fig. 17.2).

   The advantages of using hemihepatic vascular inflow control are:

   (a) As the vascular inflow to the hemiliver which forms the future remnant liver is not temporarily occluded during liver resection, there is no ischaemia-reperfusion injury to this part of the liver.

   (b) As the main portal vein is not occluded, there is no portal venous congestion of the internal organs. Normally the main portal vein can be occluded for up

Fig. 17.1 Professor Masatoshi Makuuchi

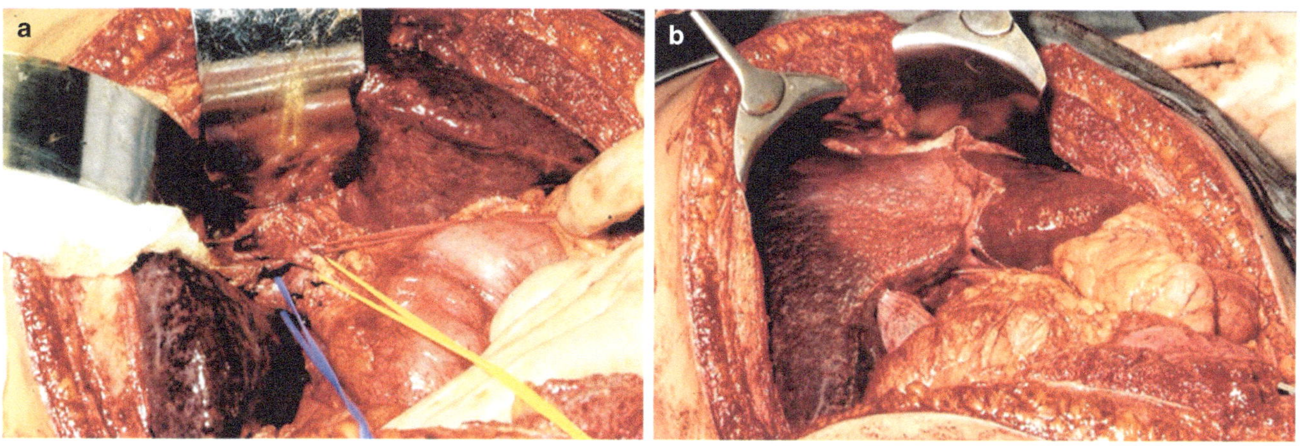

Fig. 17.2 (**a**) Dissect out the branches in the right hepatic pedicle. (**b**) Line of demarcation between right and left livers

W. Y. Lau, *Applied Anatomy in Liver Resection and Liver Transplantation*, https://doi.org/10.1007/978-981-16-0800-1_17

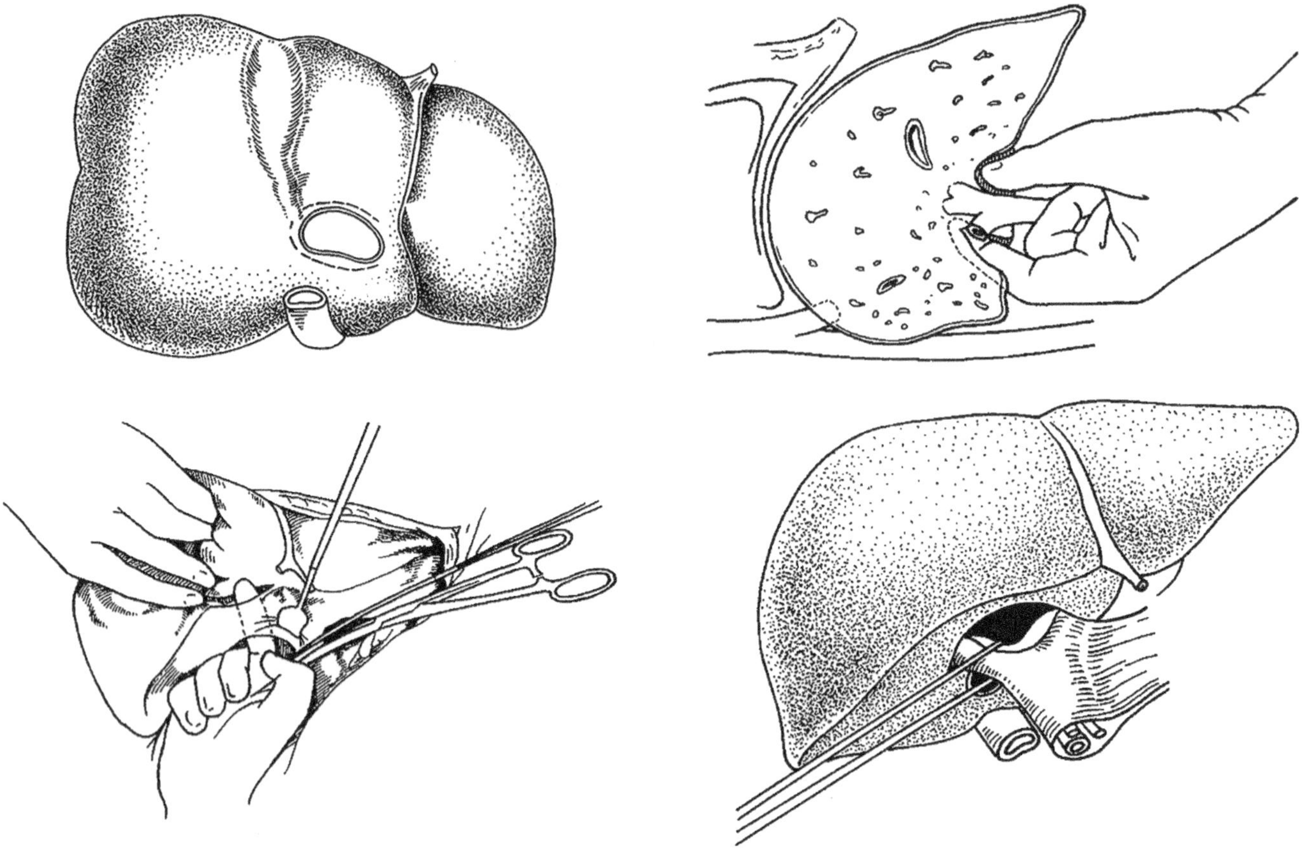

Fig. 17.3 Intrahepatic Glissonian approach by Launois and Jameison

to 1 h, in hemihepatic vascular inflow control, there is no such a time limit.

(c) Patient's hemodynamic can be more stable with hemihepatic vascular inflow control.

The disadvantages are:

(a) Blood loss can come from the hemiliver whose vascular inflow has not been occluded.

(b) This operation requires a high technical skill.

(c) In laparoscopic surgery, the demand for technical skill is even higher.

(d) If less than a hemihepatectomy is carried out on the hemiliver whose vascular inflow has been occluded for long, ischaemia-reperfusion injury can still happen to this part of the liver.

2. The Open Surgery with Intrahepatic Glissonian Approach by Launois and Jameison (Fig. 17.3)

This approach, when used in open surgery, has been described in detail in Sects. 4.3, 13.2 and 15.3.

In laparoscopic liver resection, Topal et al. (2007), Cho et al. (2007) and Machado et al. (2008, 2009) have reported the use of this approach to control either the right or the left hepatic pedicle by establishing a tunnel through the liver tissues just above the junction of these two pedicles. Following this, either right/left hemihepatic vascular inflow occlusion can be carried out laparoscopically. The advantage of this approach is liver parenchymal transection usually takes longer in laparoscopic surgery. As prolonged continuous Pringle's manoeuvre is not advisable and, intermittent Pringle can lead to bleeding from the raw surfaces of the transected liver during hilar unclamping, continuous hemihepatic vascular inflow occlusion solves these problems. The main disadvantage is tunnelling through the liver parenchyma can result in bleeding, especially in patients with a cirrhotic liver. Even though the bleeding may not be too much and the bleeding can stop with time, it still affects the laparoscopic view once bleeding occurs.

It takes a third surgical concept to solve this problem.

3. The Lowering of the Hilar Plate

Hepp and Couinaud in 1956 first described this technique in French to dissect the left hepatic duct for left hepaticojejunostomy to relieve malignant obstructive jaundice in patients with unresectable tumours. Leslie Blumgart popularised this approach in the English-speaking world. Lau WY learned this technique from Blumgart and modified this technique into a technique to control either the right or the left hepatic pedicle (Fig. 17.4a, b), and teaches this technique in his Liver Resection Workshops carried out all over China (12 workshops in different cities in China up to mid-2015). When this technique was brought to Shanghai, a series of papers on comparison between hemihepatic and total hepatic vascular inflow occlusion was published by Lau W.Y. together with Professor Zhou W.P. and Professor Wu M.C. (Fig. 17.5).

- A prospective randomised controlled trial to compare two methods of selective hepatic vascular exclusion in partial hepatectomy. Eur J Surg Oncol 2013;39:125–130.
- Selective hepatic vascular exclusion versus Pringle manoeuvre in liver resection for tumours encroaching on major hepatic veins. Br J Surg 2012;99:937–973.
- A prospective randomised controlled trial to compare Pringle manoeuvre, hemihepatic vascular inflow occlusion, and main portal vein inflow occlusion in partial hepatectomy. Am J Surg 2011;201:62–69.
- Liver resection with relative hepatic vascular exclusion: a cohort study. Ann Surg 2009;209:624–627.

17.2 Development of Laparoscopic Lowering of Hilar Plate Approach

When the concept of open lowering of the hilar plate approach to control the right/left hepatic pedicles reached Foshan, this approach evolved and was first attempted to be carried out laparoscopically by the three surgeons (Fig. 17.6).

The laparoscopic approach is conducted using ports as shown in Fig. 17.7. After the hilar plate has been dissected and opened anteriorly, it is lowered, and a biliary bougie is then used to create a tunnel between the liver plate and the liver parenchyma on the right or the left side to control either the right or the left hepatic pedicle (Fig. 17.8).

The initial results have been reported and they are very encouraging (Laparoscopic liver resection under hemihepatic vascular inflow control using the lowering of the hilar plate approach. Chen YL, et al. Hepatobiliary Pancreat Dis Int 2014;13:508–12). The mean hemihepatic vascular occlusion time was 55 min (range 30–86 min). The mean time taken to achieve hemihepatic vascular occlusion was 24 min (range 5–45 min). The mean operative blood loss was 361 mL (range 70–1500 mL). No patient required a blood transfusion.

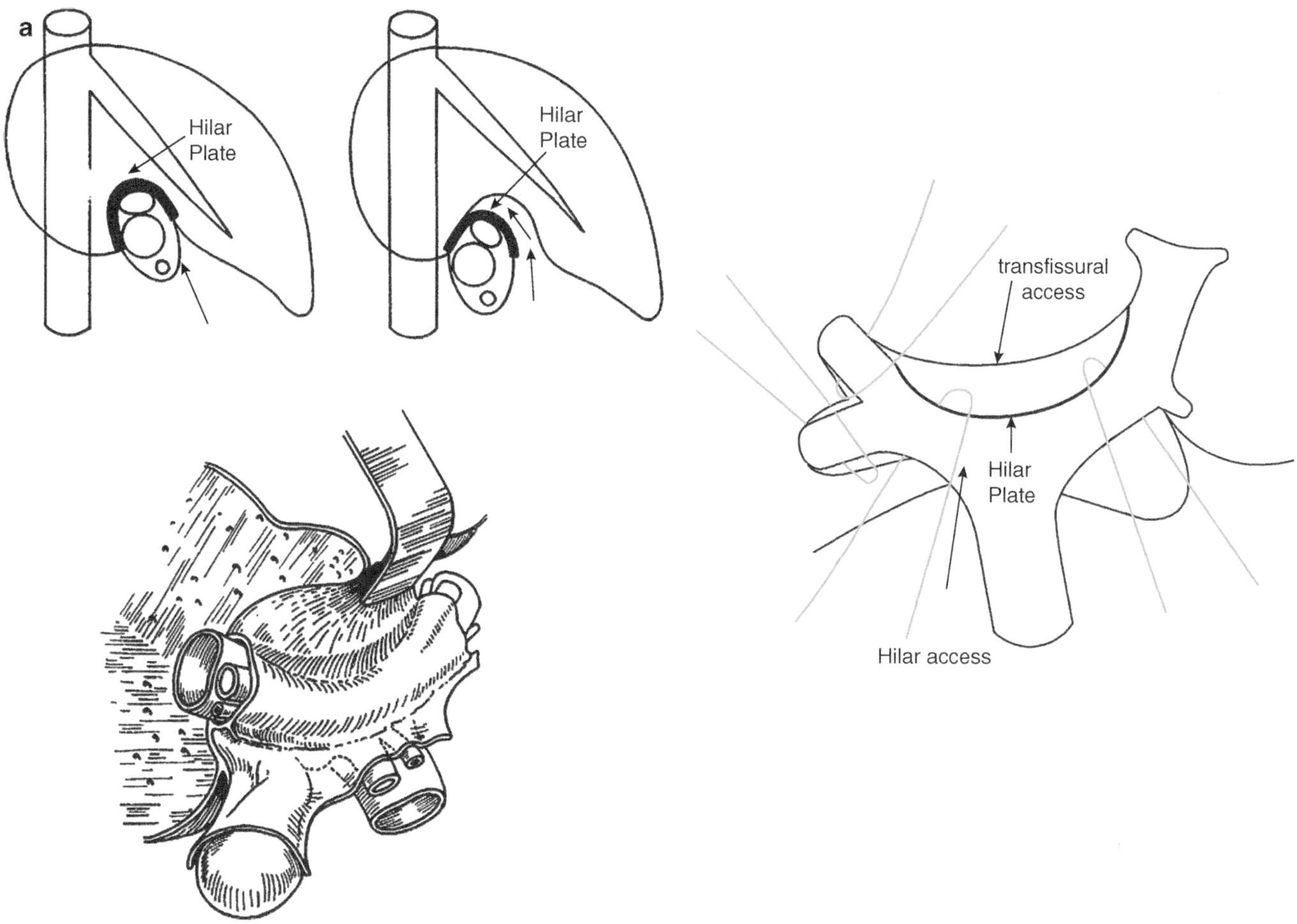

Fig. 17.4 Open surgery in lowering of the liver plate

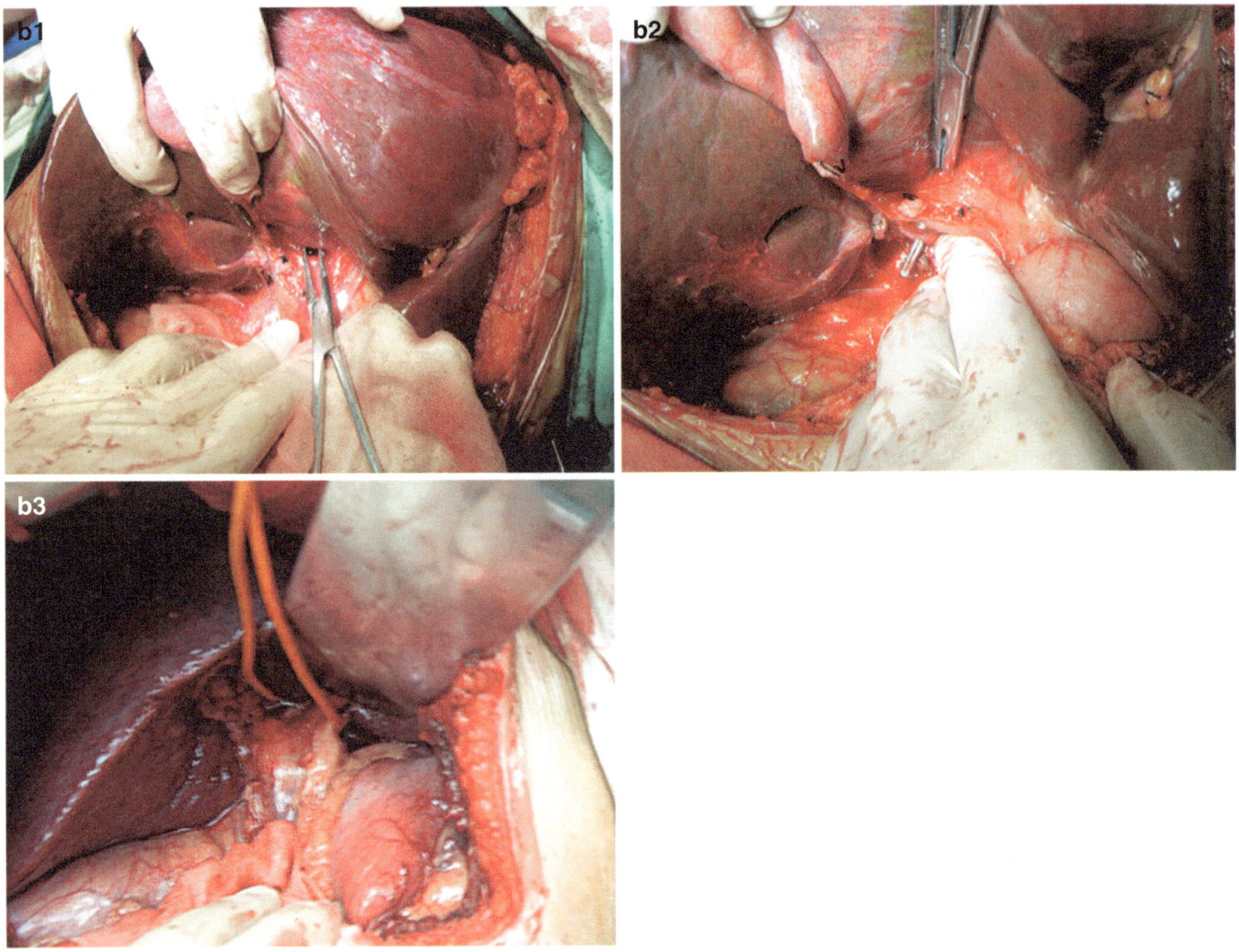

Fig. 17.4 (continued)

Fig. 17.5 (**a**) Leslie Blumgart. (**b**) Lau WY. (**c**) Zhou WP. (**d**) Wu MC

Fig. 17.6 (**a**) Zhen ZJ. (**b**) Chen HW. (**c**) Chen YJ

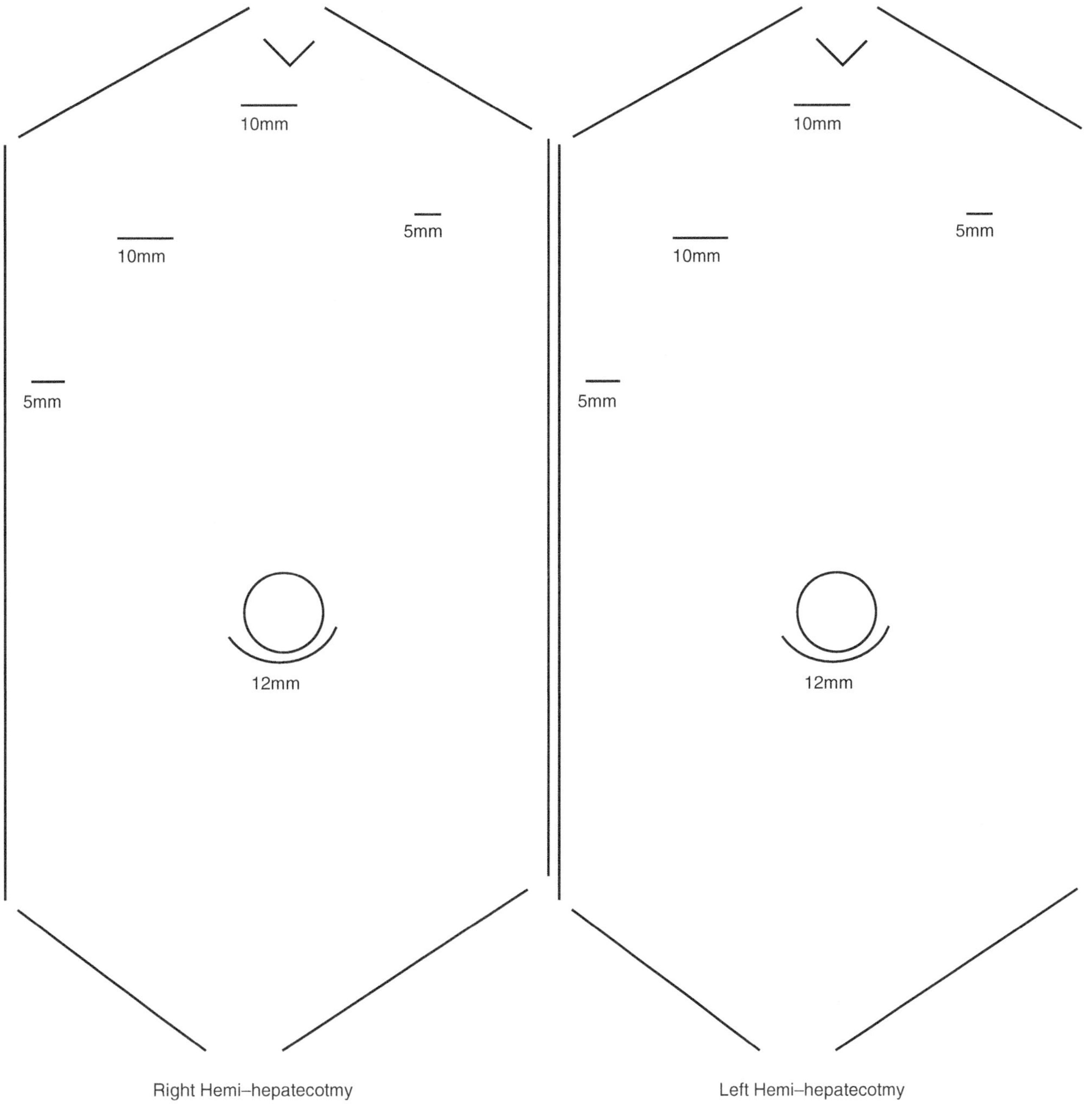

Fig. 17.7 Ports used in the laparoscopic lowering of liver plate

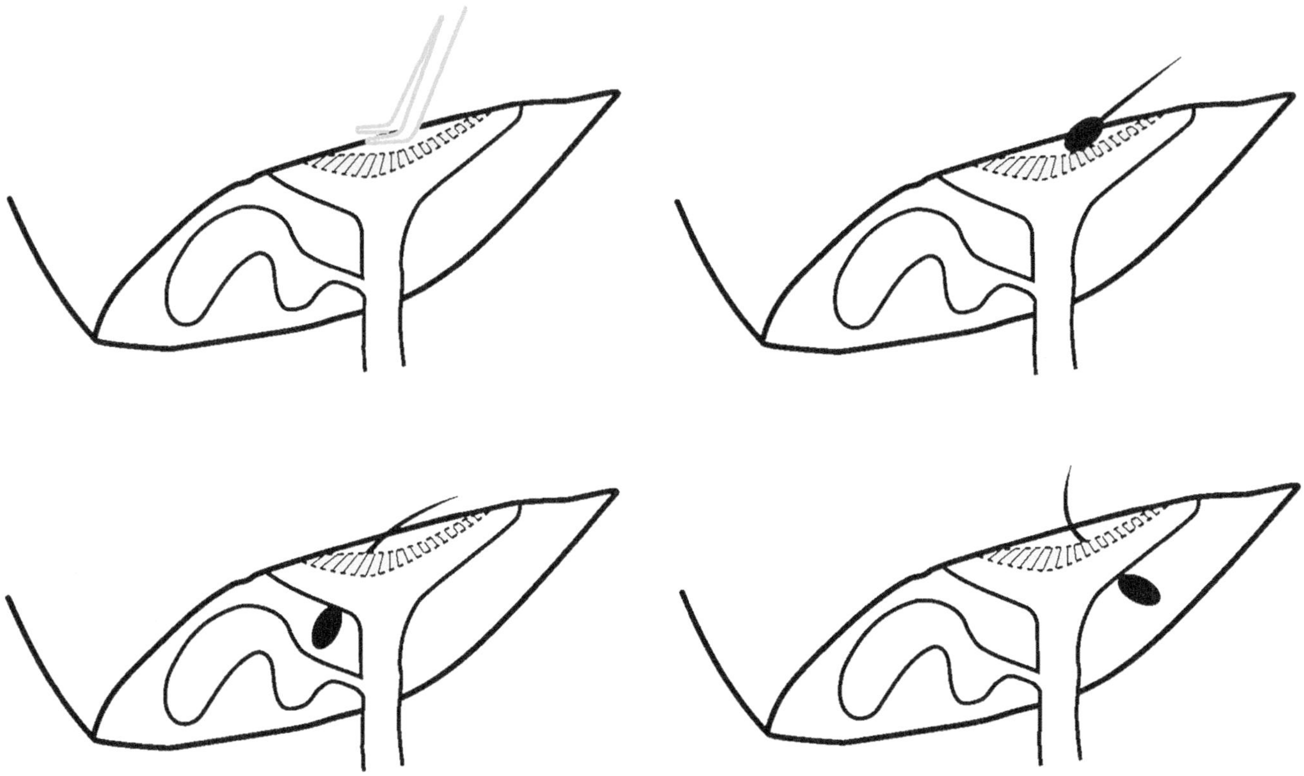

Fig. 17.8 Use of a biliary bougie to create a tunnel to control right/left pedicle

17.3 Conclusion

This laparoscopic lowering of liver plate approach to control either the right or the left hepatic pedicle for hemihepatic vascular inflow occlusion is a new technique. The reason why I include this operation into this book is because this technique has a lot of advantages. It should therefore be popularised so that it can be applied more often in laparoscopic liver resectional surgery.

Further Reading

Blumgart LH, Hann LE. Chapter 1: Surgical and radiologic anatomy of the liver and biliary tract. In: Blumgart LH, Fong Y, editors. Surgery of the liver and biliary tract, vol. 1. 3rd ed. London: W.B. Saunders; 2000. p. 3–33.

Cho A, Asano T, Yamamoto H, et al. Laparoscopy-assisted hepatic lobectomy using hilar Glissonean pedicel transection. Surg Endosc. 2007;21:1466–8.

Hepp J, Couinaud C. Approach to and use of the left hepatic duct in reparation of the common bile duct. Presse Med. 1956;64:947–8. (French)

Jameison G, Launois B. Chapter 2: Liver resection and liver transplantation: the anatomy of the liver and associated structures. In: Jameison GG, editor. The anatomy of general surgical operations. Edinburgh: Elsevier; 2006. p. 8–23.

Launois B, Tay KH. Chapter 19: Intrahepatic Glissonian approach. In: Lau WY, editor. Hepatocellular carcinoma. Singapore: World Scientific; 2008. p. 429–46.

Machado MA, Makdissi FF, Surjan RC, et al. Laparoscopic resection of left liver segments using the intrahepatic Glissonian approach. Surg Endosc. 2009;23:2615–9.

Makuuchi M, Mori T, Gyuvén P, et al. Safety of hemihepatic vascular occlusion during resection of the liver. Surg Gynecol Obstet. 1987;164:155–8.

Topal B, Aerts R, Penninckx F. Laparoscopic intrahepatic Glissonian approach for right hepatectomy is safe, simple, and reproducible. Surg Endosc. 2007;21:2111.

Robotic Liver Surgery

18

18.1 Development of Robotic Surgery

The use of robots in surgery has come about within the past 25 years. Its introduction has revolutionised the field of minimally invasive surgery. It was developed to overcome the disadvantages of conventional laparoscopic surgery. Well-known advantages of the robotic system over conventional laparoscopic surgery include improved vision via three-dimensional view, magnification, tremor suppression and flexibility of instruments allowing precise operating techniques in a variety of complex procedures in general surgery. These features allow the surgeon to perform delicate tissue dissection and precise intracorporeal suturing. The main drawback of robotic surgery is the associated cost. At the current stage of development, the benefits of the robotic approach in liver surgery have not yet been defined, and its technique has not been standardised. Every new technique should undergo a reliable assessment of safety and efficacy before it is adopted as a clinical standard. Only a small number of studies have been reported though the technique of robotic-assisted laparoscopic (robotic) hepatectomy has been available in the medical literature. It should be emphasised that proper use of robotic hepatectomy requires four conditions: (1) appropriate patient selection; (2) following the principles of open liver surgery; (3) specific expertise and training in both liver and laparoscopic surgery; (4) familiarisation with the robotic machine and appropriate knowledge of its potential dangers so that appropriate measures can be taken to prevent inadvertent visceral injury by the robotic arms because of a total loss of tactile feedback.

18.2 The Components of the Robotic System

The da Vinci surgical system (Intuitive Surgical, Sunny Valley CA, USA) is the only commercially available therapeutic robotic system in the market at this moment. This is a master-slave system rather than a true autonomous robot. The surgeon sits at a console remote from the patient, controlling the camera and robotic arms which are docked through laparoscopic ports. The da Vinci surgical system consists of three parts: (1) a patient cart with four interactive robotic arms, a vision cart and a surgeon console. The patient cart holds up to three instruments, and one 3D camera being connected to the surgeon console via a cable. The central arm holds the camera system, and the three lateral arms hold the surgical instruments. The multiple robotic arms and distal instrument joints allow seven degrees of freedom, as the human wrist in open surgery, which is named as 'EndoWrist' (Fig. 18.1).

The EndoWrist instruments are multiple-use because they are sterilisable and reusable for a defined number of procedures. The vision cart of the robot is placed adjacent to the patient cart. There are two camera control units in the vision cart. Two image synchronisers and a focus controller provide the surgeon sitting at the console with a high-quality three-dimensional view of the surgical field with adjustable magnification. The surgeon console consists of a binocular viewer

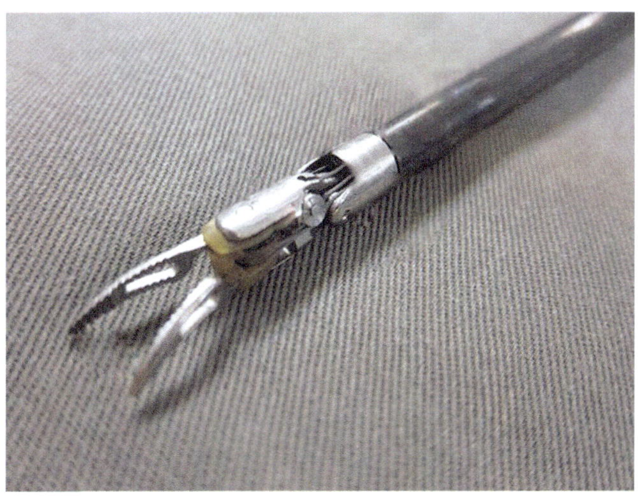

Fig. 18.1 Seven degrees of freedom of robotic instrument arm, Endowrist

W. Y. Lau, *Applied Anatomy in Liver Resection and Liver Transplantation*, https://doi.org/10.1007/978-981-16-0800-1_18

Table 18.1 Advantages and disadvantages between the robotic and conventional laparoscopic approaches

Conventional laparoscopic approach	Robot-assisted laparoscopic approach
Advantages	
• Well-developed technique with more scientific evidence	
• Less expensive	• Three-dimensional stereoscopic view with magnification
	• Camera stability
	• Improved dexterity with seven degrees of freedom
	• Facilitate anastomoses
	• Elimination of fulcrum effect
	• Elimination of physiologic tremors
	• Motion scaling
	• Improved ergonomics for surgeons
	• Allow telesurgery
Disadvantages	
• Loss of tactile feedback	
• Two-dimensional view	
• Compromised dexterity with limited degrees of motion	
• Fulcrum effect	• Separation surgeon from the operating field
	• High costs (Robotic system, maintenance cost, start-up)
	• Absence of tactile feedback
	• New technology with limited scientific evidence

of the 'Insite' vision system, the instrument controllers, a series of foot control pedals, and the computer hardware and software.

18.2.1 Pros and Cons of the Robotic Approach Versus the Conventional Laparoscopic Approach

A summary of the advantages and disadvantages of the robotic approach versus the conventional laparoscopic approach is shown in Table 18.1.

With these advantages, the robotic approach may add additional benefits to the minimally invasive liver surgery, such as (1) to overcome difficulty in laparoscopic suturing for bleeding in the hepatic parenchyma (Fig. 18.2); (2) to perform hilar dissection, and vascular pedicle control (Fig. 18.3); (3) to perform lymph nodes dissection (Fig. 18.4); (4) to perform hepatectomies requiring biliary-enteric reconstruction (Fig. 18.5); (5) to shorten learning curve; and (6) to extend indications of laparoscopic liver surgery. However, its real benefits still need more studies to evaluate.

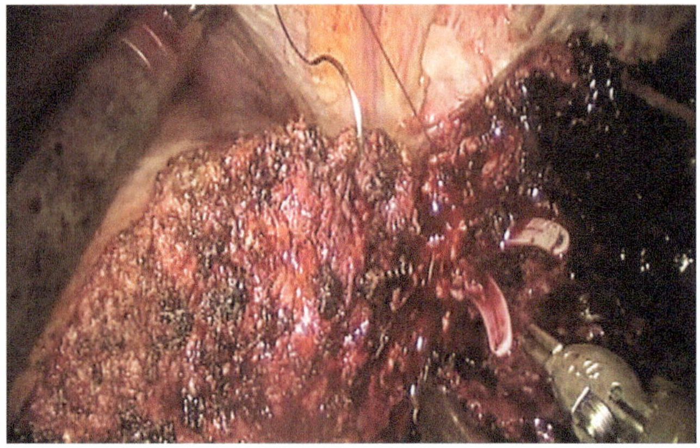

Fig. 18.2 Suturing of bleeding hepatic parenchyma

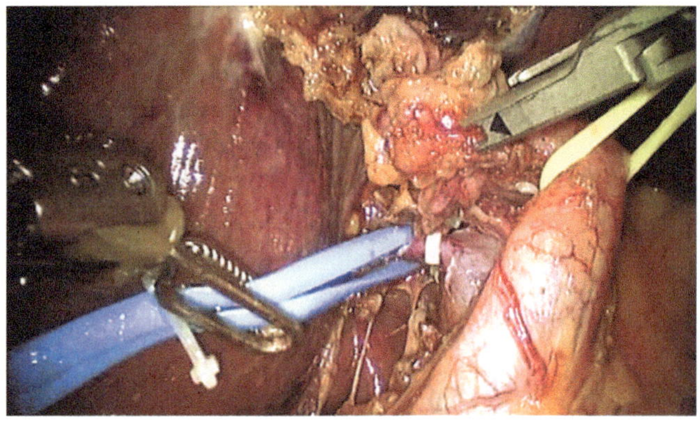

Fig. 18.3 Hemi-vascular inflow control

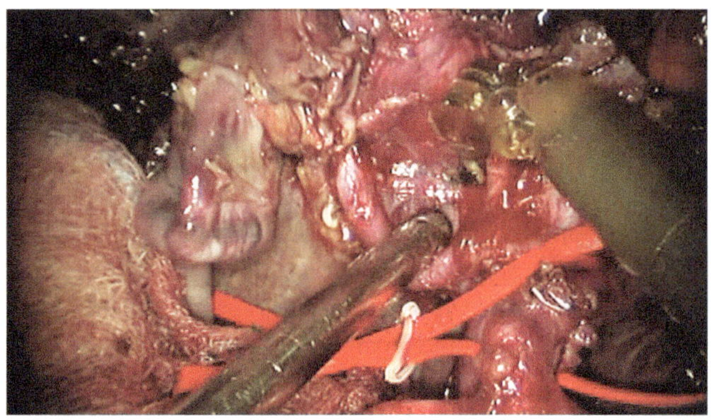

Fig. 18.4 Porta lymph nodes dissection

18.3 Robotic Hepatectomy

18.3.1 Perioperative Outcomes After Robotic Hepatectomy

Clinical trials comparing robotic hepatectomy with open hepatectomy are not available yet. No randomised clinical trial exists. A number of case series and nonrandomised clinical trials comparing the robotic technique with the conventional laparoscopic technique for hepatectomies are available (Lai et al. 2011; Ji et al. 2011; Giulianotti et al. 2011a; Berber et al. 2010; Packiam et al. 2012; Tsung et al. 2014; Spampinato et al. 2014; Tranchart et al. 2014; Wu et al. 2014; Lai et al. 2013). The liver transection techniques and energy devices used in the various series are different in these studies (Fig. 18.6).

Ho et al. reported a systematic review of 19 series, which included 217 patients (Ho et al. 2013). The most commonly performed procedures were wedge resection and segmentectomy. Right hepatectomy has been performed in a few specialised centres. The conversion to open surgery and complication rates were 4.6% and 20.3%, respectively. The most common reason for conversion to open surgery was unclear tumour margin. Intra-abdominal fluid collection was the most frequent morbidity. The mean operating time was 200–507 min. The mean intraoperative blood loss was 50–660 mL, with a tendency toward increased blood loss in series, which included major hepatectomies. The mean postoperative hospital stay was 5.5–11.7 days. Based on these nonrandomised studies, the data suggested that robotic and laparoscopic hepatectomies showed no differences in intraoperative and post-operative outcomes as measured by intraoperative blood loss, blood transfusion rate, morbidity rate, mortality rate, R_0 resection rate and length of hospital stay (Berber et al. 2010; Packiam et al. 2012; Tsung et al. 2014; Spampinato et al. 2014; Tranchart et al. 2014; Wu et al. 2014; Lai et al. 2013). The robotic approach tended to have longer operating times in these studies. Comparison of operating times between the different studies is often difficult because of the lack of uniformity in the definition. Some authors referred to a 'total operating time' to include the 'robot set-up and docking time', whereas others referred to a 'procedure time', a separate 'system time' (from positioning the robot over the patient to disconnection of the robot) and a 'dissection time' (surgeon's active time at the console). Others calculated the time from 'induction of anesthesia to incision' or from 'incision to extubation'. The benefits of robotic hepatectomy have been shown in several studies (Tsung et al. 2014; Tranchart et al. 2014; Wu et al. 2014; Lai and Tang 2014; Casciola et al. 2011; Giulianotti et al. 2011b; Lai et al. 2012). Both Tsung et al. and Wu et al. showed the robotic surgical system facilitated the completion of hepatectomies by the totally minimally invasive approach and allowed a higher proportion of major hepatectomies (Tsung et al. 2014; Wu et al. 2014). The case series of Lai et al., Casciola et al. and the nonrandomised studies of Tranchart et al. showed the robotic surgical system facilitated resection of tumours located in the superior and posterior liver segments (Tranchart et al. 2014; Lai and Tang 2014; Casciola et al. 2011).

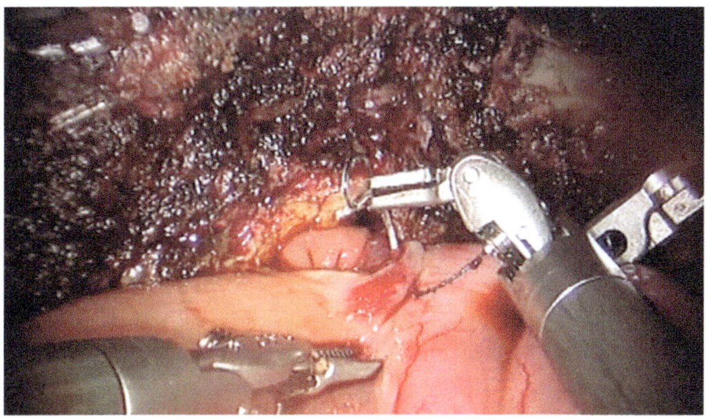

Fig. 18.5 Right hepatico-jejunostomy

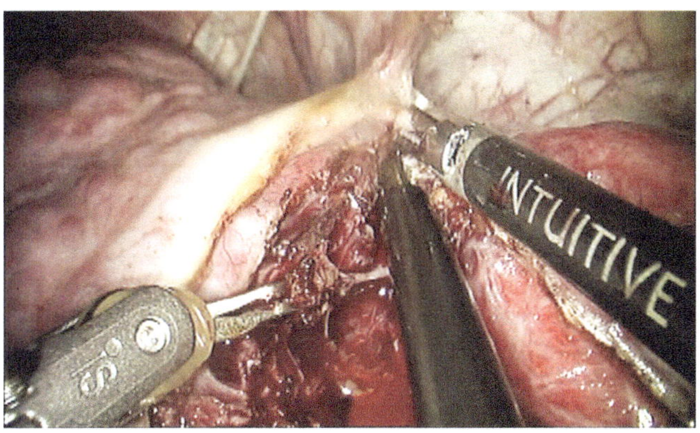

Fig. 18.6 Liver parenchymal transection with ultrasonic dissector and Harmonic scalpel

18.3.2 Oncological Outcomes After Robotic Hepatectomy for Malignancies

At present, available survival data on robotic hepatectomy for liver malignancies in the medical literature are very limited. Difficult learning curves, adequate resection margins, tumour seeding, metastases to wounds, and long-term outcomes are the major concerns in robotic surgery for malignancies. Mainly hepatocellular carcinoma (HCC) and colorectal liver metastases (CRLM) have been included, and there have been very few cases of non-colorectal liver metastases and cholangiocarcinoma which have been included in these series. No port-site recurrence has been reported. However, specific survival data have been very limited. Only Lai et al. reported on the short-term outcomes in a homogeneous series of robotic hepatectomy for HCC. Forty-one consecutive patients with HCC underwent 42 robotic liver resections (Lai et al. 2013). Five resections were carried out for recurrent HCC. In 23.8%, the operations were hemi-hepatectomies. The mean operating time and blood loss were 229.4 min, and 412.6 mL, respectively. The R_0 resection rate was 93%. The hospital mortality and morbidity rates were 0% and 7.1%, respectively. The mean hospital stay was 6.2 days. The 2-year overall and disease-free survivals were 94% and 74%, respectively. In the study by Choi et al., there was no recurrence in 13 patients with HCC at a median follow-up of 11 months (Choi et al. 2012). In the study by Giulianotti et al., 17 patients with hepatic malignancies (HCC, $n = 1$; CRLM, $n = 11$; Non-colorectal liver metastases, $n = 4$; Hepatoblastoma, $n = 1$) were reported (Giulianotti et al. 2011b). Among those patients with CRLM, 9 of 11 were alive and disease free at a mean follow-up of 36 months. One patient with HCC was alive and disease free 6 months after surgery. In Casciola et al., after a mean follow-up of 25.1 months, among 19 patients with malignant tumours (HCC, $n = 3$; CRLM, $n = 14$; Clear cell renal metastasis, $n = 1$; gallbladder carcinoma metastasis, $n = 1$), one patient with HCC died because of tumour progression (Casciola et al. 2011); two patients with CRLM died because of tumour progression after discontinuation of the adjuvant chemotherapy with hepatic, nodal and peritoneal relapse; three patients with CRLM were alive but with the disease.

Based on the current evidences, the role of the robotic approach for hepatic malignancies is still unclear because of the uncertainty of the short and long-term results and the fear of compromising oncological resection margins. A controlled trial is needed to compare the long-term oncologic results of robotic hepatectomy with those of open surgery.

18.4 Future Development

Robotic surgery has its disadvantages, such as high costs and a long set-up time when compared with open and laparoscopic surgery. In the future, the set-up time may be reduced with increasing experience with robotic surgery. The high costs of robot-assisted procedures are likely to decrease with upcoming competitors' surgical robotic systems. The robotic surgical system can be used in a more efficient way by the following measures:

- Establish a high-volume centre
- Create a specialised robotic surgery unit
- Train dedicated operating theatre staffs
- Reduce the number of disposable instruments per operation
- Reduce set-up time
- Shorten learning curve with the help of expert surgeons at the training period
- Increase multidisciplinary use and overall annual use of the robotic system

18.5 Conclusion

The robotic surgical approach offers an alternative to the conventional laparoscopic hepatectomy for minimally invasive hepatectomies. Unlike other procedures, robotic hepatectomy requires a team approach which should include a highly skilled laparoscopic surgeon at the patient's side to manage the complex instruments and techniques. Comparative data for hepatectomies showed that the operating time tends to be longer with the robotic approach, and morbidity, mortality and post-operative hospital stay were similar. Robotic hepatectomy may have benefits over laparoscopic hepatectomy in major hepatectomies and resection of tumours in technically difficult liver segments. The current medical literature lacks sufficient data on the oncological outcomes after robotic hepatectomy for liver malignancies. Trials are required to evaluate the benefits of robotic hepatectomies, to define indications for robotic liver surgery and to provide a cost-benefit analysis for both patients and institutions. The future implementation of robotic hepatectomy will depend on the advantages that it can provide over conventional laparoscopic or open surgery.

References

Berber E, Akyildiz HY, Aucejo F, et al. Robotic versus laparoscopic resection of liver tumours. HPB (Oxford). 2010;12:583–6.

Casciola L, Patriti A, Ceccarelli G, Bartoli A, Ceribelli C, Spaziani A. Robot-assisted parenchymal-sparing liver surgery including lesions located in the posterosuperior segments. Surg Endosc. 2011;25:3815–24.

Choi GH, Choi SH, Kim SH, Hwang HK, Kang CM, Choi JS, Lee WJ. Robotic liver resection: technique and results of 30 consecutive procedures. Surg Endosc. 2012;26:2247–58.

Giulianotti PC, Coratti A, Sbrana F, et al. Robotic liver surgery: results for 70 resections. Surgery. 2011a;149:29–39.

Giulianotti PC, Sbrana F, Coratti A, et al. Totally robotic right hepatectomy: surgical technique and outcomes. Arch Surg. 2011b;146:844–50.

Ho CM, Wakabayashi G, Nitta H, et al. Systematic review of robotic liver resection. Surg Endosc. 2013;27:732–9.

Ji WB, Wang HG, Zhao ZM, Duan WD, Lu F, Dong JH. Robotic-assisted laparoscopic anatomic hepatectomy in China: initial experience. Ann Surg. 2011;253:342–8.

Lai EC, Tang CN. Robot-assisted laparoscopic partial caudate lobe resection for hepatocellular carcinoma in cirrhotic liver. Surg Laparosc Endosc Percutan Tech. 2014;24:e88–91.

Lai EC, Tang CN, Yang GP, Li MK. Multimodality laparoscopic liver resection for hepatic malignancy—from conventional total laparoscopic approach to robot-assisted laparoscopic approach. Int J Surg. 2011;9:324–8.

Lai EC, Tang CN, Li MK. Robot-assisted laparoscopic hemi-hepatectomy: technique and surgical outcomes. Int J Surg. 2012;10:11–5.

Lai EC, Yang GP, Tang CN. Robot-assisted laparoscopic partial hepatectomy for hepatocellular carcinoma: short-term outcome. Am J Surg. 2013;205:697–702.

Packiam V, Bartlett DL, Tohme S, et al. Minimally invasive partial hepatectomy: robotic versus laparoscopic left lateral sectionectomy. J Gastrointest Surg. 2012;16:2233–8.

Spampinato MG, Coratti A, Bianco L, et al. Perioperative outcomes of laparoscopic and robot-assisted major hepatectomies: an Italian multi-institutional comparative study. Surg Endosc. 2014;28:2973–9.

Tranchart H, Ceribelli C, Ferretti S, et al. Traditional versus robot-assisted full laparoscopic partial hepatectomy: a matched-pair comparative study. World J Surg. 2014;38:2904–9.

Tsung A, Geller DA, Sukato DC, et al. Robotic versus laparoscopic hepatectomy: a matched comparison. Ann Surg. 2014;259:549–55.

Wu YM, Hu RH, Lai HS, Lee PH. Robotic-assisted minimally invasive partial hepatectomy. Asian J Surg. 2014;37:53–7.

Liver Transplantation 19

19.1 Cadaveric Liver Transplantation

19.1.1 Liver Procurement

The operation is often performed with simultaneous removal of other organs for transplantation usually, the heart and kidneys, sometimes also the lungs and pancreas, and occasionally the small bowel. Before the operation, the various teams should pre-plan the operation in order to adapt the technique to the need of the various teams.

19.1.2 Technique for a Haemodynamically Stable Patient

Exposure is obtained using a median sternotomy and a midline laparotomy down to the pubis. The round, falciform and left triangular ligaments are taken down. Hepatic arterial anatomy is delineated by inspection of the gastrohepatic ligament, and porta hepatis and the lesser omentum to identify aberrant arterial anatomy. The right colon and the small intestinal mesentery are mobilised and reflected towards the donor left upper quadrant. The inferior mesenteric vein is identified, controlled and slung. The infrarenal aorta is then exposed at the bifurcation with proximal and distal control obtained. The intestines are returned to the left lower abdomen. The distal common bile duct is circumferentially dissected, distally ligated and transected. The gallbladder is then opened and flushed out through the choledochostomy (Fig. 19.1). After reflection of the left liver medially, control of the supra-coeliac aorta is obtained by dividing the diaphragmatic crura. The donor is ready for heparinization and cannulation. Heparin of 300 units/kg body weight is given.

The aorta is cannulated just above the aortic bifurcation. A strong ligature is tied distally to occlude the distal aorta. A more proximal ligature is tied to fix the cannula. The inferior mesenteric vein is cannulated and tied in position. The donor is exsanguinated by opening the suprahepatic vena cava or distal inferior vena cava. The aorta is cross clamped in the supra-coeliac location and the arterial and portal circulations are flushed with University of Wisconsin (UW) solution via the two cannulae placed previously. The abdominal cavity is packed with ice slush for topical cooling (Fig. 19.2a, b).

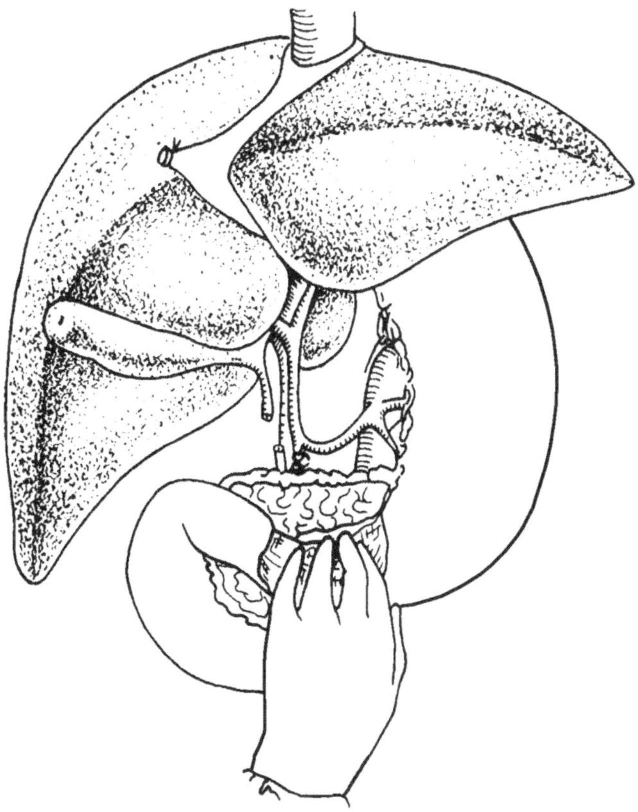

Fig. 19.1 The common bile duct is divided. The gallbladder is opened and flushed

After satisfactory cooling, the diaphragm is dissected around the suprahepatic inferior vena cava. Portal dissection is then indicated to delineate the arterial anatomy and trace the arterial inflow back to the aorta. The gastroduodenal artery is divided while dissecting the common hepatic artery down to the coeliac trunk. The left gastric artery and the splenic artery are divided. Next, the portal vein is dissected

W. Y. Lau, *Applied Anatomy in Liver Resection and Liver Transplantation*, https://doi.org/10.1007/978-981-16-0800-1_19

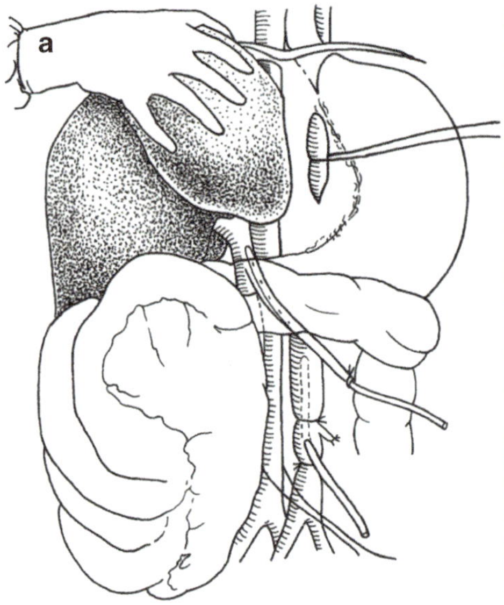

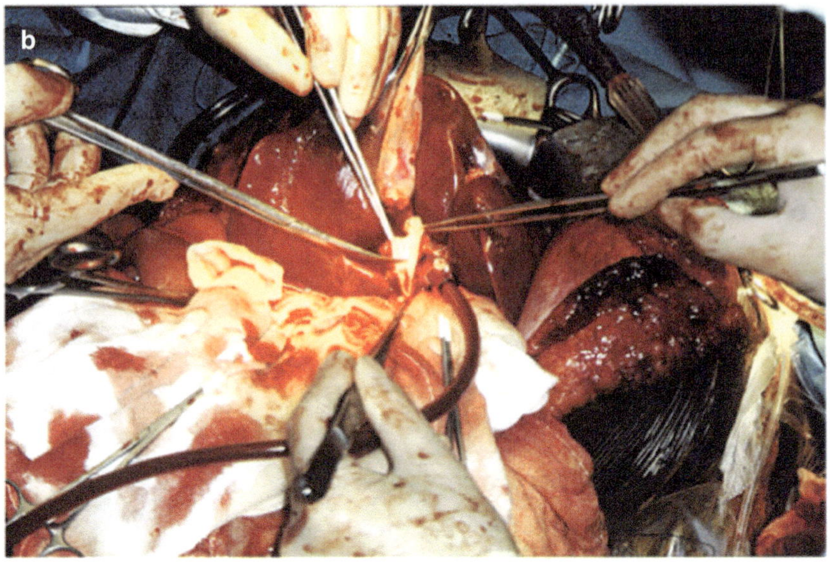

Fig. 19.2 (a) Cannulation of the aorta and the inferior mesenteric vein before cold perfusion. (b) Operative photograph showing cannula in inferior mesenteric vein. The common bile duct was being divided

and divided near the pancreas. Modification is necessary if procurement of the pancreas is being simultaneously performed. A search for replaced or accessory right hepatic artery is also undertaken through identification of the superior mesenteric artery and its proximal branches. Preservation of this artery, if present, is essential. Excision of the coeliac trunk with a Carrel patch of the aorta is performed. The inferior vena cava is divided above the renal veins. All tissues and diaphragm between the right kidney and liver are then divided, and the hepatic graft is removed from the donor and packed in ice.

19.1.3 Rapid Procurement Technique

Rapid procurement technique is required in a non-beating heart donor or in a patient with sharp drops in donor haemodynamics during procurement. The operation starts with cannulation of the distal aorta and the inferior mesenteric vein and the preparation of a section of the supra-coeliac aorta so that clamping can be carried out without further dissection of the hepatic pedicle. Once perfusion of the splanchnic organs is satisfactory, the procurement procedure is similar to those described in the conventional technique. There have been several modifications to the rapid procurement technique. In particular, a simpler technique of multiple donor organ retrieval by perfusion via the aorta only has been shown to be equally effective. In this technique, interruption and cannulation of the portal vein are not needed. Effective liver perfusion occurs via the aorta and the hepatic artery, but also via the portal vein after the fluid has transversed the intestinal circulatory bed (Fig. 19.3).

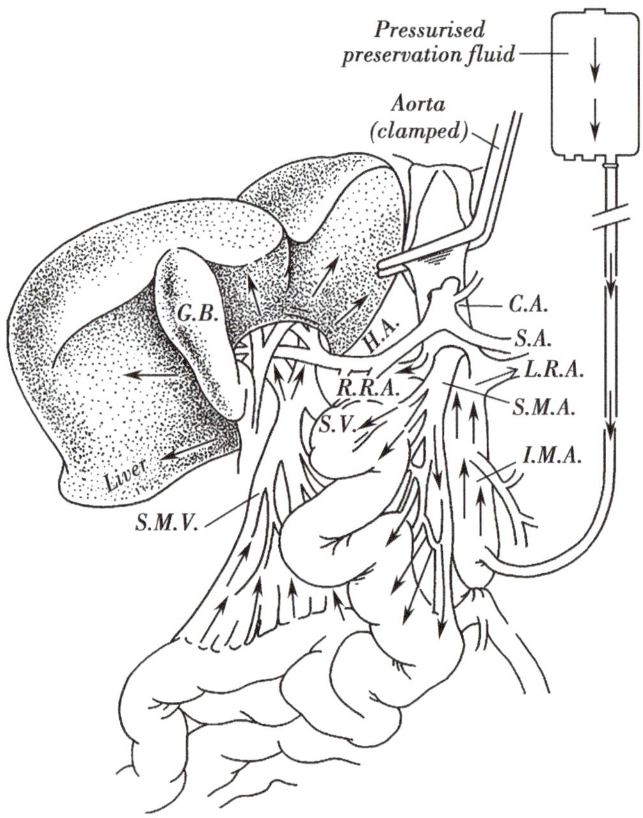

Fig. 19.3 The rapid infusion of cold solution into the aorta alone cools the liver via the hepatic arteries, but also through the portal vein after the fluid has traversed the intestinal circulation. *CA* celiac artery; *SA* splenic artery; *HA* common hepatic artery; *SMA* superior mesenteric artery; *SMV* superior mesenteric vein; *LRA* left renal artery; *RRA* right renal artery; *SV* splenic vein; *GB* gallbladder

19.1.4 Bench Surgery

The final preparation of the donor graft is done in ice cold preservative fluid and it consists of removal of the diaphragm and preparation of the suprahepatic inferior vena cava. The adrenal gland is removed, the adrenal vein ligated and the infrarenal inferior vena cava prepared. The portal vein is dissected up to the bifurcation and cannulated. The coeliac trunk is dissected up to the gastroduodenal artery with all unnecessary branches ligated. During this phase, any arterial reconstruction and conduits are performed. All the vascular cuffs are prepared for anastomosis (Fig. 19.4).

There are many anomalies in the hepatic artery (Fig. 19.5).

While most of the anomalies of the hepatic artery can be solved by ligation of the unnecessary branches and then using a Carrel patch of the aorta, the anomalies which requires special preparation during bench surgery is the anomaly (D) in Fig. 19.5 with the left hepatic artery from the left gastric artery and the anomaly (C) with the right hepatic artery from the superior mesenteric artery.

There are two ways to solve this problem. The most commonly used option is to trim the Carrel patch of the aorta as shown in Fig. 19.6a to form two patches. The two Carrel patches are sutured together as shown in Fig. 19.6b with interrupted 5.0 prolene stitch. The distal stump of the superior mesenteric artery is then used for anastomosis in the recipient.

The alternative technique is shown in Fig. 19.7a, b. The right hepatic artery is divided from the superior mesenteric artery with its Carrel aortic patch. This right hepatic artery is anastomosed to the stump of the splenic artery or to the gastroduodenal artery. The discrepancy between the size of the vessels determines which artery stump the right hepatic artery should be anastomosed. The Carrel patch of the aorta can be anastomosed to the recipient.

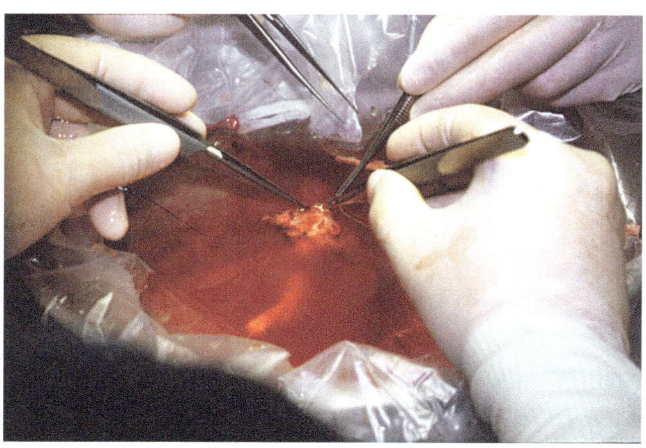

Fig. 19.4 Bench surgery

19.1.5 Recipient Hepatectomy

This is the most demanding part of liver transplantation. Previous surgery, portal vein thrombosis, coagulopathy and portal hypertension further increase the degree of difficulty of this operation.

Most surgeons use a bilateral subcostal incision with an upper midline extension. The umbilical, falciform and left triangular ligaments are taken down for maximal exposure. The porta hepatis is dissected and the right, middle and left hepatic arteries are ligated and divided. The cystic duct is divided to allow circumferential dissection of the common hepatic duct which is divided high up in the hilum. The portal vein is then skeletonized back to just above the confluence of the splenic and superior mesenteric vein. If veno-venous bypass is used, the portal vein is cannulated and bypass instituted (Fig. 19.8).

In patients where veno-venous bypass is not used, the portal vein can simply be clamped proximally, ligated in the hilus and divided.

With the portal vein either cannulated or clamped, exposure for the infrahepatic dissection and circumferential control of the inferior vena cava is easily obtained. At this stage, the retrohepatic inferior vena cava is mobilised out of the retroperitoneum from the left side. The right triangular ligament is taken down and the retroperitoneal inferior vena cava dissected from the right side. The adrenal vein is ligated. The dissection frees the retrohepatic inferior vena cava up above the hepatic veins to allow application of the infrahepatic and suprahepatic IVC clamps. The recipient's liver is then sharply excised with care taken to leave cuffs of inferior vena cava above and below the liver. The recipient's liver is removed from the patient (Fig. 19.9).

19.1.6 Recipient Implantation of Donor Liver

Orthotopic implantation of the donor's liver starts with anastomosis of the suprahepatic inferior vena cava and the infrahepatic inferior vena cava, and utilisation of adequate venous cuffs is crucial for these anastomoses. The posterior wall is sutured with a running 3.0 monofilament suture. The anterior layer is sutured externally with a continuous technique. During the anastomosis, ice cold Hartmann solution is flushed through the catheters in the portal vein and the hepatic artery in the donor graft to keep it cool and to flush out all UW solution.

The portal anastomosis is next undertaken. If veno-venous bypass is used, interruption of the portal circuit is followed by removal of the portal cannula. The donor and the recipient portal veins are anastomosed using an end-to-end anastomose with an everting cuff technique with 60 prolene. The incorporation of a growth factor or the use of interrupted stitches in the anterior row is essential to prevent anastomosis stenosis. Before the final stitch is tied, all air within the portal vein is driven off by filling the veins with heparinized saline, the infrahepatic IVC clamp, the portal vein clamp and the suprahepatic IVC clamp are taken off.

The hepatic artery is then anastomosed using 60 prolene. The branch patch technique is commonly used to facilitate anastomosis.

Biliary continuity is established by the choledochocholedochostomy or by choledochojejunostomy. The aim is to establish a tension-free anastomosis (Fig. 19.10).

19.1.7 Piggy-Back Technique

The first step of the recipient hepatectomy is identical to the traditional technique with dissection of the hepato-duodenal ligament. The portal vein is isolated as far as the bifurcation and the right portal vein is encircled with a ligature. Once the division of the right triangular ligament is complete, the hepato-caval ligament is divided. The liver is raised, and the short hepatic veins are ligated and divided, proceeding in a caudal-cranial direction, mobilising the liver from the anterior surface of the retrohepatic vena cava as far as the junction of the right hepatic vein. The left side of the caudate lobe is completely freed of its ligaments and the peritoneum which covers the left edge of the caudate lobe is divided. The left liver is exposed. Short hepatic veins running from this side of the liver into the vena cava are ligated and divided.

The liver is only attached to the trunk of the three hepatic veins. The right hepatic vein is clamped with an angle clamp and simultaneous ligation of the right portal branch is done to avoid venous congestion in the right liver. The right hepatic vein is divided and sutured with 40 prolene. The common trunk of the middle and the left hepatic vein is clamped, and the main trunk of the portal vein is divided. The liver is removed. The orifice of the middle and left hepatic vein is modelled by dividing the septa and extending the incision on the vena cava inferiorly by 1 cm, thus making a triangular shape opening. Caval anastomosis is made with

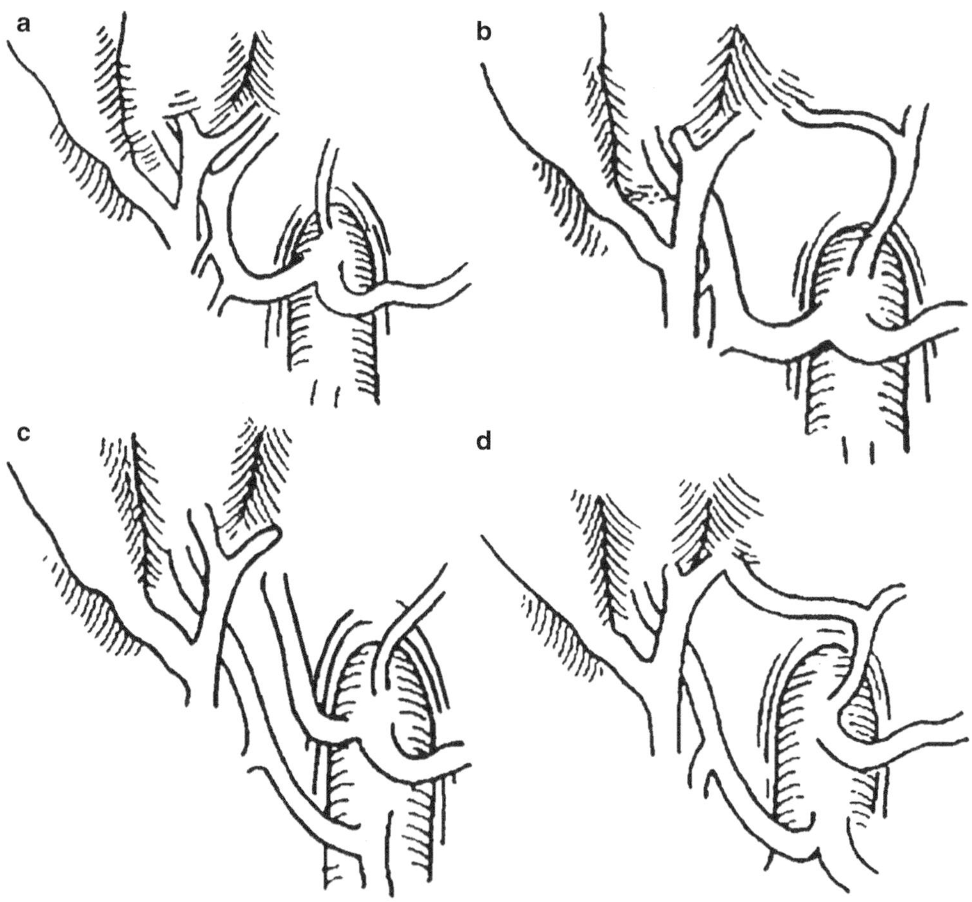

Fig. 19.5 Anomalies in the hepatic artery. (**a**) Prevailing pattern: common hepatic artery from coeliac axis. (**b**) Left hepatic artery from left gastric artery. (**c**) Right hepatic artery from superior mesenteric artery. (**d**) Left hepatic artery from left gastric artery, right hepatic artery from superior mesenteric artery. (**e**) Common hepatic artery from superior mesenteric artery. (**f**) Common origin of coeliac and superior mesenteric artery; (**g**) Right and left hepatic arteries from aorta. (**h**) Left hepatic artery from aorta; right hepatic artery from coeliac

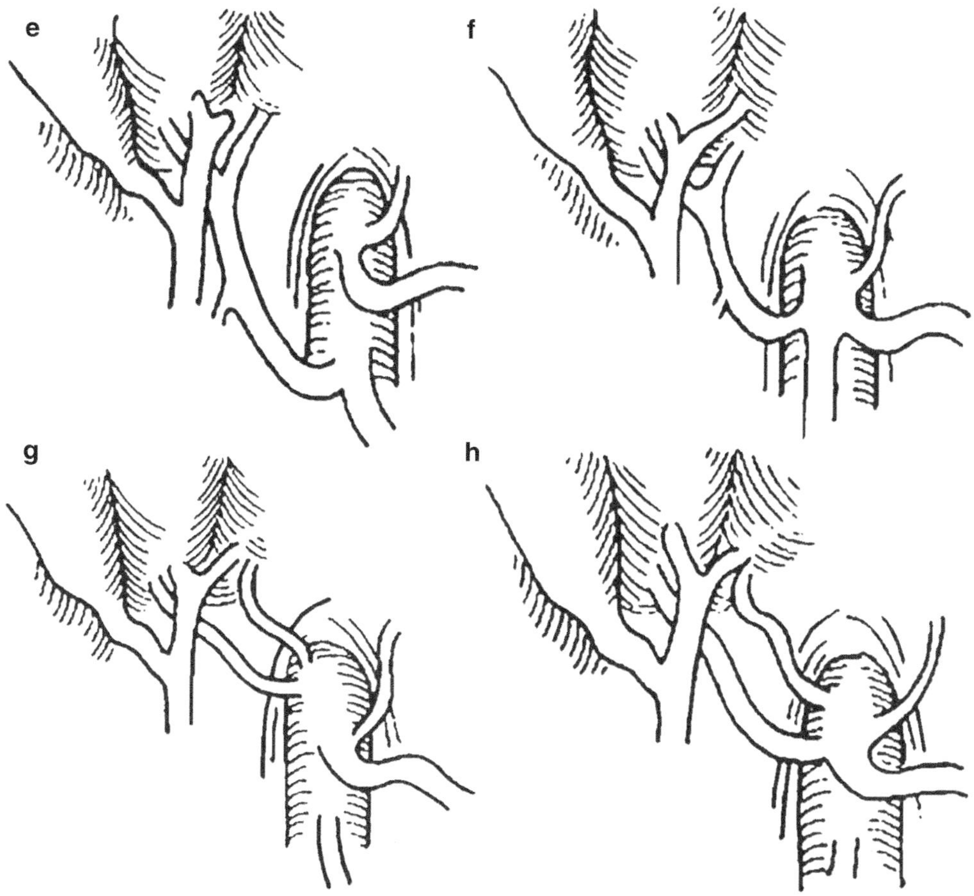

Fig. 19.5 (continued)

prolene 30 as described in conventional transplantation. Irrigation of the liver through the portal vein with cold Hartmann solution is done. The subhepatic stump of the vena cava is closed with a vascular stapler.

The rest of the procedure follows exactly the conventional technique, with portal anastomosis, followed by revascularization of the graft, arterial and biliary anastomosis.

19.2 Reduced-Grafts Liver Transplantation

When a patient can benefit from transplantation of a whole liver, this is always the operation of choice. However, in small adults or children, the availability of a donor of the same size is limited. The use of reduced-size liver transplantation solves this problem.

In 1984, both Professor Bismuth in France and Professor Broelsch in Germany almost concurrently performed the first reduced-size liver transplantation.

The reduction technique is based on the segmental anatomy of the liver, according to Couinaud. The principle is to reduce the size of the allograft while preserving all the important hilar structures and vena cava for subsequent reanastomosis during graft implantation. The left lateral section (segments 2 and 3) is the most often used reduced-size liver graft using the piggy-back technique for paediatric patients (Fig. 19.11). Either the left hemiliver (Fig. 19.12) or the right hemiliver (Fig. 19.13) can be employed in bigger or adult recipients. In cases with gross size mismatch, a single segment may also be used. The reduction is performed during the 'bench work' preparation and is best carried out next to the recipient operating room as the correct reduction can often be assessed exactly only after visual comparison of the size of the donor graft and the recipient hepatic fossa. For the use of the left lateral section, the right hepatic artery, the right portal vein and the right bile ducts are ligated close to the bifurcation. Dissection of the liver parenchyma just to the right of the falciform ligament proceeds as for usual liver resection. The right and the middle hepatic veins are divided close to the inferior vena cava and closed with a vascular suture. The raw surface of the liver is then sprayed with fibrin glue. Similarly, the left hemiliver (segments 2, 3 and 4), or the right hemiliver (segments 5, 6, 7 and 8) can be used. The implantation operation is otherwise similar to that with a full-size graft. However, in reduced-size liver transplantation, an adult liver is cut down to a smaller portion. The

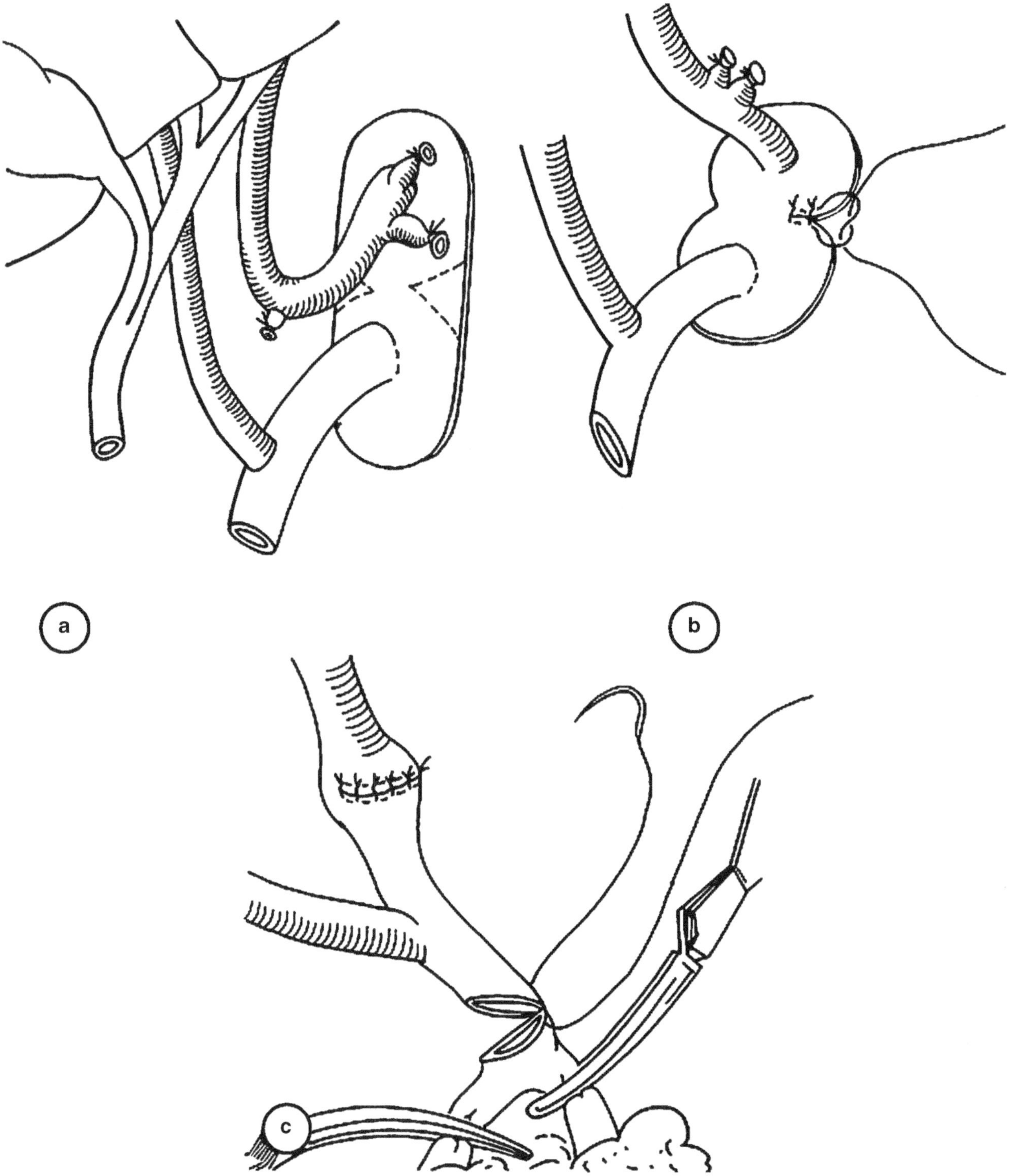

Fig. 19.6 Solution to solve the problem arising from anomalies (**c**) and (**d**) in Fig. 19.5

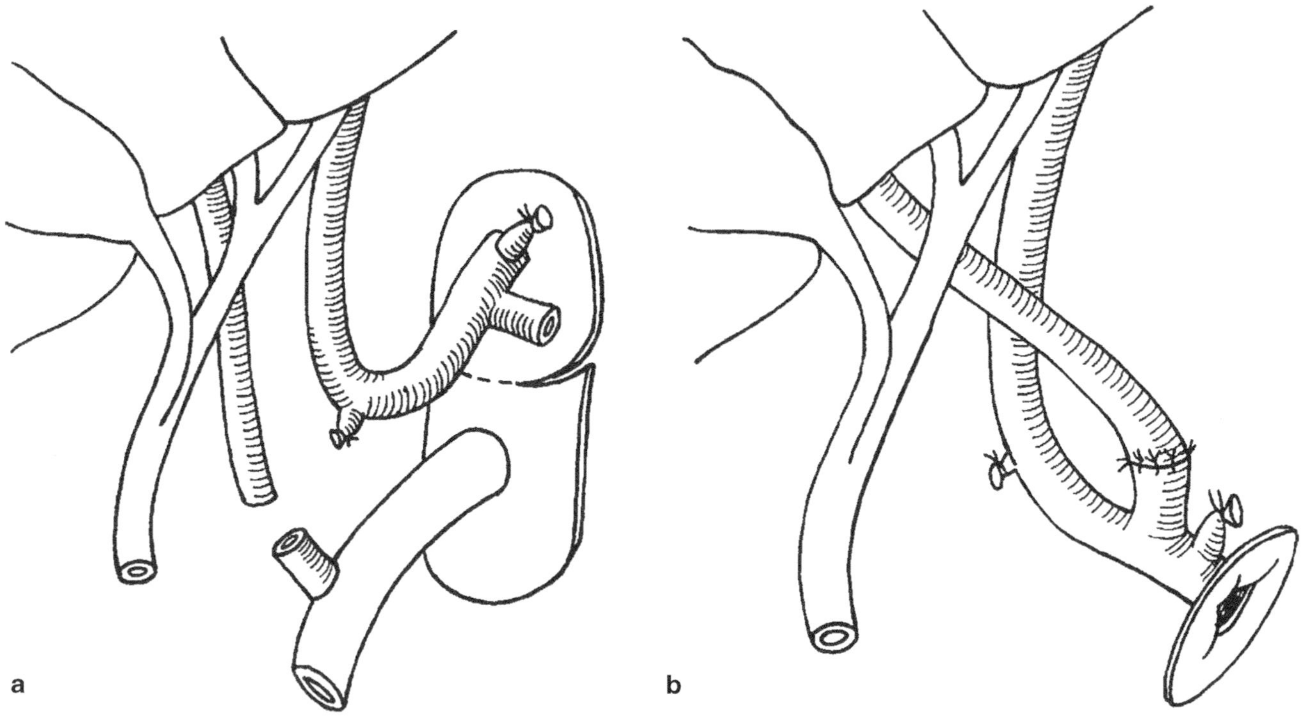

Fig. 19.7 (**a**, **b**) Alternative to solve the problem arising from anomalies (**c**) and (**d**) in Fig. 19.5

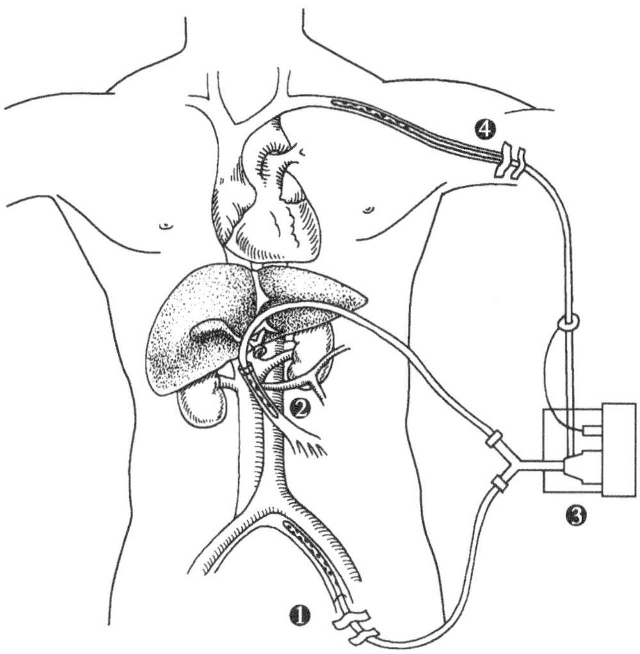

Fig. 19.8 Veno-venous bypass

remaining liver is discarded. The current graft shortage has prompted the development of split liver transplantation where one donor liver is split into two grafts, thereby increasing the number of grafts available for two recipients.

19.3 Split Liver Transplantation

In 1988, Professor Pichlmayr successfully split an adult liver into two, and the parts were transplanted into two patients (Fig. 19.14).

The initial results showed survival was inferior, and there was a high incidence of complication. Apart from the inevitable initial learning curve, it is undoubtedly a more difficult operation. Also, the selection of poor risk recipients can partly explain the poor results. Recently, much better results have been reported.

Before a split should take place, there are requirements of the donor and the liver graft. The donor should be haemodynamically stable. In general, liver surgeons prefer a donor of over 60 kg of weight and age under 50 years. The donor should be an optimal donor, and the liver should be macroscopically normal. The two recipients should be matched in size to the respective parts of the liver. The hospital must have the organisational arrangements in place to allow two liver transplantations to take place with a reasonable cold ischaemic time. This usually refers to a local or a near regional donor. Often, in a split transplant programme, there is a second hospital catering for paediatric transplants. Finally, the vascular and biliary anatomy of the donor's liver must allow safe splitting of the liver.

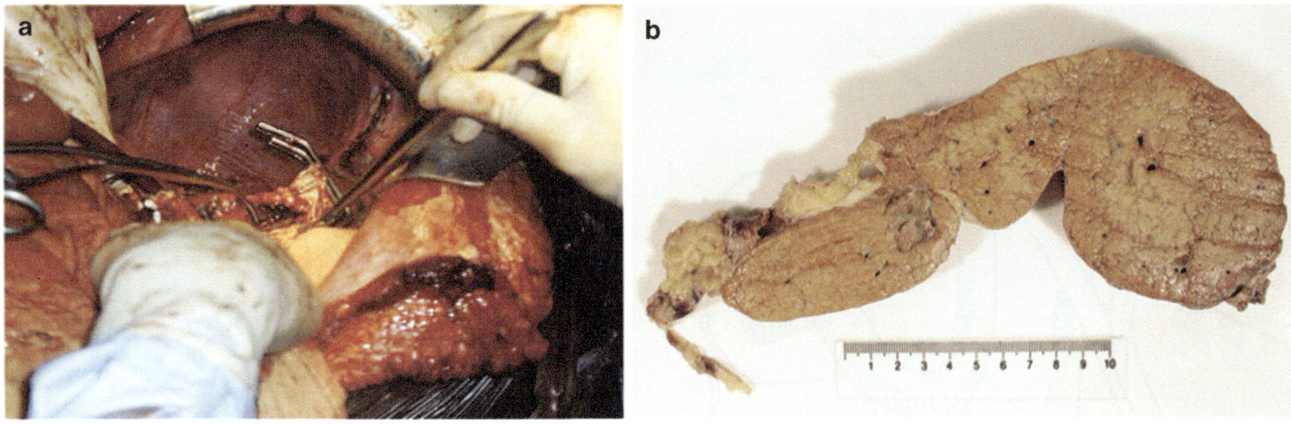

Fig. 19.9 (a) Recipient liver resected from recipient. Note a big space left behind with a cross clamp across the divided inferior vena cava. (b) Ideal patient with hepatocellular carcinoma to receive liver transplantation. Note small shrunken liver and a small tumour (arrow)

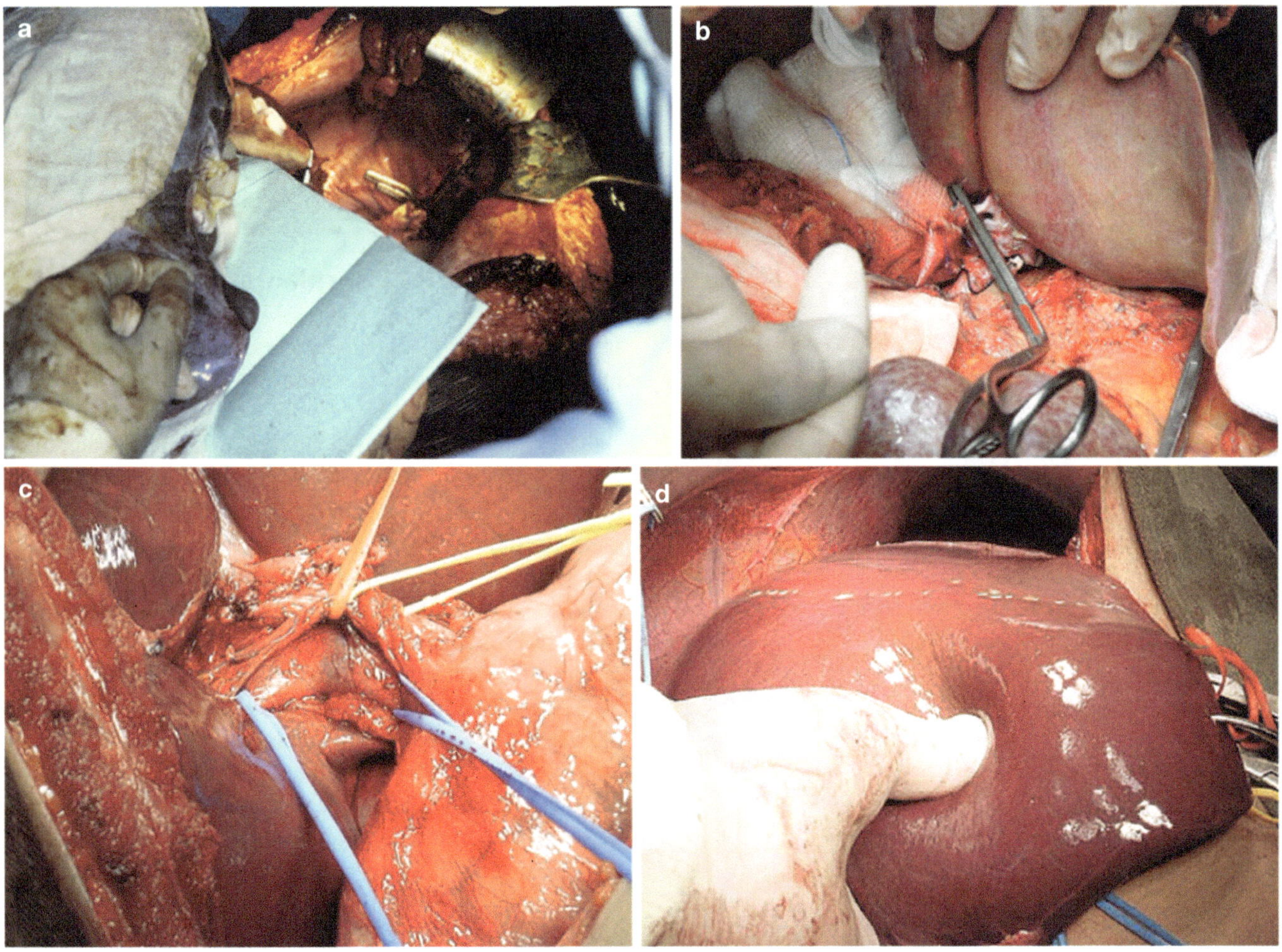

Fig. 19.10 Recipient implantation of donor liver. (**A**) Suprahepatic inferior vena cava anastomosis. (**B**) Anastomosis of portal vein. Note donor liver still not yet perfused. (**C**) Completion of portal vein, hepatic artery and bile duct anastomoses. (a) Hepatic artery anastomosis. (b) Bile duct anastomosis. (c, d) Portal vein anastomosis. (**D**) Donor liver reperfused

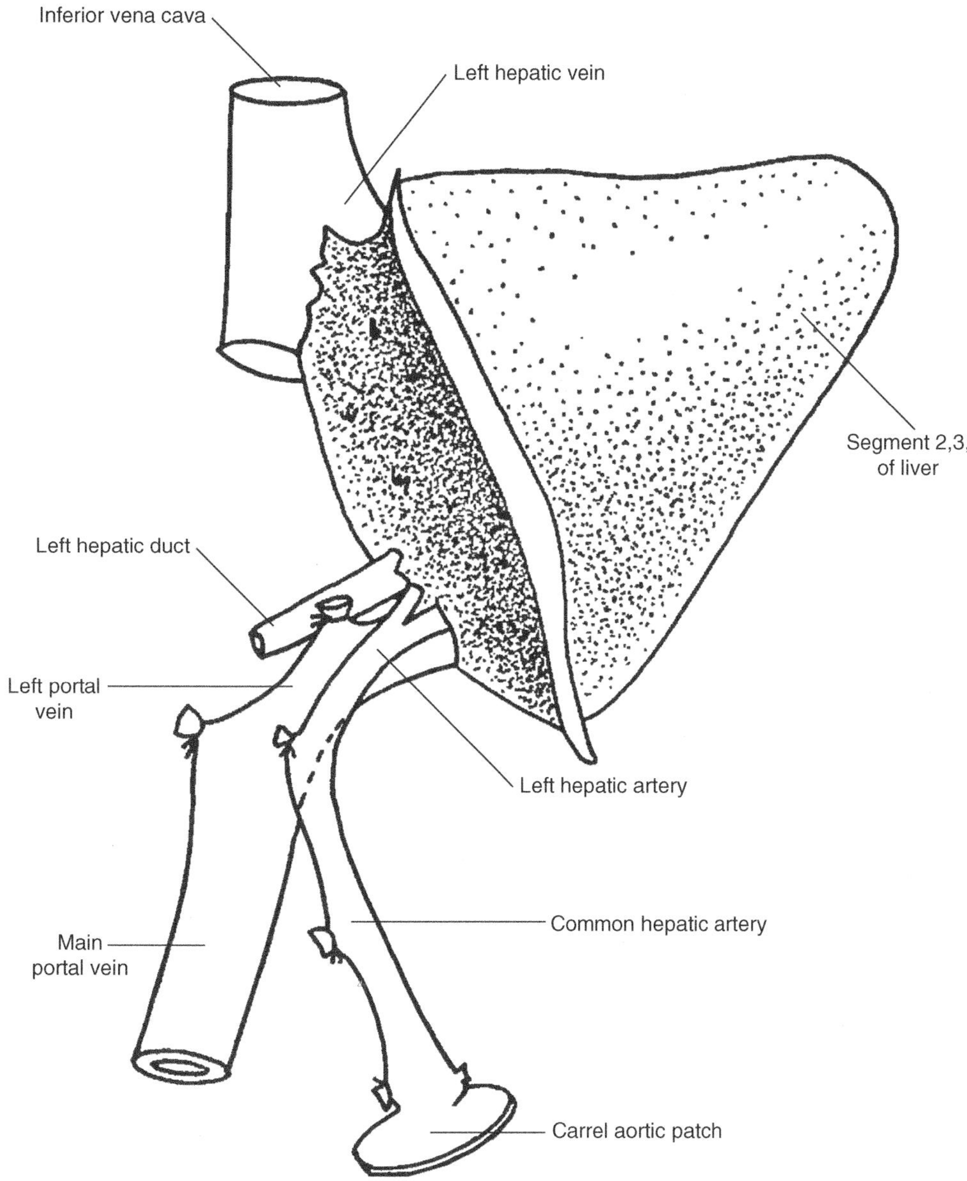

Fig. 19.11 Reduced size liver transplantation using the left lateral section of the liver

The right hemiliver usually retains the vena cava, the common bile duct and the portal trunk. The right branch of the hepatic artery is usually anastomosed with 70 prolene to the donor iliac artery to make implantation easier. The left hemiliver retains the common hepatic artery with the coeliac trunk and a Carrel's aortic patch. The left branch of the portal vein is usually anastomosed with a segment of the donor iliac vein. The junction of the middle and left hepatic veins is fashioned to obtain a wide single trunk. The short left hepatic duct can be anastomosed to a loop of the jejunum.

The actual splitting of the liver can be done inside the body of the cadaveric donor with a beating heart. The alternative is to split the liver in the bench after cooling and whole organ harvesting.

Absolute contraindications to splitting are the absence of a portal vein bifurcation and atrophy of the left liver. Back table angiography to identify hepatic artery anomalies has been suggested particularly when the left hepatic artery arises from the left gastric artery, or the right hepatic artery arises from the superior mesenteric artery. The ideal hepatic arterial anatomy for splitting is simple division of the common hepatic artery, and cholangiography has also been recommended to detect usual biliary duplications.

19.4 Living Donor Liver Transplantation (LDLT)

The success of reduced-size liver transplantation provided the surgical foundation of using part of the liver from living donors for living donor liver transplantation (LDLT). The left lateral section of the liver was used successfully first (Fig. 19.15)

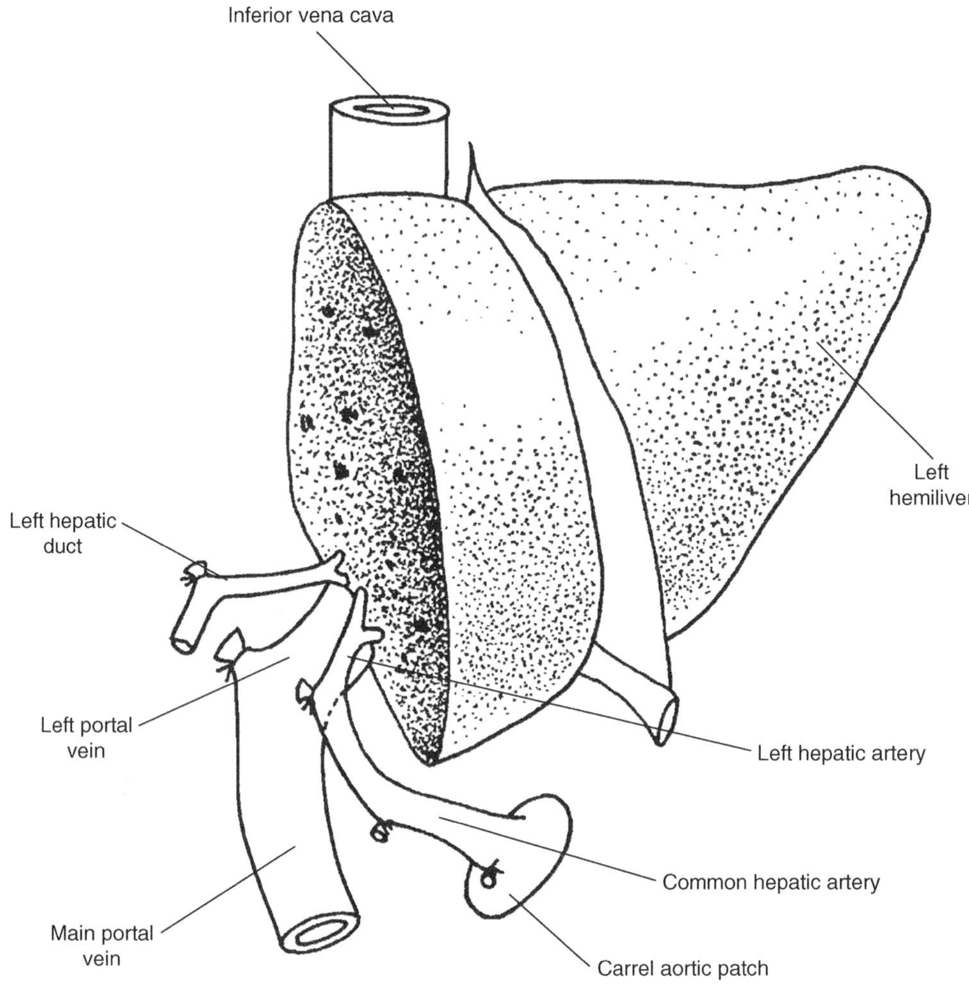

Fig. 19.12 Reduced size liver transplantation using the left hemiliver

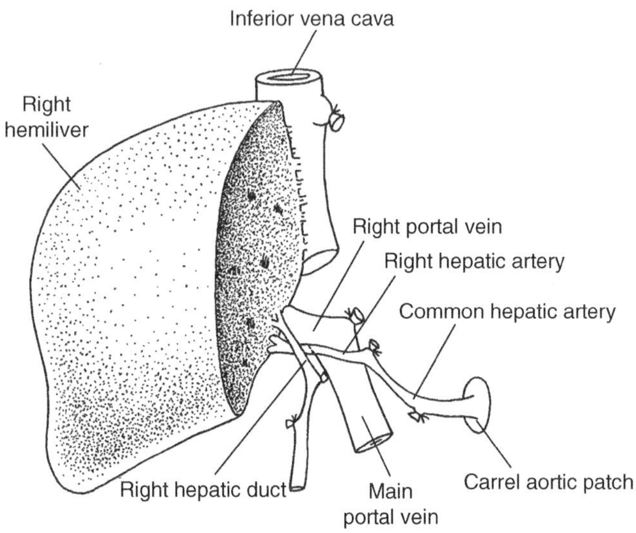

Fig. 19.13 Reduced size liver transplantation using a right hemiliver

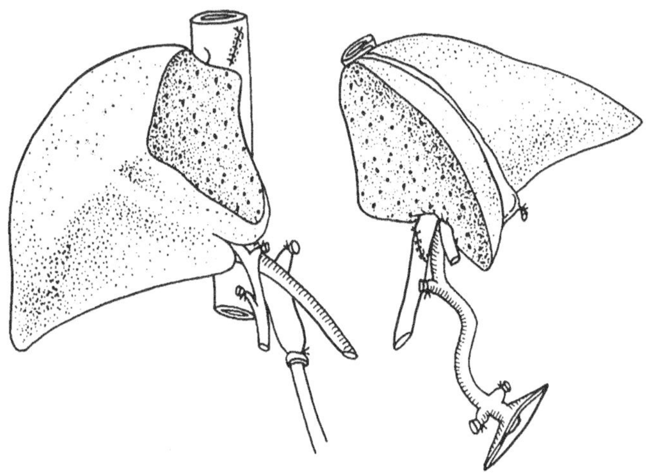

Fig. 19.14 Split liver grafts

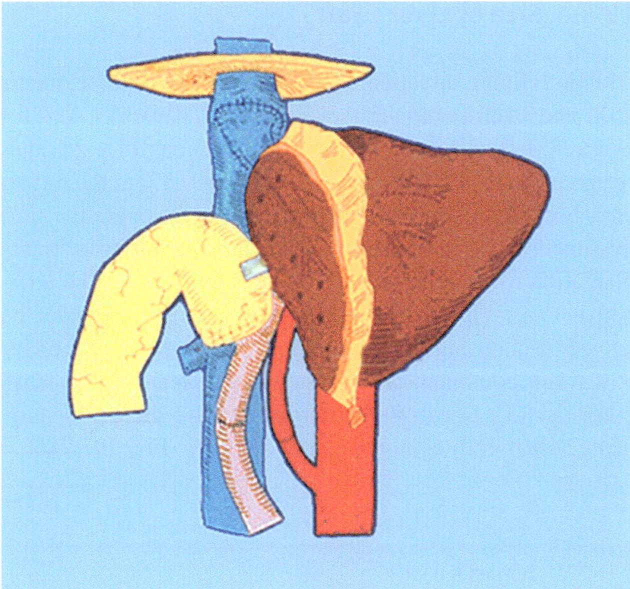

Fig. 19.15 Living donor liver transplantation using the left lateral section

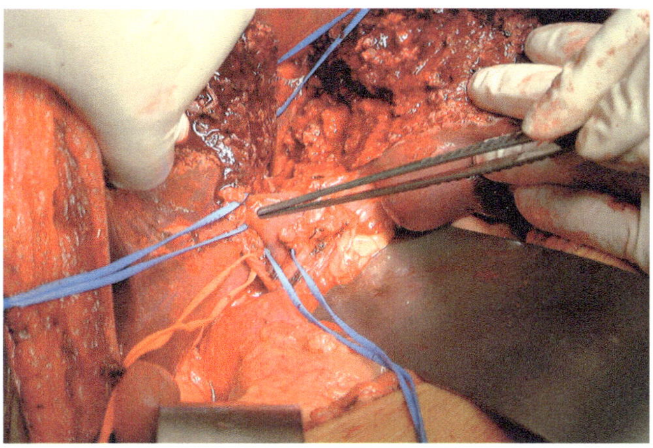

Fig. 19.16 Living donor liver transplantation. Donor hepatectomy using the right hemiliver. Note the liver had been split along the midplane. Right portal vein, right hepatic artery and right hepatic vein all slung and were ready to be transected (Right hemiliver had been mobilized with all the short hepatic veins to the right liver divided). Sling (a) right hepatic vein; sling (b) right portal vein; sling (c) right hepatic artery; sling (d) main portal vein

followed by the left hemiliver, then the right hemiliver (Fig. 19.16). LDLT was first reported in two patients by Raia et al. in 1989. Both recipients died of medical complications shortly after the procedure. The first successful LDLT was reported by Strong from Australia on a child using the left liver of the mother in 1989. LDLT has since been increasingly used, particularly in countries where procurement from brain-dead patient is prohibited by law and in countries where the cadaveric organ donation rate is low.

19.4.1 Selection of LDLT Donors

Eligible donors are usually between the ages of 18 and 55 years. Acute or chronic medical illness, especially liver disease, is excluded by a detailed history and physical examination. The standard requirements include ABO blood group matching, normal liver function and the same negative viral serology in the donor, as in the cadaveric liver transplant setting. Donors must be satisfactory on psychological assessment. All potential donors should then undergo volumetric computed tomography screening to assess the liver volume and to exclude unsuspected intraabdominal pathology. Anomalous vascular incompatible with the donation is excluded by hepatic angiography. Endoscopic retrograde cholangiopancreatography (ERCP) is performed only if there is doubt about the anatomy or adequacy of the biliary tract. Parents of children undergoing transplantation for Alagille syndrome are a good example of this. Liver biopsy may be necessary to ascertain the quality of donor's liver in cases such as suspected fatty infiltration. Obviously any one of the above investigation, especially the invasive ones, carries a small but real risk to the donor. Magnetic resonance imaging in the places of angiography and cholangiography are being used more frequently to replace some of the more invasive investigations.

19.4.2 Potential Advantages of LDLT

The advantages offered by LDLT include:

1. An ideal liver graft in which the liver graft is expected to function immediately because of the selection of the good donor.
2. The ability to schedule the operation electively, allowing maximum preparation of the recipient.
3. The recipient does not need to put on the waiting list to wait for a cadaveric graft. This is important in recipients with hepatocellular carcinoma because the tumour can grow despite the use of anti-cancer therapy during the waiting period.
4. The recipient can be designated, receiving a graft from his/her close relative, thus patients with hepatocellular carcinoma which is beyond the Milan Criteria can still be transplanted.
5. There may be a theoretical immunologic advantage of receiving a living-related organ, as suggested by the lower incidence of steroid-resistant rejection compared with cadaveric liver transplantation.

19.4.3 Disadvantages of LDLT

Despite the impressive results of LDLT, considerable debate persists concerning donor safety. Data about the peri-surgical complication and long-term outcome of living liver donors are sparse. However, extrapolation from the data concerning liver resection for benign and malignant disease suggests that the mortality rate in non-cirrhotic liver resection should be 0–0.5% in experienced hands. Clearly, all forms of LDLT are subject to varying degrees of complication. Despite good intentions and experienced hands, there can be no argument that a definite risk of death to the donor exists and has been reported. Death and major complications to a living donor are bad, as the operation is 'unnecessary' to that normal individual.

19.4.4 Size of Liver Graft

Although living donation of a left lateral section (segments 2, 3) and a left hemiliver (segments 2, 3, 4) works well for children, left side graft are rarely large enough for an adult recipient. Indeed, a liver graft of less than 0.8% of the recipient's body weight has a high risk of early post-operative graft dysfunction. As a consequence, adult-to-adult liver transplantation using right-sided grafts has been used. The alternative to provide a large graft to the recipient cannot be solved by choosing a large-size donor, and a small size recipient as this combination of donor and recipient is very rare to come by. S.G. Lee from Korea tried to solve this problem by using a dual graft coming from two donors (Fig. 19.17).

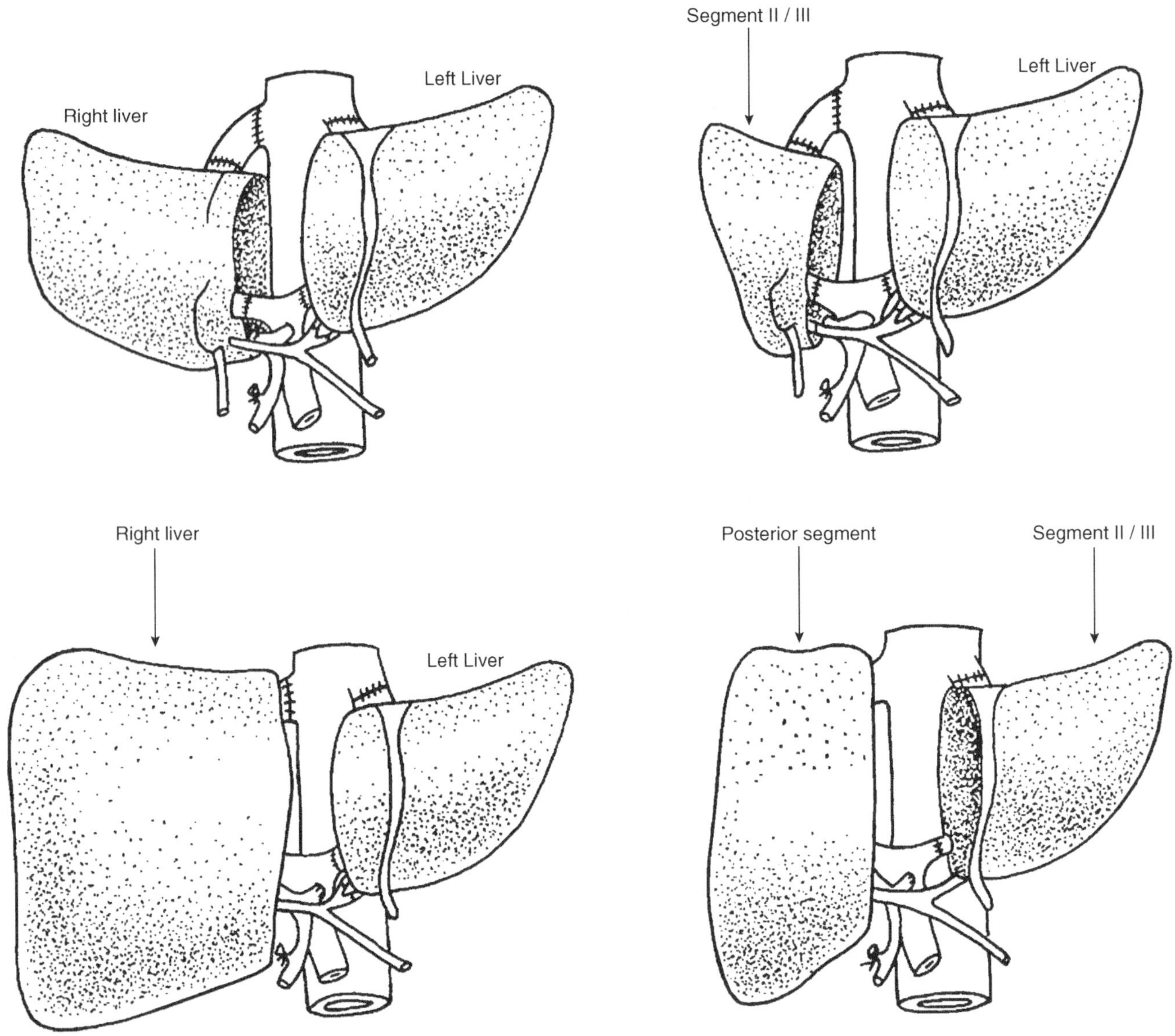

Fig. 19.17 Types of dual grafts

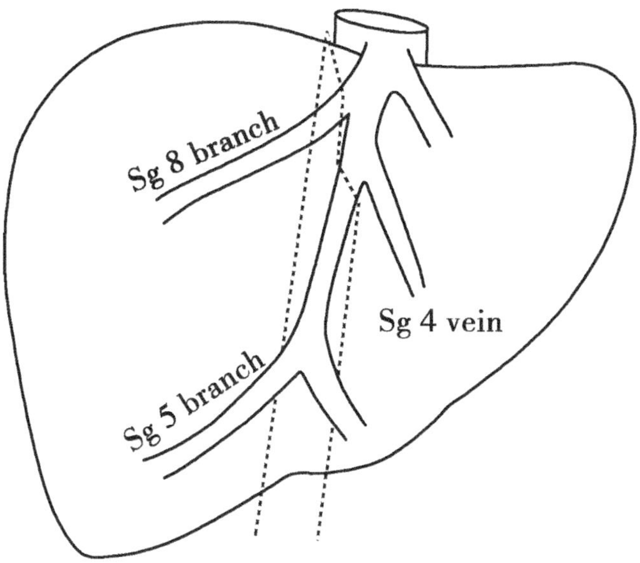

Fig. 19.18 Controversies in the use of the right hemiliver for LDLT

19.4.5 Controversies in the Use of Right Hemiliver for LDLT

There are two major controversies in the use of the right hemiliver in LDLT:

1. Whether the middle hepatic vein should be included in the graft (Fig. 19.18). If the middle hepatic vein is included in the graft, care must be taken to preserve the segment 4 vein or the liver segment 4 left in the donor's body after the operation will turn dusky in colour because of venous congestion. The advantages of leaving the middle hepatic vein to the graft are the graft size is bigger, and the segments 5 and 8 have better venous drainage after implantation. However, this is done at a higher risk to the donor.
2. The second controversy is if the middle hepatic vein is not included in the graft, should the segments 5 and 8 branches be reconnected to the recipient's inferior vena cava via vein grafts to provide better venous drainage to the donors' segments 5 and 8. The disadvantages of using vein grafts are technical difficulties and the added complexity to the surgery.

19.5 Auxiliary Liver Transplantation

The early attempts in placing a liver graft in the abdominal cavity but leaving the recipient's whole liver was met with poor results, and this procedure has been abandoned.

Recently, there has been a renewed interest in the use of auxiliary liver transplantation (ALT) in patients with fulminant or sub-fulminant hepatitis for the following reasons:

1. While immunosuppressive agents are effective in preventing rejection, the most frequent cause of death after transplantation is from sepsis as a result of immunosuppression. It is, therefore, good to have a treatment that can end up with discontinuation of immunosuppression.
2. Complete regeneration of the native liver of patients transplanted with an auxiliary graft for acute liver failure is possible;
3. In patients with hepatitis B virus-related fulminant liver failure, there is a frequent disappearance of viral infection;
4. Technical improvements making the ALT safer.

19.5.1 Techniques of Auxiliary Liver Transplantation

I have no intention to discuss the details of the operation.

In the use of a right graft, the recipient undergoes a right hepatectomy and the right graft is put into the orthotopic position (Fig. 19.19).

Similarly for a left graft, the recipient undergoes a left hepatectomy and a left graft is put into the orthotopic position (Fig. 19.20).

Preliminary data on this procedure showed 68% of the patients had complete regeneration of the native liver, while incomplete regeneration with areas of obvious fibrosis or extensive fibrosis was observed in 14% and 18% of the patients, respectively. Thirteen of the 19 survivors had been withdrawn from all immunosuppression.

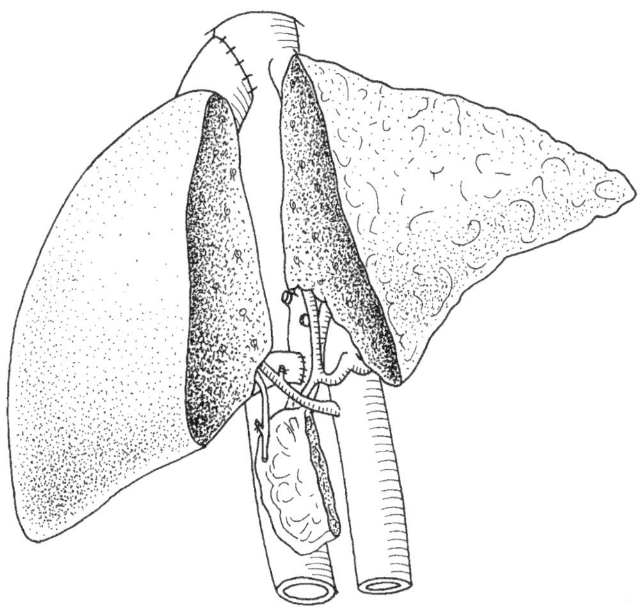

Fig. 19.19 Auxiliary liver transplantation using a right graft after recipient right hepatectomy

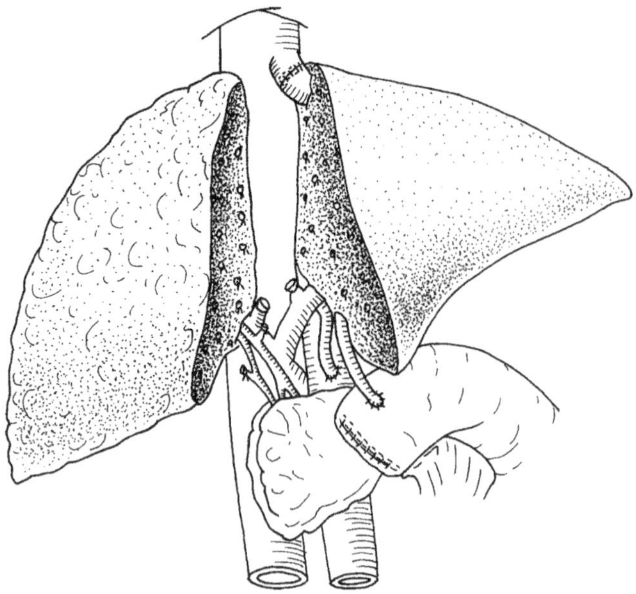

Fig. 19.20 Auxiliary liver transplantation using a left graft after recipient left hepatectomy

Further Reading

Fan ST. Living donor living transplantation. Hong Kong, China: Takungpao Publishing Co. Ltd.; 2007.

Köckerling F, Schwartz. Liver surgery. Operative techniques and avoidance of complications. Heidelberg: J.A. Barth; 2001.

Mazziotti A, Cavallari A. Techniques in liver surgery. London: Greenwich Medical Media; 1997.

20 Associating Liver Partition and Portal Vein Ligation for Staged Hepatectomy (ALPPS)

The abbreviation of this operation is ALPPS, not ALPS (a famous mountain in Europe).

20.1 Establishment of ALPPS

The concepts which finally led to the establishment of ALPPS were developed in stages by different surgeons. In 1980 Professor Makuuchi used portal vein embolization (PVE) to induce hypertrophy of the contralateral liver. In 2000 Professor Adam used two-staged hepatectomy to treat multiple liver tumours. In 2004, Professor Jaeck used wedge resections for left liver tumours, PVE of right portal vein followed by extended right hepatectomy after the left liver had hypertrophied. In 2007, Professor Clavien used a staged operation to treat multiple liver tumours: Stage 1 wedge resections of left liver tumours + right portal vein ligation; Stage 2, after the left liver had hypertrophied, extended right hepatectomy was carried out. At around the same time, Professor Kiammanesh used staged hepatectomy + branched portal vein ligation to treat multiple liver tumours.

For all these operations, the wait for the contralateral hemiliver to hypertrophy to a size adequate for a second-stage resection is usually 2–3 months. During this period of wait, the liver tumours might have progressed to an unresectable stage. The scene is set for the establishment of ALPPS.

In 2007, Professor Hans Schlitt (Fig. 20.1) operated on a patient with hilar cholangiocarcinoma. Intraoperatively he found the volume of the left lateral section of the liver (segments 2 and 3) was very small. He decided to carry out palliative left hepaticojejunostomy.

Fig. 20.1 Professor Hans Schlitt

For technical reasons, he transected the liver along the right border of the falciform ligament to expose the left hepatic duct. After the left hepaticojejunostomy, he decided to ligate the right portal vein with the aim to result in hypertrophy of the left lateral section. Driven by curiosity, on post-operative day 8, he carried out a computed tomography and found the left lateral section of the liver to become immensely hypertrophied. A second-stage extended right hepatectomy was carried out. This started the first ALPPS—a creation merely by chance.

W. Y. Lau, *Applied Anatomy in Liver Resection and Liver Transplantation*, https://doi.org/10.1007/978-981-16-0800-1_20

20.2 International Uptake of ALPPS

This procedure first appeared as a poster put up at the European-African Hepato-Pancreato-Biliary Association Conference in 2011 by a group from Mainz, Germany, who operated on three patients. The unnamed procedure was then taken up by an Argentina group of surgeons who reported its application to three cases that were previously considered unresectable owing to inadequate future liver remnant volumes. In 2012, Schmitzbauer et al. published a series of 25 patients in the Annals of Surgery. A subsequent editorial in the Journal by deSantibanes and Clavien proposed a name—Associating Liver Partition and Portal Vein Ligation for Staged Hepatectomy (ALPPS). This procedure has since elicited much discussion and debate in the global hepatobiliary community.

In 2012, an international ALPPS registry was created. At the end of 2014, there were more than 430 ALPPS procedures registered from 75 centres around the world on this website: www.alpps.net/q=registry.

20.3 Conventional ALPPS (Fig. 20.2)

First operation on Post-op Day 6 CT	Second operation on Post-op Day 7
– Excise all tumours in the left lateral section	
– Ligate right portal vein	
– Transect liver along the right side of falciform ligament	
– Cholecystectomy	
– Operative cholangiogram through the cystic duct	
– Suture any bile leak site	
– Introduce two abdominal drains	– After the left lateral section hypertrophied
	– Divide right hilar pedicle to carry out extended right hepatectomy

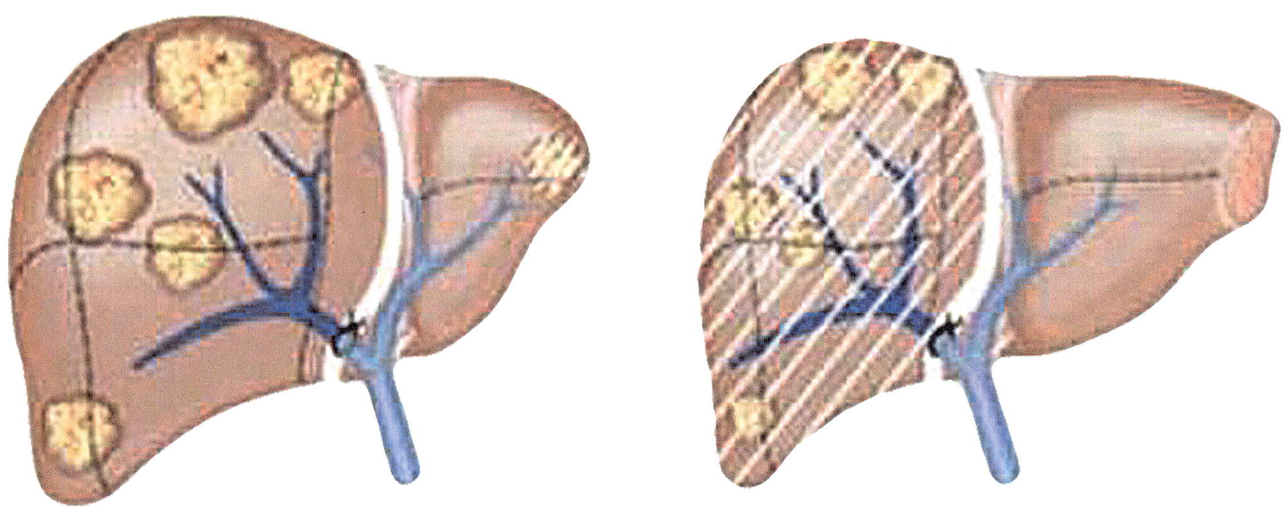

Fig. 20.2 Conventional ALPPS

The first operation consists of excising all tumours in the left lateral section, ligation of the right portal vein, transection of the liver along the right side of the falciform ligation, cholecystectomy and then carrying out operative cholangiogram through the cystic duct stump. Any bile leak site should be carefully sutured. The operation finishes by closing the abdominal wall after two abdominal drains have been put in to drain the right and left sides of the liver (Fig. 20.3a, b).

On post-operative days 6–7, computerised tomography is carried out (Fig. 20.4).

The second operation is usually carried out on days 7–14 after the first operation, when the left lateral section has hypertrophied to an adequate volume. The second operation consists of an extended right hepatectomy (Fig. 20.5a, b).

20.4 Deviations from the Conventional ALPPS

Deviation here means a modification in a part of the surgical procedure.

20.4.1 Deviation 1

The time interval between the first and the second operations exceeds the normal interval of 7–14 days.

The main reasons for carrying out the second operation within the period of 7–14 days are:

1. Bile leak after the first operation is common. A delay in the second operation can result in sepsis and death.
2. Liver segment 4 is ischaemic, a delay in the second operation can lead to necrosis, liver abscess and sepsis.
3. Early fibrinous adhesions are easier to separate than when the adhesions have become vascularized, which can result in bleeding.

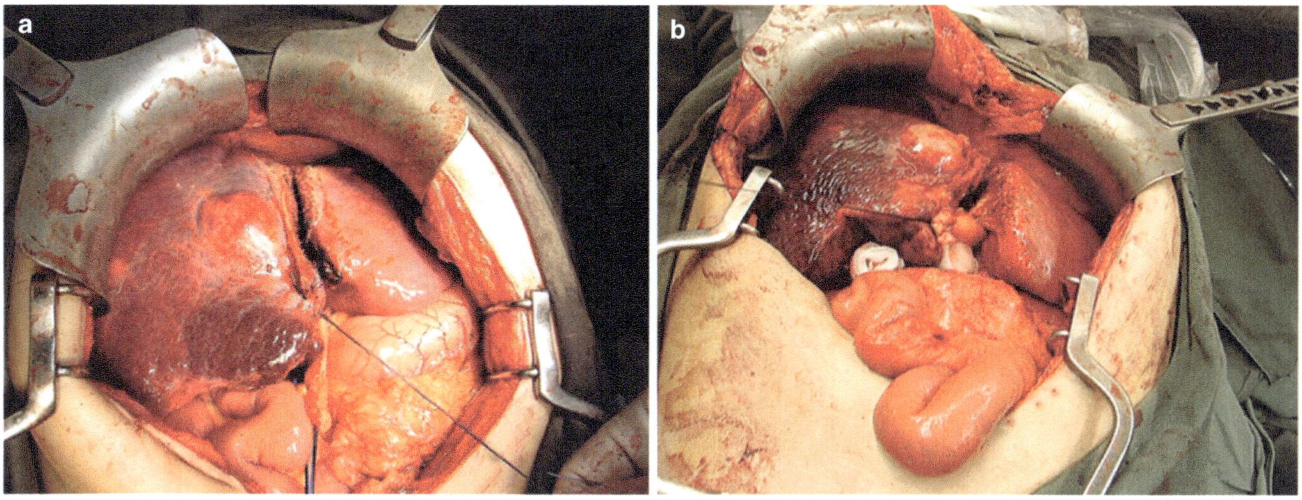

Fig. 20.3 (**a**) First operation with transection of liver on the right of falciform ligation. (**b**) Please note ischaemia in segment 4 liver

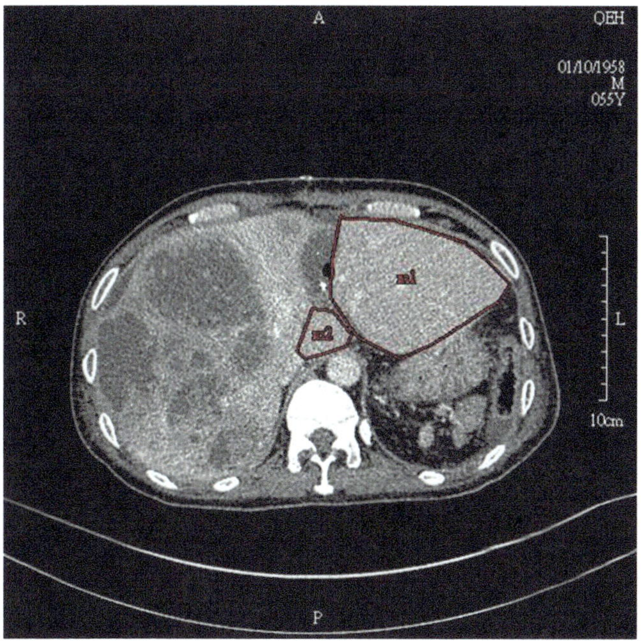

Fig. 20.4 CT on post-operative day 6. The left lateral section hypertrophied from 26.9% (283 mL) to 37.4% (532 mL)

4. This time is required for adequate hypertrophy to occur. Occasionally, the second operation has to be delayed to wait for liver hypertrophy to become adequate.

20.4.2 Deviation 2

As bile leak is common after the first operation, some surgeons use a plastic bag to wrap around the right liver and to put in a drain which leads outside of the patient's body to drain fluid out.

Merits

1. Reduces the chance of fluid collection with subsequent infection and septic complication.
2. Reduces the degree of adhesions in the second operation.

Demerits

1. A plastic bag is a foreign body that increases the chance of infection.
2. Even if the future liver remnant does not hypertrophy enough, a second operation still needs to be carried out to remove the plastic bag. For this reason, Machado and his associates recommended using biological glue instead of a plastic bag.

20.4.3 Deviation 3

Some surgeons advocated resection of liver segment 4 in the first operation aiming to decrease the chance of developing necrosis and liver abscess. Most surgeons now believe that resection of liver segment 4 is unnecessary.

20.4.4 Deviation 4

Dokmak and Belghiti advocated in addition to ligating the right portal vein, also ligate the right hepatic duct so that liver hypertrophy can occur even better. Unfortunately, bile leak after the additional ligation of the right hepatic duct

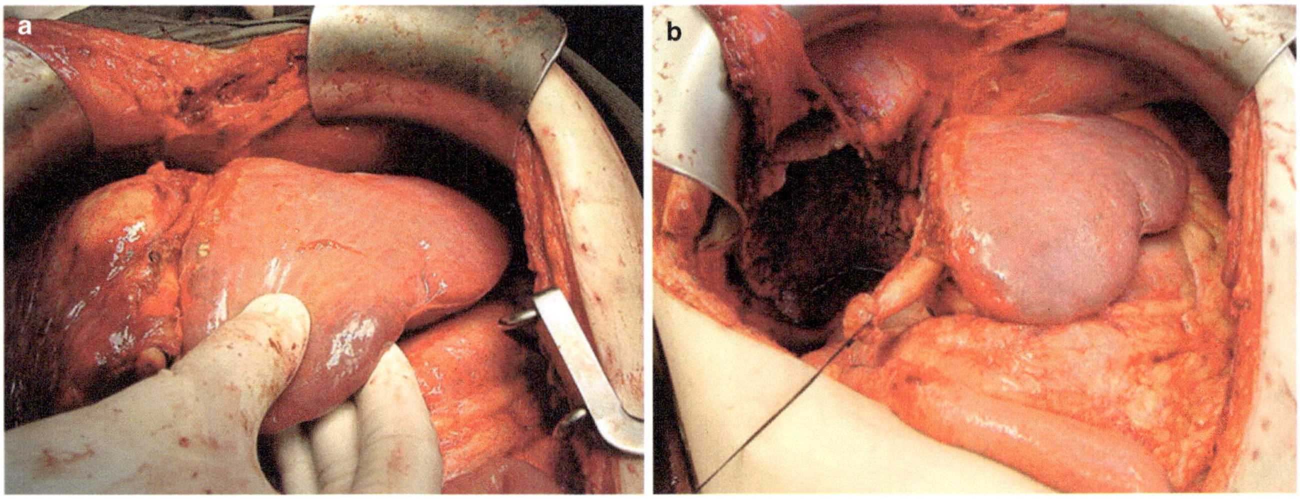

Fig. 20.5 (**a**) Please note hypertrophy of the left lateral section at the second operation. (**b**) After extended right hepatectomy

happened in 87.5% of patients, and thus this ligation should not be carried out.

20.4.5 Deviation 5

Using no touch technique.

A group of surgeons from Hamburg advocated the use of no touch technique in ALPPS, and named the procedure 'hybird ALPPS'.

The first operation consists of using the anterior approach to transect the liver on the right side of the falciform ligament. The surgeon should avoid dissecting the porta hepatis and the right liver.

After the first operation, PVE is carried out on the right portal vein. After the left lateral section has hypertrophied to an adequate volume, extended right hepatectomy is carried out as a second operation.

Merits

1. Reduces adhesions in the second operation.
2. Reduces the chance of spreading tumours cell into the bloodstream during the first operation.

Demerit: Technically more difficult in the first operation.

20.4.6 Deviation 6

To keep the middle hepatic vein intact.

In the conventional ALPPS, many surgeons divide the middle hepatic vein during the first operation. Hernandez-Alejandro et al. advocated to keep the middle hepatic vein intact, thus avoiding any venous congestion in liver segments 4, 5 and 8, and reduces the chance of ischaemia, necrosis and bile leakage.

20.4.7 Deviation 7

Using a liver tourniquet.

Robles Campos and his associates advocated in open surgery the use of a Vicryl suture which serves as a tourniquet around the liver. After tightening of the suture and using intraoperative ultrasound to ensure there is no blood flow into the right liver, the abdomen is then closed. When the future liver remnant has hypertrophied to an adequate volume, the second operation, which consists of a right hepatectomy is carried out.

The main advantage of this operation is the first operation requires less technical skill.

20.4.8 Deviation 8

Radiofrequency ablation for liver transection and ligation of the right portal vein in the first operation.

After the future liver remnant has hypertrophied adequately, a second operation is carried out.

20.4.9 Deviation 9

Laparoscopic ALPPS which can be

– Hand-assisted	in the first operation
or	or
– Total laparoscopic	in the first and second operations

Professor Cai has reported the laparoscopic use of a liver tourniquet in the first operation.

20.5 Modification of ALPPS

A modification is a change in the whole operation.

20.5.1 Modification 1: Left ALPPS (To Preserve the Right Liver)

The first operation consists of ligation of the left portal vein, multiple wedge resections of liver tumours in the right liver and transection of the liver through the mid plane (between the right and left hemilivers). After the right liver has hypertrophied enough, the second operation consists of left hepatectomy (Fig. 20.6).

20.5.2 Modification 2: Right ALPPS (To Preserve the Central Liver Segments 4, 5, 8)

The first operation consists of ligation of the posterior sectional branch of the right portal vein, left lateral sectionectomy, multiple wedge resections of liver tumours in the right anterior section and segment 4 and liver transection through the plane between right anterior and posterior sections. After the middle liver segments have adequately hypertrophied, the right posterior sectionectomy is then carried out as a second-stage operation (Fig. 20.7).

Fig. 20.6 Modification 1. Left-sided ALPPS to preserve right hemiliver

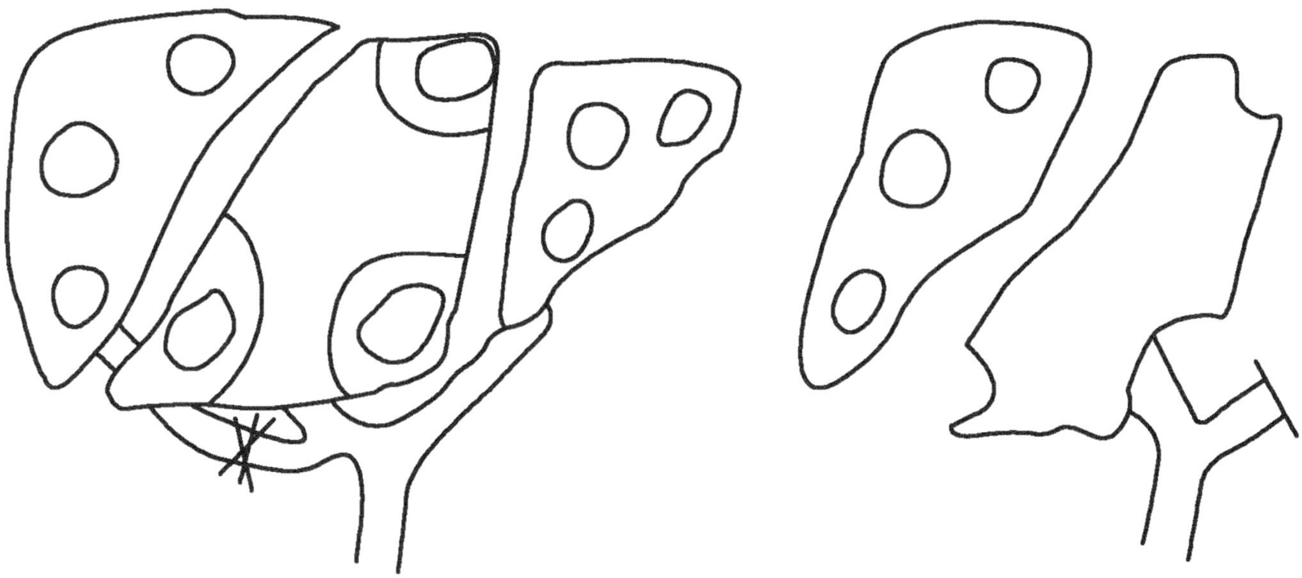

Fig. 20.7 Modification 2. Right side ALPPS to preserve liver segments 4, 5, 8

20.5.3 Modification 3: To Preserve Left Liver (Segments 2, 3, 4)

This modification can be used in patients with a huge liver tumour in the right liver and with a small future liver remnant in the left liver. By using ALPPS, the second stage of resectional surgery hopefully can be carried out in about 1–2 weeks instead of the usual waiting time of 4–8 weeks for adequate hypertrophy of the FLR to occur using either portal vein embolization or portal vein ligation.

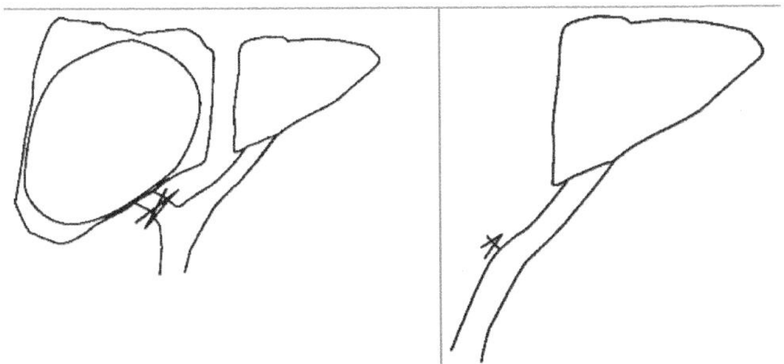

First operation	Second operation
– Ligation of right portal vein	
– Transection of mid plane of liver between the right and left hemiliver	– Right hepatectomy

20.5.4 Modification 4: Salvage ALPPS

When right/left portal vein embolization results in inadequate liver hypertrophy, salvage ALPPS can be done by transecting the liver along the mid plane of the liver between the right and left hemilivers. After adequate liver hypertrophy, a right or left hepatectomy can be carried out as a second-stage operation.

20.6 Indications for ALPPS

ALPPS can be considered in any liver resectional surgery when:

1. In a normal liver, the future liver remnant is <30%.
2. In an abnormal liver (obstructive jaundice, fatty liver, cirrhosis or after chemotherapy) the future liver remnant is <40%.

The usual indications are:

1. Colorectal liver secondaries
2. Hepatocellular carcinoma
3. Extrahepatic cholangiocarcinoma
4. Neuroendocrine tumour with liver secondaries
5. Klatskin tumour

ALPPS has been successfully carried out in a patient with multiple hepatocellular carcinoma with underlying fibrotic liver with portal vein tumour thrombosis (Cavaness et al. 2013) and in a patient with hepatitis B-related hepatocellular carcinoma, which arose from a fibrotic liver (Chia et al. 2014).

20.7 Contraindications to ALPPS

1. Unresectable liver tumours.
2. Unresectable primary tumours.
3. Unresectable extrahepatic metastasis.
4. Serious portal hypertension.
5. R_0 resection cannot be achieved.
6. Serious pre-morbid conditions that preclude any patients to receive surgery.
7. High anaesthetic risk patients.

20.8 Merits of ALPPS

1. Rapid hypertrophy of future liver remnant. The hypertrophy can be 74–87% of the future liver remnant in 7 days after the first ALPPS operation.

   Reasons for rapid liver hypertrophy are:

   (a) After ligation of the right portal vein, the portal blood can only flow through the left portal vein into the liver. The additional liver parenchymal transection prevents the development of intrahepatic collateral blood vessels.

   (b) Stimulating factors and growth factors. Ligation of the right portal vein results in stimulating and growth factors to flow all into the remaining left liver. These factors stimulate the future liver remnant to hypertrophy. Unfortunately, these factors can also stimulate liver tumours to grow faster.

   (c) Transection of the liver resulting in local damage which results in increase in stimulating factors to induce hypertrophy.

   A retrospective review by Schadde et al. in 2014 stated 'the rate of hypertrophy in the future liver remnant after ALPPS is 11 times faster than after portal vein embolization or portal vein ligation, e.g. 34.8 versus 3 mL/day'.
2. ALPPS allows the patient to become more tolerant to complications, because:

   (a) The portal venous flow increases tremendously into the remaining liver remnant resulting in early and rapid liver hypertrophy.

   (b) The part of the liver which is going to be resected is still supplied by the hepatic artery. This dual blood supply to the liver assists the future liver remnant to

carry out the necessary liver functions as well as to hypertrophy at the same time.
3. The second ALPPS operation can usually be done within 1–2 weeks of the first operation, thus.
   (a) Fibrinous adhesions can be dealt with easily.
   (b) The chance of tumour progression or metastasizing before the second operation is small.
   (c) The patient can receive chemotherapy earlier after ALPPS.
4. ALPPS allows in the first operation simultaneous resection of a primary tumour, or resection of any tumour in the future liver remnant.
5. In hilar cholangiocarcinoma, portal vein embolization requires a decision to be made before surgery as to whether to preserve the right or the left liver. In ALPPS, this decision can be made at operation.

20.9 Reasons for Unsatisfactory Hypertrophy of the Future Liver Remnant

The reasons can be:

1. The liver remnant has been affected by cirrhosis, fibrosis or obstructive jaundice.
2. A phenomenon which is similar to 'the small for size syndrome' after liver transplantation.
3. Effects of chemotherapy or radiotherapy on the liver.

20.10 Demerits of ALPPS

The common criticisms on ALPPS are:

1. High operative mortality and morbidity rates.
2. Lack of long-term oncological results.
3. Violate the 'no touch' principle in oncological surgery. Some surgeons call the procedure 'all touch technique'. Theoretically the chance of intraoperative manoeuvres spreading tumour cells is very high.
4. Data on ALPPS in cirrhotic patients are scarce.

20.11 The Most Updated Information on ALPPS

1. An international website for the registration of ALPPS was established in 2012. Up to the end of 2014, there have been 430 ALPPS operations from 75 countries being registered.

In 2015, a systematic review on ALPPS was published in the International Journal of Surgery by Bertens et al. (International Journal of Surgery 2015;13:280–7). This systematic review included 59 articles on ALPPS, in 13 of these articles, there were more than 5 cases of ALPPS making a total of 419 cases. The remaining 46 articles included 1–2 cases. Thus the total number of cases reviewed in this systematic review was around 500 cases. The results showed:

- Patient age ranged from 50 to 70 years with a median of 54–67 years.
- Most ALPPS procedures were carried out on patients with colorectal liver secondaries, 79–100% of these patients had received chemotherapy before ALPPS.
- The standardised future liver remnant was 0.19–0.27.
- The interval between the first and the second operation was 9–14 days.
- The liver remnant had hypertrophied 61–93%.
- Ninety-five percent to hundred percent of patients were able to go onto the second operation.
- The R0 resection rates were 86–100%.
- The operative morbidity rate reported in the biggest series of ALPPS (n = 202 patients) was 28% for Calvien-Dindo classification over IIIB.
- The 90-day operative mortality
  - Initially reported by Schnitzbauer AA et al. in 2012 was 12%.
  - Could be 22–29% in small series reported by Cai et al. in 2014 and Li et al. in 2013.
  - Zero percentage has been reported by Sala et al. in 2012, Hermandez-Alejandro et al. 2014, Alverez et al. 2013, Oldhafer et al. 2014.
- In general, the best results were obtained in patients with colorectal liver secondaries.
- ALPPS for patients with hilar cholangiocarcinoma and gallbladder carcinoma resulted in higher mortality and morbidity rates because of the need to carry out lymphadenectomy.
- ALPPS for non-colorectal liver secondaries is a bad prognostic factor.
- The long-term oncological results showed:
  - For all patients who underwent ALPPS
    Median survival at 180 days 73–95%
    at 1 year 46–60%
  - For colorectal liver secondaries
    Tumour free survival 1 year 59%
    2 years 41%
    Overall survival 6 months 86–100%
    2 years 59%

20.12 Comparison Between ALPPS and the Conventional Two-Staged Operation Using Portal Vein Embolization PVE/Portal Vein Ligation PVL

The results are best shown in Tables 20.1 and 20.2.

In a reply to a Letter to the Editor, Aloia stated, 'after portal vein embolization or portal vein ligation, liver hypertrophy was 34% when compared with ALPPS it was 77%'.

Table 20.1 ALPPS versus conventional two-staged operation using PVE/PLV, a retrospective study by Shindol et al. (2013)

	PVE/PVL	ALPPS
n	144	25
Liver hypertrophy	62%	74%
Complications	33%	40%
Bile leak	5.8%[a]	24%[a]
Sepsis	0.7%[a]	20%[a]
Reoperation	2.9%[a]	28%[a]
Liver related mortality	5.7%	12%
Unable to go to definitive liver resection	27.8%	0%

[a]Significant difference

Table 20.2 ALPPS versus conventional staged operations using PVE/PVL

	PVE/PVL	ALPPS
n	83	48
90-Day operative mortality	6%	15%
Liver failure	9%	13%
R_0 resection	66%	83%
Recurrence/12 months	52%	54%

A retrospective study by Schadde et al. (2014)

Thus after PVE/PVL, liver hypertrophy was slower and was less than after ALPPS.

20.13 Conclusions

ALPPS is a technically demanding operation. It is a new operation and it is a very controversial operation. Short-term studies reveal that ALPPS can increase the R_0 resection rate for advanced liver tumours, but at a price of high operative and morbidity rates. The mid-term oncological results are beginning to surface, and the results are acceptable. However, there is little data on the long-term oncological results after ALPPS. There is also little data on liver hypertrophy after ALPPS in cirrhosis.

Further Reading

Bertens KA, Hawel J, Lung K, et al. ALPPS Challenging the concept of unresectability—a systematic review. Int J Surg. 2015;12:280–7.

Chia NH, Lai ECH, Lau WY. Associating liver partition and portal vein ligation for a patient with hepatocellular carcinoma with a background of hepatitis B related fibrotic liver. Int J Surg Case Rep. 2014;5:1077–81.

Zhang GQ, Zhang ZW, Lau WY, et al. Associating liver partition and portal vein ligation for staged hepatectomy (ALPPS): a new strategy to increase resectability in liver surgery. Int J Surg. 2014;12:437–41.

Printed by Printforce, the Netherlands